作品2019
中国建筑设计研究院作品选
SELECTED WORKS 2019 OF CHINA ARCHITECTURE DESIGN & RESEARCH GROUP

中国建筑设计研究院 编

中国建筑工业出版社
CHINA ARCHITECTURE & BUILDING PRESS

PREFACE

Putting a big stack of manuscript of CADG collection on my desk, Mr. Zhang Guangyuan, the former director of architectural culture center of CADG, told me that over 100 projects had been included in this collection, which means that CADG's projects have achieved new heights in both quantity and quality, marking another year of "bumper harvest".

Flipping through the meticulously prepared pictures of the projects, I have been informed of the architects and engineers' respect for nature, cities, culture and life, as well as their positive values based on both rational thinking and emotional consciousness, which have guaranteed the high quality of CADG's projects. So I proposed several outstanding architects for descriptions on their focus of research and practice, sharing our design philosophies with readers to reach consensus that will benefit both our projects and clients.

The global pandemic, the volatile international situation, the prevalence of economic recession and the waning of globalization have all made 2020 a tough year. However, I hope architects and engineers can always proceed with optimism and good will, accomplishing our tasks in a positive and pragmatic way, for the value of high-quality buildings lies not only at the moment, but also in the days to come, and our efforts will pay off as blessings for the future.

Looking forward to our next CADG collection and a better world in the future!

Cui Kai
Sep. 9, 2020

SELECTED WORKS 2019

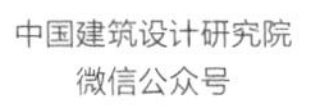

中国建筑设计研究院
微信公众号

中国建筑设计研究院
网站

地址：北京市车公庄大街19号
邮编：100044
电话：010-88328888
传真：010-68349921
网址：www.cadg.com.cn
邮箱：jy@cadg.cn

ADDRESS: No.19, Chegongzhuang Street, Beijing, P. R. China
ZIPCODE : 100044
TELEPHONE: 8610-88328888
FAX: 8610-68349921
WEBSITE: www.cadg.com.cn
E-MAIL: jy@cadg.cn

序

设计院新的作品集那厚厚的书稿又摆在了面前，院建筑文化传播中心张广源老主任笑眯眯地告诉我，这本书收录了几年来的一百多项作品，虽然还有几个值得期待的作品也快建成了，但不能再等了，否则这本书就太厚了。言下之意，这几年作品的数量和质量相比以往又有了提升，又是一个“丰收年”。

翻看着书稿，一幅幅精美的项目图片，不仅表现了我院各专业设计师辛勤工作的成果，从更深层次看，还表达了我们对自然、对城市、对文化、对生活的尊重和热忱，反映出充满正能量的价值观；而这种基于理性思维和感性自觉的价值观成为我们中国建筑设计研究院在建筑创作方面多出上品的重要保证。因此我提议让院里几位建筑创作带头人分别针对各自不同的研究和实践方向写一段文字，让读者在翻阅作品集时能够更清晰地了解我们的思考和立场，从作品的欣赏达到价值观的共识，进而推动新的设计、新的工程顺利前行。也许这正是这本作品集的意义所在。

2020年注定是不平凡的一年，疫情的世界性传播、国际关系的暖冷交替、经济衰退的普遍存在、全球化的大退潮，都使我们对未来多少感到迷茫。但我以为，为人居环境服务的建筑师应该永远保持乐观的心态和美好的憧憬，以善意面对出现的种种情况，以积极的态度和务实的方法完成业主交给我们的任务。因为建筑的价值并不仅仅在当下，越好的建筑，它的价值就越长久。我们为建筑而努力工作，就是为未来而祈福！

期待未来的世界更美好，下一本作品集更精彩！

2020.9.9

在生态文明建设的新时代，我们在每项工程中践行因地制宜的绿色生态理念。

在生态园林中，我们尊重自然地貌，吸纳"巧于因借、精在体宜"的造园思想，挖掘传统营建智慧创造现代景观建筑；在边远地区中，我们顺应自然气候，以当地材料和低技措施表达自然和地域文化特点；面对既有建筑时，我们尊重原有场所精神，以"少拆除、多利用、高活力、可再生"为原则，以"微介入"为策略，将废弃城市空间改造为生长型城市社区，使绿色生态思想与既有建筑改造融合在一起；在密集的城市环境中，我们尊重城市微环境，将大体量建筑纳入街区肌理，实现绿色生态的城市更新。

绿色设计不是技术和设备的堆砌。我们将"绿色生态"理念作为最基本的设计思维，使人们能够通过建筑空间感受绿色、享受生态。

——景 泉

Sustainability

In the new era of ecological civilization, we are implementing sustainab concepts that suit to local conditions.

We build ecological gardens with respect for natural landforms and traditio buildings, highlight the glamor of remote areas with local materials and low-te approaches, renovate existing structures with less demolition and more utilizati with an approach featuring "micro-intervention", integrate large-scale buildin into their densely occupied neighborhood to facilitate sustainable urban renewa To us, sustainable design goes beyond the application of technology a equipment. It's our fundamental mindset for designing buildings, whe sustainability can be perceived.

—Jing Qua

半个世纪前，库恩提出了"范式转移"理论："一个稳定的范式如果不能提供解决问题的适当方式，它就会变弱，从而出现范式转移"。中国的城市发展正在验证着这一理论。2010 年以前的很长时间里，中国的城市发展几乎就是意味着不断扩张与膨胀。而最近这个十年里，这种"范式"正在悄悄地"转移"：一方面建筑增量不再依靠"楼盘"的形式，简单扩张城市边界，更多地呈现出重大事件激发的城市拓展，有的离散为城市边界的新生活样板，成为"边缘优势"的反中心化例证；另一方面，既有城市发展区也在进行着存量提升与品质再造，不仅对老城进行着"驱弊兴利"式的迭代，也在迎接发展区新旧动能的转换节点。

中国建筑设计研究院呈现的诸多与城市发展相关的建筑案例，强调的不是为创新而创新，而是为新问题而创造新的应对方法。从这一点看，它们已不再是一般的"范式转移"，而是试图在新的社会发展结构中寻求对范式的再次约定。

——曹晓昕

Urban Development

According to the theory of paradigm shift, put forward by Thomas Sammual Ku half a century ago, if a stable paradigm fails to solve problems appropriate it will weaken and lead to paradigm shift. The theory can be verified by urban development in China, which means relentless expansion before 20 However, the paradigm began to shift in the past decade: on one hand, simple extension of urban boundary has been replaced by expansion trigge by significant events; on the other hand, the developed urban regions are go through optimization and iteration.

For CADG's projects relevant to urban development, innovation means n approaches to solve new problems. From this point, the projects are no lon typical paradigm shift, but re-arrangement of paradigm in a new structure social development.

—Cao Xiao

我们希望在设计中通过对当地文化的自觉回应，使我们的建筑真正地融入城市，通过可识别的形式语言来彰显城市的个性，获得市民的自我认同感和归属感，但绝非形式化的仿古守旧。立足于本土文化的创新应首先回归到对建筑本体逻辑和空间逻辑的真实反映。

同时，文化要获得真正的生命力，需要融入真实的现代生活。建筑创作对文化的回应不仅要注重形式语言的创新，更要关注市民的真正需求。我们希望营造具有开放性、共享性的场所，获得市民的广泛参与，激发城市的应有活力。

建筑需要延续我们的文化基因，更应承载当代的城市生活。从中国院最近几年的设计实践中，大家可以看到这些年来我们的探索与尝试，以及我们所坚持的价值取向。

——任祖华

Cultural Life

We aim for the integration of buildings and cities through the spontanec response to local culture, so that a sense of identity for local people will created. And this land-based innovation should reflect the logic of spaces in first place without falling into the trap of formalism or simple adherence to past.

Furthermore, the authentic vitality of culture lies in real life, requiring architectu design to focus on the real needs of people. As a result, we aim to present op spaces to motivate public participation and bring out the vigor of cities.

Built for inheritance of cultural mems and accommodating modern life, CAD works in recent years are just reflections of our values and principles.

—Ren Zuh

科技创新

科技创新在当下的建筑设计语境中变得越来越重要。一方面，在传统的文
、居住、办公等建筑中，计算科学的快速进步为设计提供了更便利的手段，
对新型结构、绿色、装配式等技术的发展，产生了更多建筑空间、形态和性
上新的可能性；另一方面，大量前沿企业也有越来越多的和科技相关的建造
动，包括集群性质的实验研发，大规模数据存储处理等前所未有的空间和功
需求，同样在行为、心理和文化上提供了新的设计视角。

我们的相关设计实践，也从这两个方面进行了探索和尝试，一方面用不断
现的“新技术”提高建筑各方面的性能，努力使建筑作为产品和作品具有更
的品质；另一方面，面对新型的科技建造需求，尤其是科研人员对环境、使
模式等的不同需求，用更体贴的“新建筑”来为科研服务。这也是我们对待
技创新的态度：科技为人服务，创新使生活更美好。

——徐 磊

Scientific & Technological Innovation

Scientific & Technological innovation is playing an increasingly significant role in architectural design. On one hand, the rapid progress of computing technology has offered more advanced design methods for buildings, while new opportunities emerge with the development of new structures, green technologies and pre-fabrication, etc.; On the other hand, a large number of cutting-edge enterprises are embracing building practices relevant to new technologies, so that we look at behavior, psychology and culture in new perspectives.

In response, our design practice are improving building qualities with new technologies and more consideration to people's demands, especially those from scientific researchers. Every new building we present reflects our belief that scientific & technological innovation makes our life better.

—Xu Lei

有机更新

高速增量的城镇化进程逐渐放慢脚步，如何面对并解决粗放发展阶段积累
问题，把关注点从规模转向质量，成为未来中国城市发展的一个主题。

城市是一个复杂的、多维系统交织的整体。从源起到兴盛，从兴盛到衰落，
从衰落走向复兴，在这样往复更迭的历史进程中，城市的特征和魅力沉淀下
。那些转换了用途的场所，改变了功能的建筑，在更长的生命周期里，带着
去的特征，讲述着今天的故事，为城市增添了可以阅读的厚度，城市的发展
程，也由此与人们的生活和情感紧密地交织在一起 。

我们追求一种渐进式的、生长式的、混搭式的、修补完善式的改造状态。
重城市记忆和既有环境，珍视时间痕迹和历史信息，关注日常状态和生活需
力求塑造更加和谐友善、可以辨识、持续发展的城市环境，让生活于其中
人们都能感受到善意、尊重和关照。

——柴培根

Integrated Renewal

With the slowdown of the rapidly incremental urbanization process, the shift from scale to quality, as well as the solutions to remaining problems of previous periods, have become the focus of future urban development.

As a complex and multi-dimensional system, a city goes through rounds of evolvements and thus obtains its own features and charm. Some places, with their functions changed in history, are available for people's "reading" of both the past and the present, strengthening the bond between people and the city.

We're pursuing a renewal approach featuring gradual, growing, hybrid and patching process with respect for history, context and daily life, so that people will feel cared and respected in such a city.

—Chai Peigen

地域特色

理性与诗意

对地域特色的理解，体现在更深层次的对地理气候环境特征的理解，对地
传统文化内涵的尊重，对当下社会发展的回应上，让设计兼具了地域性和创
性。基于此，建筑表达将脱离模仿性、装饰性建筑语言，转变为形体、空间、
构、材料、色彩这些基本要素对场地地理气候特征和文化特质作出的理性回
。这些理性而又充满智慧的设计策略让建筑在高原大漠等特殊场地中呈现出
种整体的诗意表达。

现实与实现

偏远地区落后的施工技术能力和低廉投资的现实条件并未束缚建筑师的设
理想，创造性地对地方材料和传统技艺的应用，即使最普通的砌块和最廉价
原竹，在高原和深山都可以成为助力理想实现的神来之笔。

——于海为

Local Characteristics

Rationality & Poetry

The understanding of local characteristics is reflected in the deeper understanding for geographic and climatic features, the respect for local culture and the response to social development, so that the design will be both land-based and innovative. Without simple imitation or excessive decorations, the design will be a rational response to fundamental elements, presenting a poetic expression in the wilderness.

Reality & Realization

The design ideals have not been held back by outdated construction techniques or low costs in projects in remote areas. Instead, the creative application of local materials and traditional techniques have achieved what we desired.

—Yu Haiwei

目 录　CONTENTS

北京世界园艺博览会中国馆 China Pavilion at Beijing International Horticultural Expo

地点 北京市延庆区 / 建筑面积 23,000m^2 / 设计时间 2016年 / 建成时间 2019年

方案设计 崔 愷、景 泉、黎 靓、李静威、张翼南、郑旭航、田 聪、徐松月、李晓韵、及 晨

设计主持 崔 愷、景 泉、黎 靓

建　　筑 郑旭航、田 聪、吴洁妮、吴南伟、吴锡嘉

结　　构 张淮湧、施 泓、曹永超、李艺然

给 排 水 黎 松、林建德

暖　　通 刘燕军、孙淑萍

电　　气 王苏阳、姜海鹏、沈 晋

电　　讯 刘 炜

总　　图 吴耀懿

室　　内 邓雪映、李海波、焦 亮、李 倬、林泽潭

景　　观 史丽秀、赵文斌、刘 环、路 璐、贾 瀛、李 旸、刘卓君、盛金龙、齐石茗月

中国馆是北京世园会最重要的场馆之一，如一柄温润的“玉如意”坐落于山水园林之间。巨型屋架从花木扶疏的梯田升腾而起，恢宏舒展。设计从园艺主题联系到中国的农耕文明，以层层叠叠的梯田体现传统农耕智慧。建筑的平面为半环形，南侧留出广场迎接八方来客；首层中部底层架空，形成南北贯通的通廊，与北侧妫汭湖建立了联系，并利用道路与湖面的高差关系，实现了不同标高的进馆和离馆流线。整个场馆采用单一方向的参观流线。一层展厅埋于土下，室内绿叶的软膜顶棚和深浅搭配的绿色格栅，仿佛把游客带入了森林。二层采用鱼腹式桁架屋顶，覆盖 ETFE 膜，室内光线柔和、细腻，中部的观景平台可供远眺世园会园区。地下一层的水院空间，让流水从瓦屋面跌落，形成中国传统民居“四水归堂”的奇妙景致。

As one of the most important pavilions of the Horticultural Expo in Beijing, the China Pavilion resembles a "Ruyi" located among hills and waters. Its huge roof truss seems to have grown up from the terraces with a stretched gesture, showcasing both the local horticulture and the time-honored agricultural civilization of China. The building has a semi-ring shaped plan, with a square to its south with a welcome gesture. The central part of the 1st floor is stilted, providing access to the lake on the north. The exhibition hall of the 1st floor is covered under the ground, with green granitic plasters, soft film ceilings and gratings making visitors feel as if they are in the forest; on the 2nd floor, ETFE films on the truss structure introduces soft light into the interior. On viewing platforms on the 2nd floor, visitors can overlook the Yongning Tower and Guirui Theater in the Expo Garden.

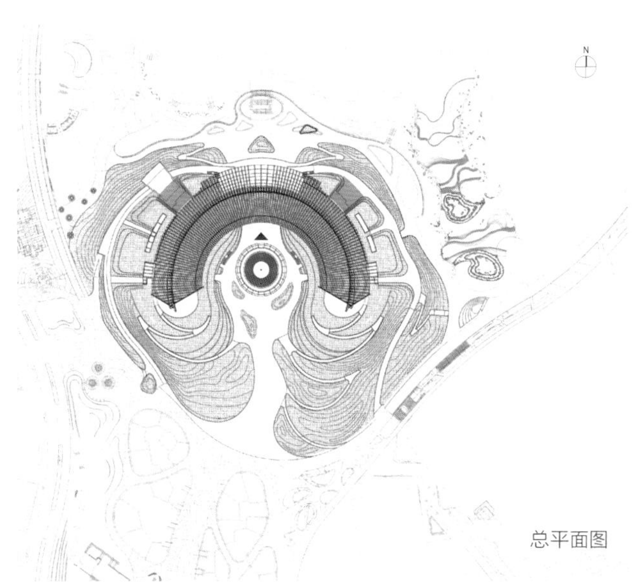

总平面图

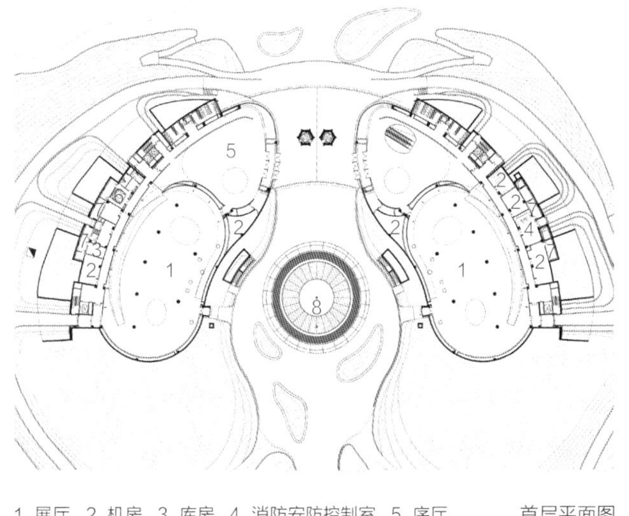

1. 展厅　2. 机房　3. 库房　4. 消防安防控制室　5. 序厅　　首层平面图

6. 贵宾接待室　7. 庭院　8. 下沉水院上空

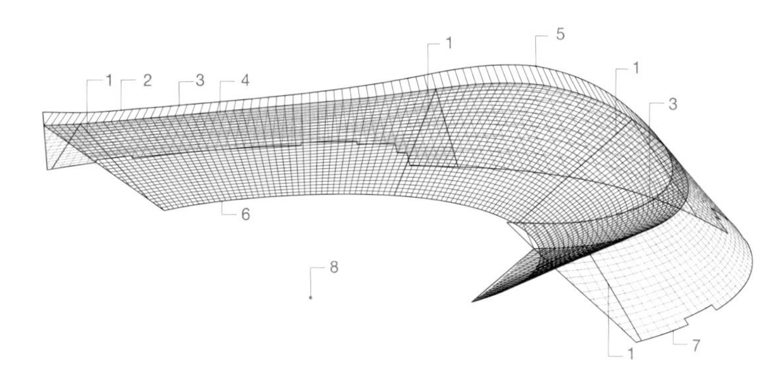

1. 室内外分界山墙
2. 屋脊挑出线
3. 最低点
4. 屋脊线
5. 最高点
6. 南侧檐口
7. 北侧檐口
8. 圆心定位点

屋架单线透视图

李季 摄

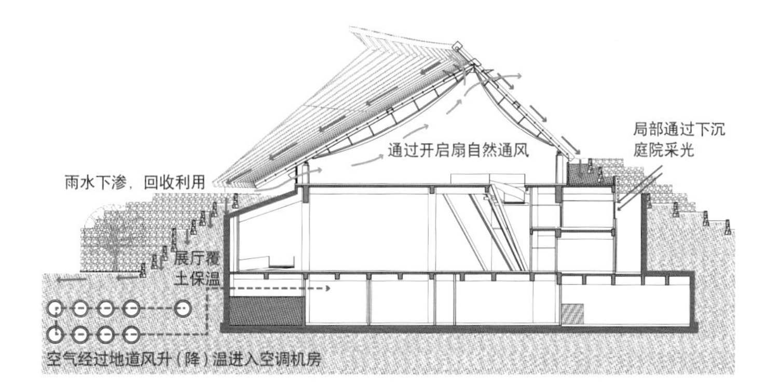

绿色设计分析图

根据延庆地区光照、降水、通风、温度等气候条件，设计选择适宜的绿色技术——展厅覆土、地道风系统、自然通风、光伏玻璃、雨水回收利用，使中国馆成为一座有生命、会呼吸的绿色建筑。

Based on local climatic conditions, appropriate green technologies, including soil covering, tunnel ventilation system, natural ventilation, photovoltaic glasses and rainwater recycling, are applied to endow the building with breath, life and sustainability.

北京奥林匹克塔 Beijing Olympic Tower

地点 北京市奥林匹克公园 / 用地面积 81,437m^2 / 建筑面积 18,687m^2 / 高度 248m / 设计时间 2005年 / 建成时间 2015年

方案设计 崔 愷、康 凯
傅晓铭、喻 弢
设计主持 崔 愷、范 重、康 凯

建　　筑 叶水清、吴 健、邢 野
结　　构 刘先明、杨 苏、彭 翼
给 排 水 郭汝艳、朱跃云
设　　备 宋孝春、王 加
电　　气 王 健、曹 磊、刘征峥
电　　讯 许 静、郭利群
总　　图 白红卫
景　　观 冯 君

位于奥林匹克公园中心区的北京奥林匹克塔，其灵感来自于自然界植物生长的形态，也寓意奥运精神生生不息。塔座部分覆土而建，缓缓升起的绿坡覆盖整个大厅，既与周边景观自然衔接，又为游客提供仰望塔顶的座席。从基座破土而出的塔身，底部采用实体围护，随着向上生长的态势，逐渐如枝叶般分叉，露出内部树枝肌理的银白色金属幕墙，与银灰色的镀膜玻璃幕墙虚实交织，让整个建筑显得轻巧起来。五个位于不同高度的塔顶如水平伸展的树冠，在空中似分似合。站在观景平台上，既可远眺北京周边的奥运景观，也能感受到超尺度的建筑和结构本身带来的震撼。

The Beijing Olympic Tower, located in the Olympic Green, symbolizes the vigorous spirit of the Olympic Games with its growing gesture. The base of the tower covered under green slopes serves as a buffering zone between the building and the landscape. The tower's body, looking as if it had grown from the soil, is enclosed at the lower parts while "branches" emerge at upper parts, adding to the lightness of the building. Standing on the viewing platform, visitors can enjoy both a panorama of the Olympic-themed scenes and the grandeur of the tower's mega-scale structure.

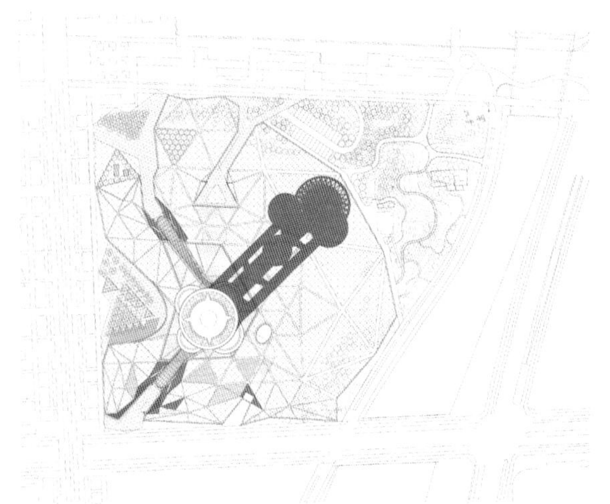

总平面图

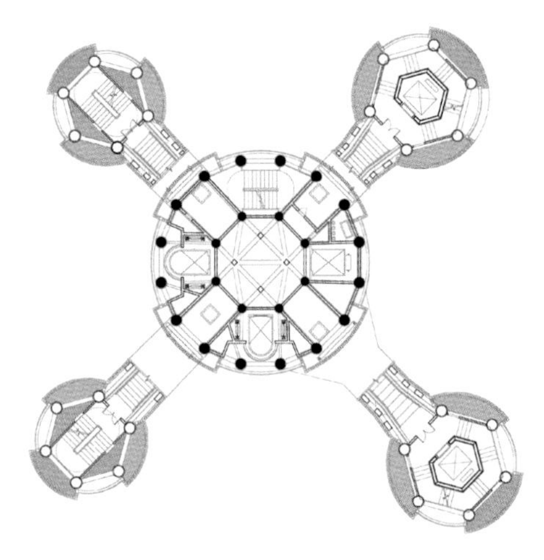

塔冠平面图

科技创新

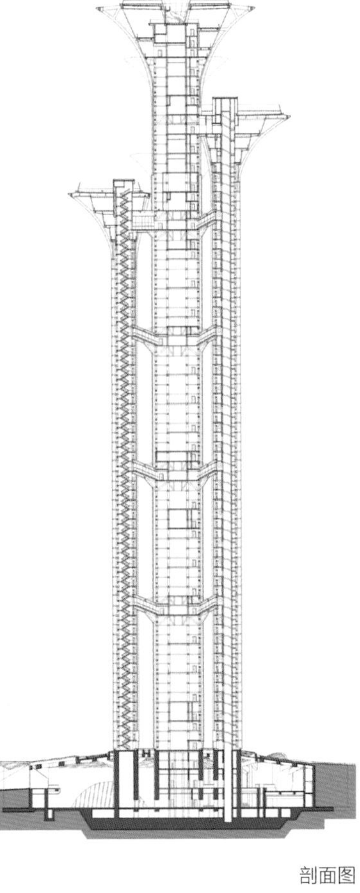

剖面图

雄安市民服务中心企业办公区 Xiong'an Civic Service Center Enterprise Office Area

地点 河北省雄安新区 / 用地面积 60,370m² / 建筑面积 36,023m² / 高度 11m / 设计时间 2017年 / 建成时间 2018年

方案设计 崔　愷、任祖华、梁　丰
　　　　 王　俊、陈谋朦
设计主持 崔　愷、任祖华

建　　筑 梁　丰、庄　彤、朱宏利
　　　　 王　俊、陈谋朦、盛启寰
　　　　 邓　超、朱　巍
结　　构 刘长松、丁路通、任乐明
　　　　 孔维伟、白倩楠、王海波
　　　　 武晓敏、张兰英、蔡玉龙
给 排 水 刘志军、李仁杰、张　越
设　　备 张英杰、张思健
电　　气 裴韦杰、杜建勋
总　　图 路建旗、李　爽、王　炜
室　　内 彭典勇、韩文文、饶　劢

雄安市民服务中心是雄安新区的第一个建设工程，由公共服务区、行政服务区、生活服务区、企业办公区四个区域组成。中国建筑设计研究院承担了其中容纳企业办公、酒店、公共服务设施的企业办公区的设计。作为雄安新区的起步项目，设计既要满足快速建造的要求，也要对未来建设提供示范，因此采用了全装配化、集成化的集装箱式建造技术，实现结构、内装、外装、设备管线全工厂生产，整体运输至施工现场统一吊装，最大限度减少建造垃圾，实现绿色施工。立面采用水泥压力板、金属网与玻璃，以集装箱为单位，通过四种标准幕墙单元的组合，形成条码化的外观，便于生产施工，也使得立面设计与模块化构造逻辑一致。办公基本单元为 1000 ~ 1200m² 的“十字”单元组合，呈现对周边环境开放的姿态，以小进深实现自然通风采光。“十字”单元可通过局部变形，相互组合，蔓延于环境之中。

As the first construction project in Xiong'an New Area, the Xiong'an Civic Service Center consists of four sectors catering to various functions. CADG has undertaken the design of the corporate office sector, which is situated on the north of the center and incorporates functions including offices, hotel and public service facilities. As the starting project of Xiong'an New Area, the building must be quickly built into an exemplary project for future construction works. As a result, the container-style building technology, featuring pre-fabrication and integration, has realized complete factory production of structural parts, interior decorative parts, exterior decorative parts and pipelines, which could be installed on site, minimizing the quantity of building waste. Cement pressure plates, metal mesh and glasses dominate the façades.

总平面图

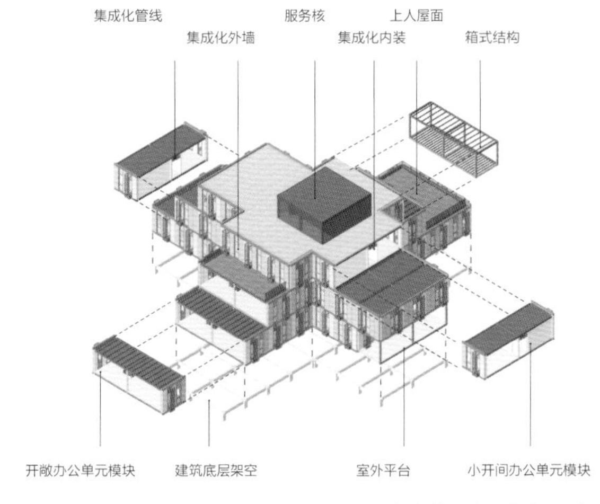

十字单元布局拆解示意

科技创新

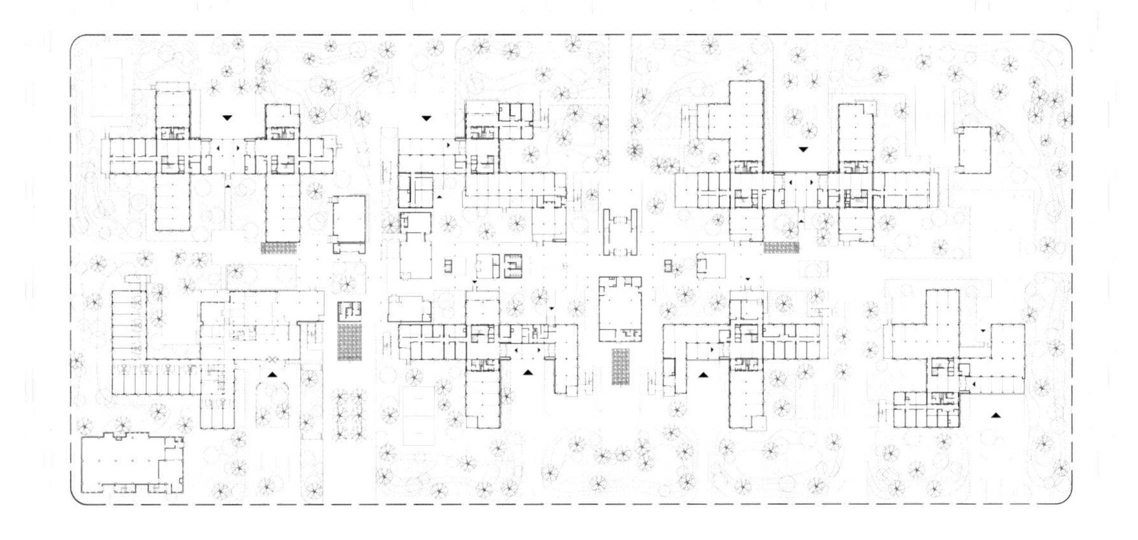

首层平面图

办公楼与酒店设计以集装箱作为基本模块单元，适合运输，便于灵活组合。商业和机房服务设施顶部则形成连续的室外平台，将各建筑连为一体。场地中部设置了公共服务平台体系，包含职工餐厅、特色餐饮、咖啡店、健身房、无人超市等功能。

In the office and hotel sectors of the center, containers are used as fundamental modules to facilitate transportation and flexible combination. Various sectors of the center are connected with an outdoor platform.

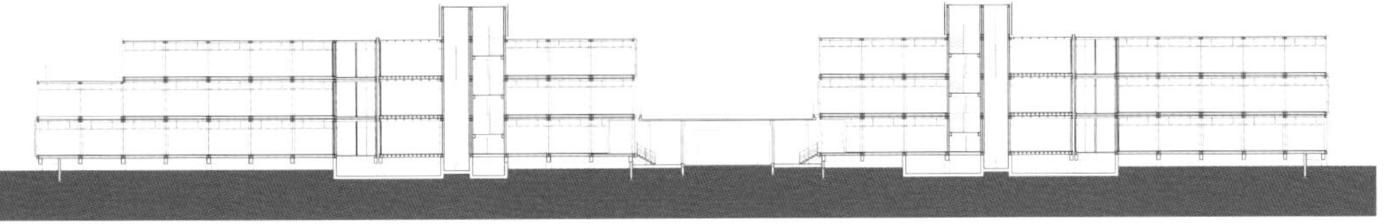

剖面图

北京城市副中心行政办公区A2项目 Project A2 in the Administrative Area of Beijing Sub-Center

地点 北京市通州区 / 用地面积 79,452m² / 建筑面积 290,611m² / 高度 50m / 设计时间 2016年 / 建成时间 2018年

方案设计 崔　愷、郑世伟、聂兆征
莫曼春、赵国璎、李慧敏
罗　云、郑伯煊、崔杨波
设计主持 刘燕辉、郑世伟

建　　筑 聂兆征、莫曼春、赵国璎
李慧敏、罗　云、郑伯煊
崔杨波
结　　构 张淮湧、何相宇、王　超
给 排 水 杨东辉、黎　松
设　　备 郑　坤、韩武松
电　　气 王苏阳、李　磊
电　　讯 张月珍、张　雅
总　　图 高　治、路建旗
室　　内 董　强、张　磊、刘力扬
景　　观 史丽秀、刘　环、李　旸

行政办公区是通州城市副中心的核心功能区之一，肩负着疏解北京城市功能的示范带头作用。其启动区包含市委、市府、人大、政协及委办局等部门办公楼，是行政办公区的重中之重。其中的A2工程设计，秉承了集约、高效、本土、人文、朴素、典雅的设计原则，力图打造能够满足当代政府需求、展现当代政府形象，具有前瞻、经济、共享、绿色特点的“千年大计”。建筑外观采用典雅、稳重、简洁、现代的风格，体现浓浓的中国风、北京味。作为绿色建筑示范工程，设计围绕人流活动、建筑空间使用特点展开，最新绿色建筑设计理念的应用贯穿项目设计始终。

As one of the major functional districts of Beijing Sub Center in Tongzhou, the administrative office district bears the exemplary role of function-relieving of Beijing, and the office buildings in the region bears great significance. Under the principle of centralization, efficiency and simplicity, the design for A2 project aims for a “millennium project” that is forward-looking, economic, open and sustainable, meeting the demands of a contemporary government with its elegant and modern style with Beijing features.

装配式混凝土外墙立面细部设计

规划采取中轴对称院落式布局形式，并结合连廊使各院落空间彼此分隔，又相互联系，兼具秩序性和层次感，不仅有利于创造良好的园林式办公空间尺度，而且与行政办公建筑端庄、严肃的气质特征呼应，形成典型的中国传统建筑空间意象。

A symmetrical layout with courtyards has been adopted, with corridors connecting or separating the courtyards. The layout is appropriate for not only the scale of a garden-style office space, but also the modest and dignified feature of administrative buildings, presenting a typical image of traditional chinese architectural spaces.

北京城市副中心行政办公区B1、B2项目 Project B1 & B2 in the Administrative Area of Beijing Sub-Center

地点 北京市通州区 / 用地面积 64,400m² / 建筑面积 172,000m² / 高度 36m / 设计时间 2015年 / 建成时间 2018年

方案设计 丁　峰、杨　凌、刘　涛
　　　　 刘　建、李　腾、张伟玲
　　　　 蒋　鑫、张柏榕
方案指导 崔　愷、徐　磊
设计主持 刘燕辉、丁　峰

建　　筑 杨　凌、张　莹
　　　　 董国升、徐　明
结　　构 孔江洪
给 排 水 王耀堂、李　伟
设　　备 孙淑萍、李　莹
电　　气 张　青、陈沛仁
电　　讯 任亚武、唐　艺
总　　图 高　治、路建旗、李　爽
室　　内 邓雪映、李　钢
　　　　 李海波、林泽潭
景　　观 史丽秀、赵文斌
　　　　 刘　环、路　璐

北京城市副中心行政办公楼B1、B2工程位于A2工程东侧，通过两组团间的道路巧妙地与A2工程的东西轴线连接；A2工程的主楼东立面作为街道底景，增加了空间层次。呼应A2工程中间主楼四角配楼的布局方式，B1、B2工程的体量采用东高西低的方式，东侧对应城市道路，西侧与A2工程低区配楼相对应，形成了均衡城市空间体量关系。每个办公组团均以两个院落空间为核心，同时为减少对院落压迫，院落角部开放，以两个L形建筑半围合而成院落。对角相套的两个院落互相连通，又以连廊、小建筑串接、分隔，产生了丰富而有序的空间变化。建筑立面遵循古典构图法则，具有得体的节奏和韵律。

Located to the east of the Project A2, the Project B1 & B2 is connected to the east-west axis of the municipal office building by a road between the two clusters. The building descends from east to west, achieving balance of volumes with surrounding buildings and roads. Each office cluster consists of two courtyards, the corners of which are open to form semi-enclosed spaces. Two courtyards sharing one corner are partly connected and partly separated by structures such as corridors and small buildings, creating spaces that are both diverse and orderly. Classic rules of composition are applied on the façades.

立面材料以灰色石材和平瓦为主，配以传统的院落空间布局，整组建筑群显得低调平和，清新典雅，京味京韵。

With their façades dominated by gray stones and tiles, the buildings take on a modest and elegant look with Beijing characteristics.

北京雁栖酒店 Beijing Yanqi Hotel

地点 北京市怀柔区 / 用地面积 437,900m² / 建筑面积 43,000m² / 高度 20m / 设计时间 2011年 / 建成时间 2014年

方案设计 美国 AECOM 公司
设计主持 刘 勤

建　　筑 宋 焱、杨丽家
结　　构 许 庆
给 排 水 赵 昕、钱江锋
设　　备 王春雷、梁 琳
电　　气 李俊民、陈双燕
总　　图 王 炜

南广场景观 史丽秀、赵文斌、刘 环

雁栖酒店位于北京雁栖湖国际会都，是北京的国宾级会议接待目的地，APEC、G20 峰会等世界级会议都曾在此举行。其规划及建筑设计在传承传统文化的基础上，探求建筑与自然的和谐共生，并满足现代高端会议活动的需求。酒店位于国际会都核心岛的东部，紧临会议中心。酒店三面环路，东临雁栖湖，其布局采用院落围合式，利用东西向场地高差，使建筑群呈现既分散又相对集中，错落又有秩序的风格，从空间上体现了中国传统院落布局的精髓，坡屋顶按现代方式处理，也继承了传统建筑的特点。

Positioned as a state-class conference center, the Yanqi Lake International Conference Resort is situated by the lake. The planning and design aims to combine the modern conference requirements and the traditional Chinese culture without interrupting the beauty of nature. The courtyard-style layout takes advantages of the height difference between east and west parts of the site and presents the essence of traditional Chinese architecture with carefully planned positions of the buildings.

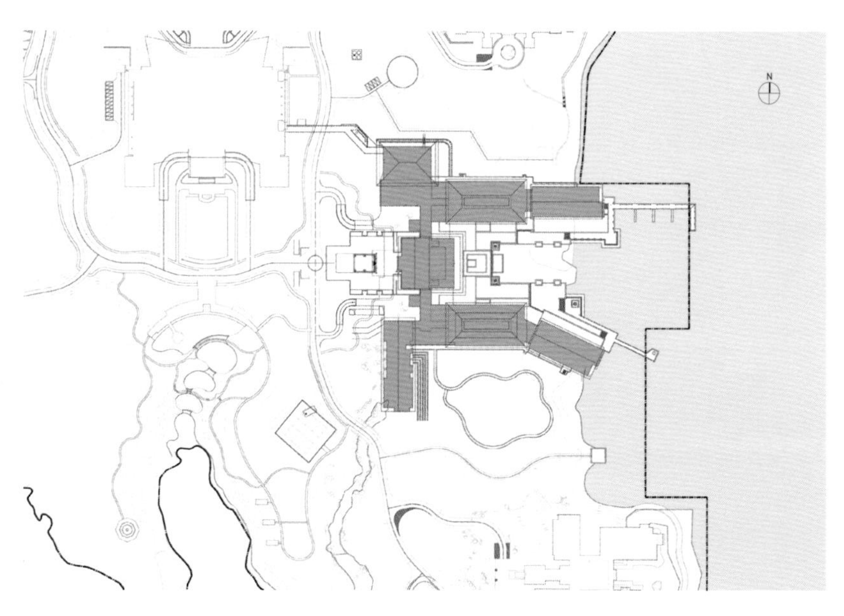

总平面图

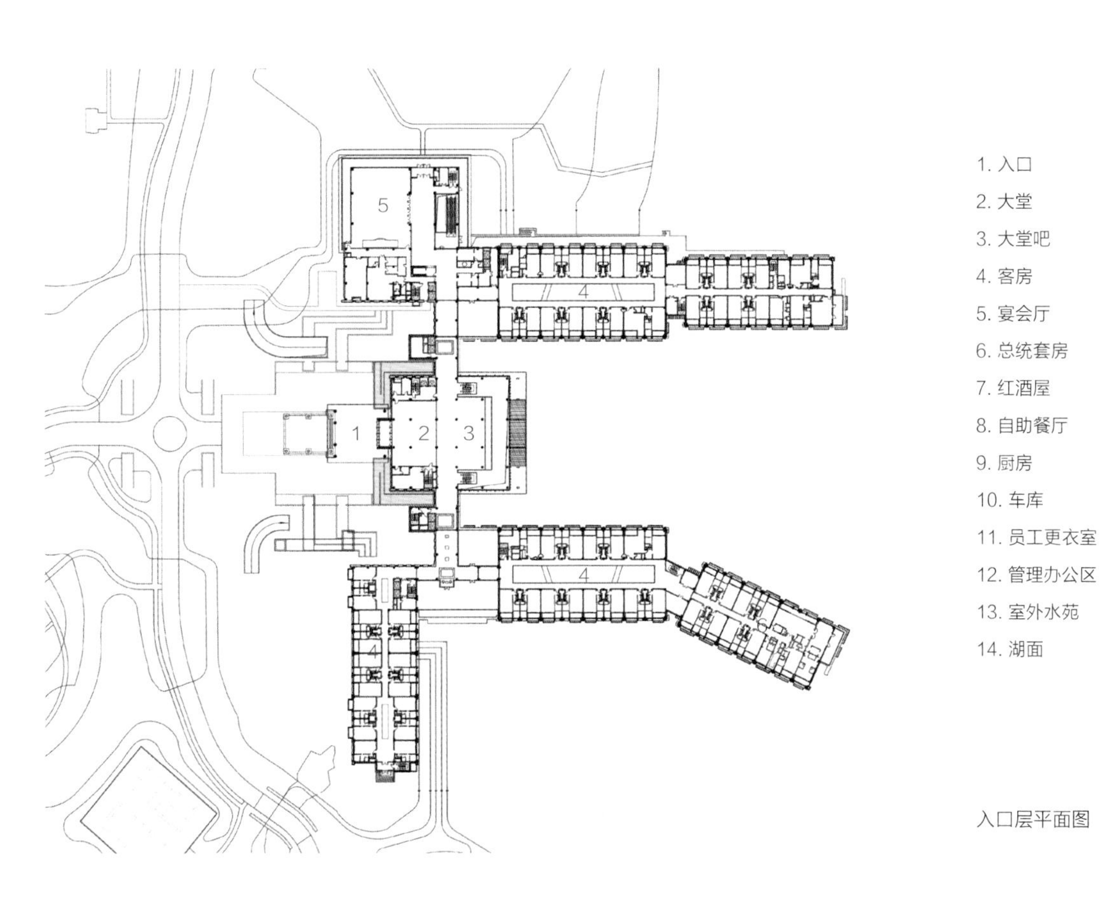
1. 入口
2. 大堂
3. 大堂吧
4. 客房
5. 宴会厅
6. 总统套房
7. 红酒屋
8. 自助餐厅
9. 厨房
10. 车库
11. 员工更衣室
12. 管理办公区
13. 室外水苑
14. 湖面

入口层平面图

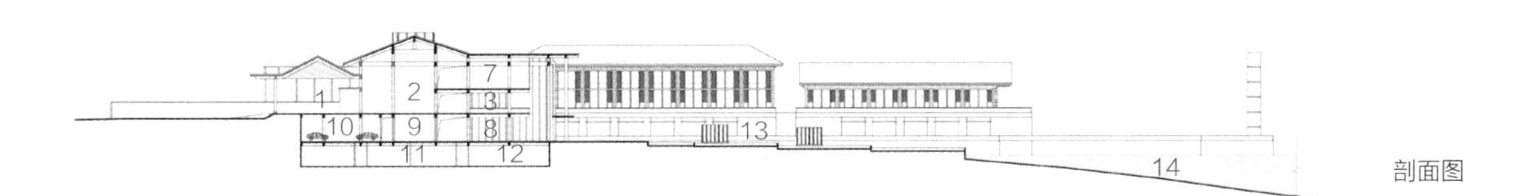

剖面图

酒店房间单元环绕合院布置，围合岛东岸的 U 形水湾，使所有客房均有良好的湖面景观，同时又保有一定私密性。内院延伸处设有游艇码头，成为整个园区水上活动的集散点。

Rooms are arranged around the courtyard and a U-shaped bay, so that each of them enjoys the view of the lake while maintaining an appropriate degree of privacy. Beyond the courtyard is a marina, serving as the hub for activities on the water.

会议中心南广场位于整个雁栖岛的入口处。其景观设计用礼仪形制营造空间格局及行为动线，用情与意延展中华文化的艺术魅力，强调了中轴对称，并按照传统宫殿建筑的空间序列形成礼仪节奏，融入了汉唐文化演绎而来的特色景观元素。

The southern square of the conference center is located at the entrance of Yanqi Island. The landscape design features a spatial sequence resembling that of traditional palaces to present ritual characteristics while integrating landscape elements derived from Han and Tang dynasties.

中国建筑设计研究院创新科研示范中心 Innovation & Scientific Research Demonstration Center of CADG

地点 北京市西城区 / 用地面积 3,578m² / 建筑面积 41,438m² / 高度 60m / 设计时间 2011年 / 建成时间 2018年

方案设计 柴培根、周　凯、李　楠
田海鸥、任　玥、杨文斌
戴天行
设计指导 修　龙、崔　愷
设计主持 柴培根、周　凯

建　　筑 田海鸥、李　颖、任　玥
结　　构 霍文营、孙海林、郭家旭
给 排 水 赵世明、赵　昕、李建业
设　　备 潘云钢、何海亮、李　嘉
电　　气 陈　琪、王　旭、林　佳
电　　讯 陈　琪、任亚武
总　　图 高　治、吴耀懿
室　　内 饶　劢、韩文文、顾大海
李　甲、曹　诚、曹　雷
景　　观 刘　环、王　婷、李　旸

创新科研示范中心的设计是一次城市有机更新的实践。在处理邻里关系的过程中，首当其冲的是解决与周边居住区的日照关系——按照高度和用地范围的限制确定最大可建设体量后，以日照最不利点为基点，阳光的移动轨迹会把最大可建设体量雕琢成一个不规则的原型。日照原型是外在的限制，而获得更大的使用空间是内在的需求，建筑的体形在这一里一外两种力量的挤压被塑造出来。

建筑的功能并不复杂，设计改变了从功能分区到使用空间的简单化处理方式，把功能转化为使用者的行为，以行为去引导场所的生成。当建筑能够包容或引发城市生活的活力时，创新楼多元复合的状态也在一定程度上改变了大型设计机构通常给人留下的刻板印象，消解了以效率为先的大开间办公环境的枯燥与单调。

两层通高的门厅串联起咖啡厅、展厅、图书区、小卖部、会议室和多功能厅等公共服务功能——这里成为创新楼公共生活的客厅。

三层到十四层是各个设计部门的办公区，由于退台的造型造成各层面积逐渐缩小，连续退叠的室外平台和室内中庭共同将大开间办公区组织到一起。中庭顺应退台的形式，成为独树一帜的内部空间。

The Innovation & Scientific Research Demonstration Center of CADG can be seen as a project of urban renewal. With the maximum volume of the center determined by height and land occupation restrictions, a final prototype was generated by both the external restrictions of daylight shading and the internal demands of more space.

With relatively simple and clear functional demands, the spaces were generated based on behaviors defined by functions, rather than defined simply by functional division. With its inclusion and stimulation of the city's vigor, the diversified status of the Innovation Center avoided the stereotype featuring the monotony of the efficiency-oriented office space with large spans.

The double-height lobby provides access to a café, an exhibition hall, a reading area, a mini supermarket, conference rooms and a multi-functional hall, serving as a lounge for public life in the Innovation Center.

Office spaces for design organizations of CADG are located on 3-14 floors. The open office area on different floors are connected and integrated by the terraced exterior platforms and the interior atrium.

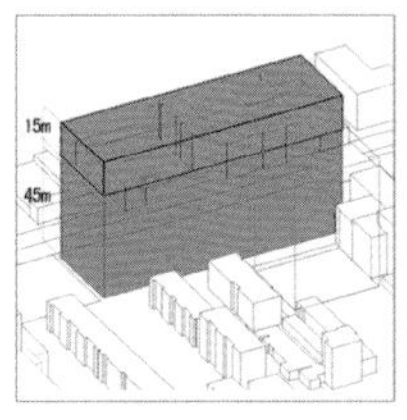
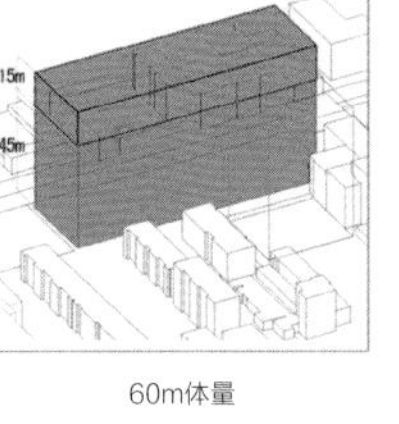

60m体量

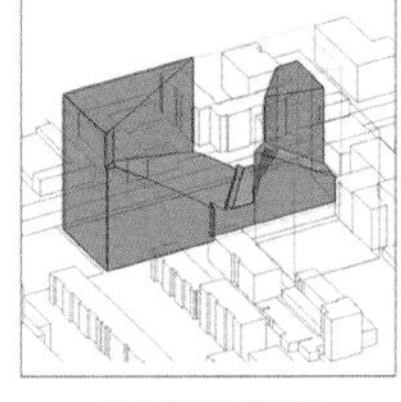

由日照条件反推形体

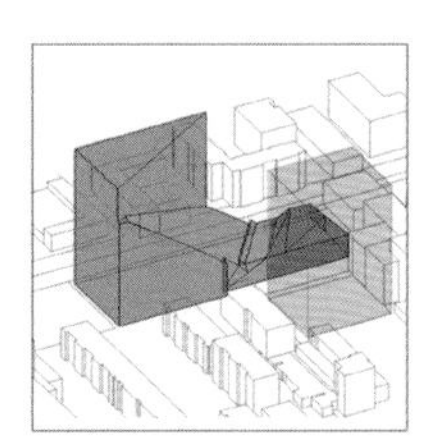

降低北侧体量以减弱与相邻宾馆的对视

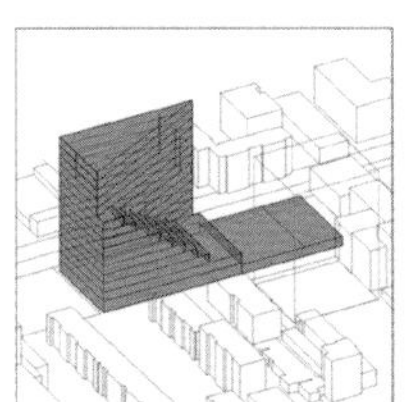

用退台表达建筑形体

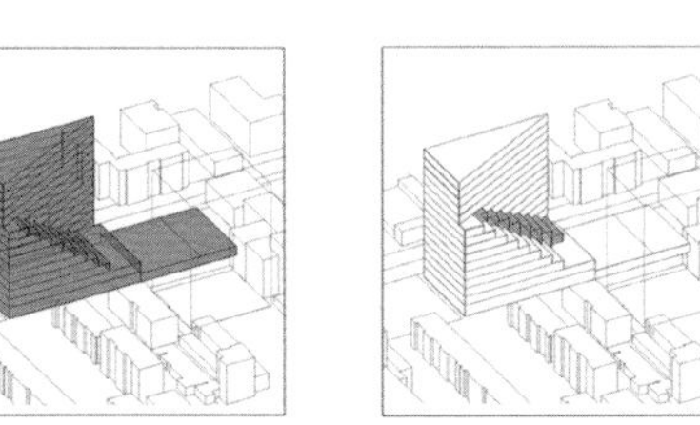

中庭加强各层联系并为大进深办公空间采光

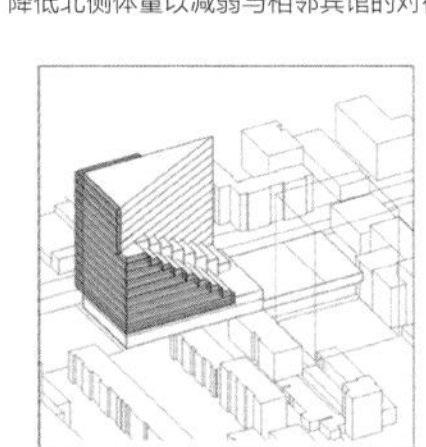

东南西侧悬挑以增加更多面积

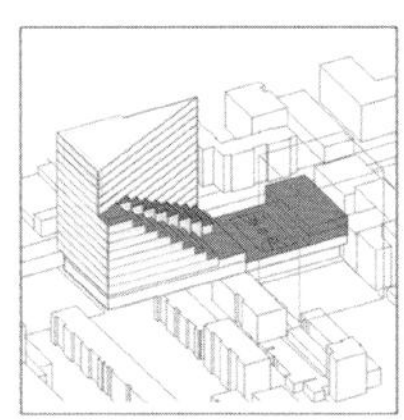

结合平台营造员工活动场所

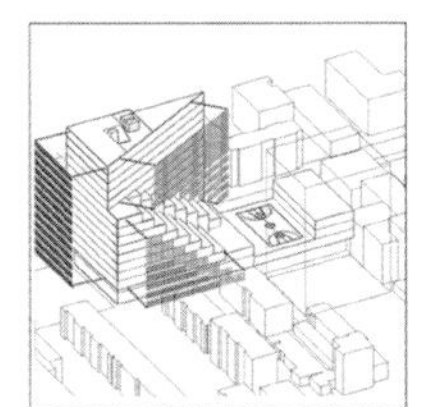

根据朝向推敲格栅形式

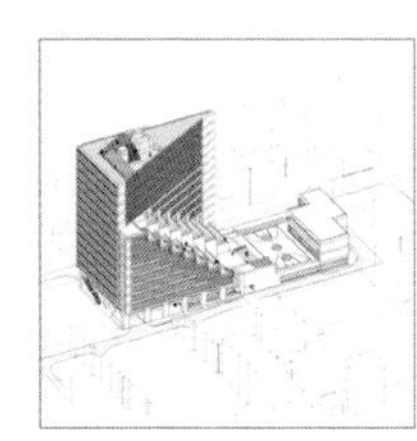

建筑整体状况

形体生成

平台、中庭、球场等场所打破了封闭的空间边界，不期而遇的交流、生活化的场景，都成为对设计行为的支撑，由此集成的场所也在整体上体现了设计企业的特征。

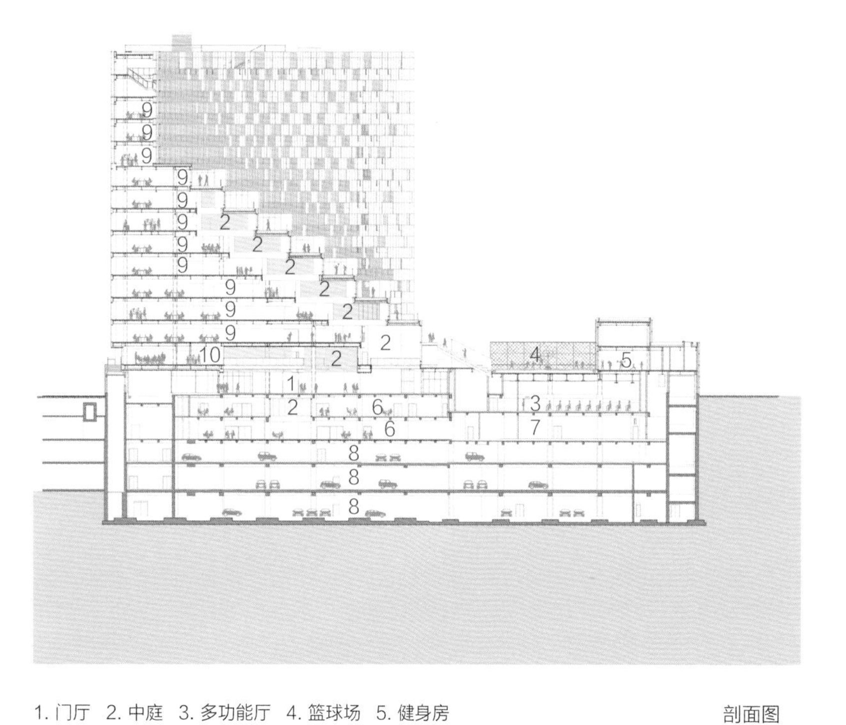

1. 门厅 2. 中庭 3. 多功能厅 4. 篮球场 5. 健身房
6. 餐厅 7. 厨房 8. 地下车库 9. 办公 10. 会议室

剖面图

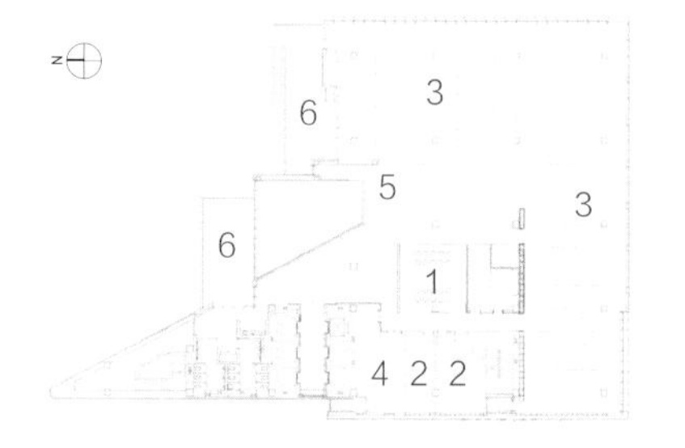

1. 会议室 2. 办公室 3. 开敞办公
4. 空调机房 5. 中庭上空 6. 露台

六层平面图

The terraces, the atrium and the basketball court have blurred the boundary of the enclosed space. The improvised communication and scenes of various lifestyles in the space could serve as the backdrop for design activities, presenting a space with authentic features of a design firm.

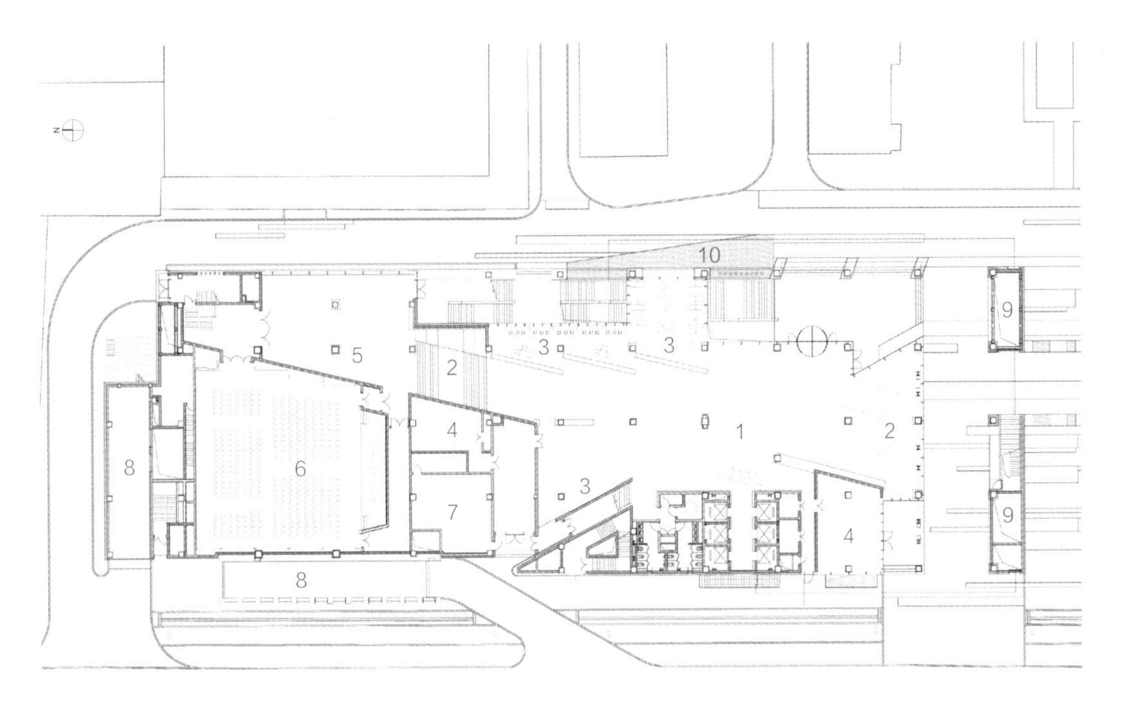

首层平面图

1. 门厅 2. 展厅 3. 休息区 4. 对外营业 5. 多功能厅前厅 6. 多功能厅 7. 安全消防控制 8. 汽车坡道 9. 通风井 10. 景观水池

南京青奥中心 Nanjing Youth Olympic Center

地点 江苏省南京市 / 用地面积 52,021m^2 / 建筑面积 481,079m^2 / 高度 315m / 设计时间 2011年 / 建成时间 2014年

方案设计 Zaha Hadid Architects
设计主持 汪 恒、任庆英、买友群
李燕云、赵丽虹、李 凌
叶 铮、刘 勤

建 筑 王 斌、任毅伟、陈 曦
杨 旭、宋 焱
结 构 刘文珽、茅卫兵
给 排 水 郭汝艳、杨东辉
石小飞、史嵘梅
设 备 孙淑萍、金 健
刘燕军、柏 文
电 气 李俊民、王为强
总 图 连 荔
经 济 祴新伦

合作设计 深圳华森建筑与工程设计顾问
有限公司

青奥中心位于南京青奥 CBD 轴线上，面朝江心洲岛，由会议中心区和双塔区两部分组成。整体设计方案延续了扎哈 · 哈迪德一贯曲线、流动的风格，建筑从屋顶到立面形成一体化的形式，简洁流畅，具有强烈的雕塑感。其中，会议中心区位于用地西北端，包含2000座大会议厅、500座音乐厅以及4个多功能厅。两栋超高层建筑与会议中心通过两处空中走廊连通。建筑的屋顶天窗、设备凹槽、侧向采光窗、通风百叶等均被统一在斜向菱形编制体系内。幕墙体系为双层一体化构造：内层为铝合金复合屋面（外墙）板，外层为干挂的具有自洁功能的 GRC 板。室内空间以流动的曲线、倾斜的墙壁、高吊的天花板和倾斜的柱子延续了整体的风格。

Located on the axis of Nanjing Youth Olympic CBD, the Youth Olympic Center consists of a conference center and twin towers. The conceptual design is typical of Zaha Hadid's style with its curves and fluidity, endowing the whole building with simplicity and sculptural beauty. The conference center consists of a 2,000-seat conference hall, a 500-seat concert hall and 4 multi-functional halls. The conference center and the super high-rise twin towers are connected with 2 skywalks. Various parts of the complex, such as the skylights, the equipment grooves, the side windows and ventilation louvers, are all integrated into a rhombus-shaped system.

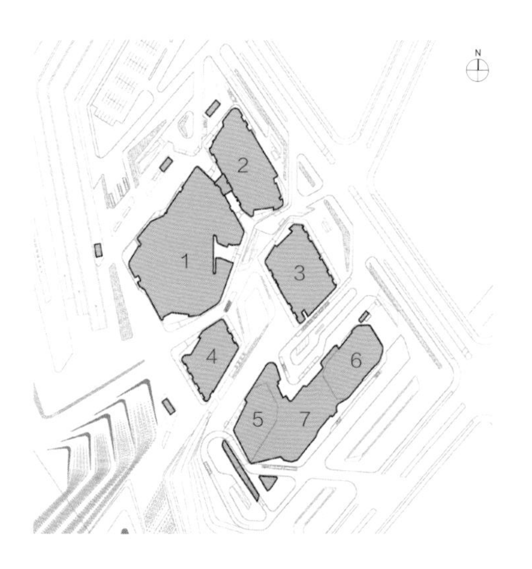

1. 会议厅
2. 多功能厅及展示厅
3. 音乐厅
4. 商业
5. 会议酒店
6. 办公及五星级酒店
7. 裙房

总平面图

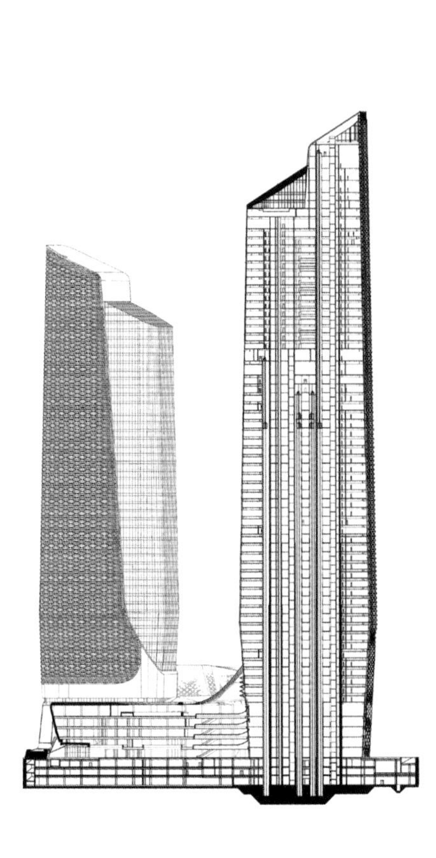

塔楼剖面图

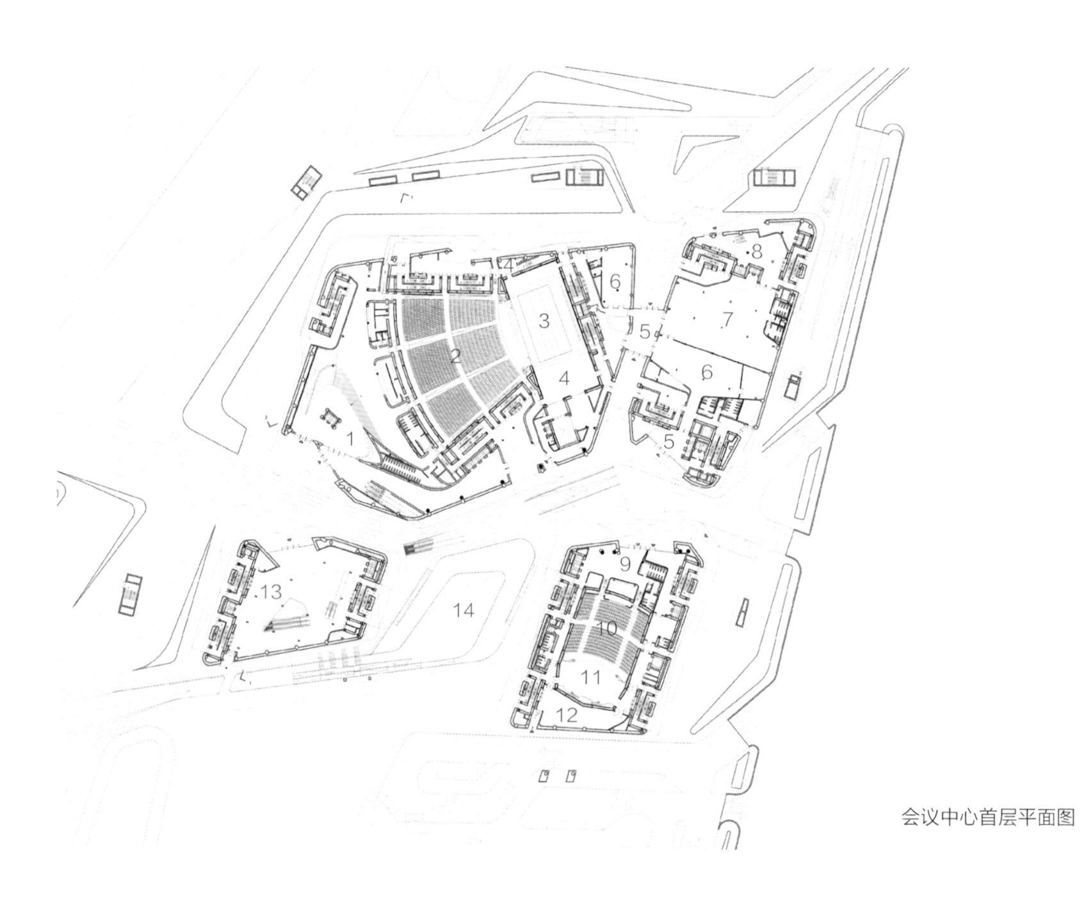

会议中心首层平面图

1. 会议厅前厅 2. 会议厅座席 3. 会议厅升降舞台 4. 会议厅侧舞台 5. 贵宾入口门厅 6. 贵宾休息室 7. 展览厅
8. 新闻发布前厅 9. 音乐厅前厅 10. 音乐厅座席 11. 音乐厅舞台区 12. 音乐厅后台 13. 商业 14. 中庭上空

北京绿地中心 Beijing Greenland Center

地点 北京市朝阳区 / 用地面积 19,770m² / 建筑面积 173,079m² / 高度 260m / 设计时间 2011年 / 建成时间 2016年

方案设计 SOM 国际建筑设计有限公司
设计主持 汪 恒、安 澎

建　　筑 孟海港、曲秉直
结　　构 范 重、彭 翼
　　　　 杨 开、杨 苏
给 排 水 王耀堂、王世豪
设　　备 刘玉春、李 莹
电　　气 曹 磊、高 洁
总　　图 高 治

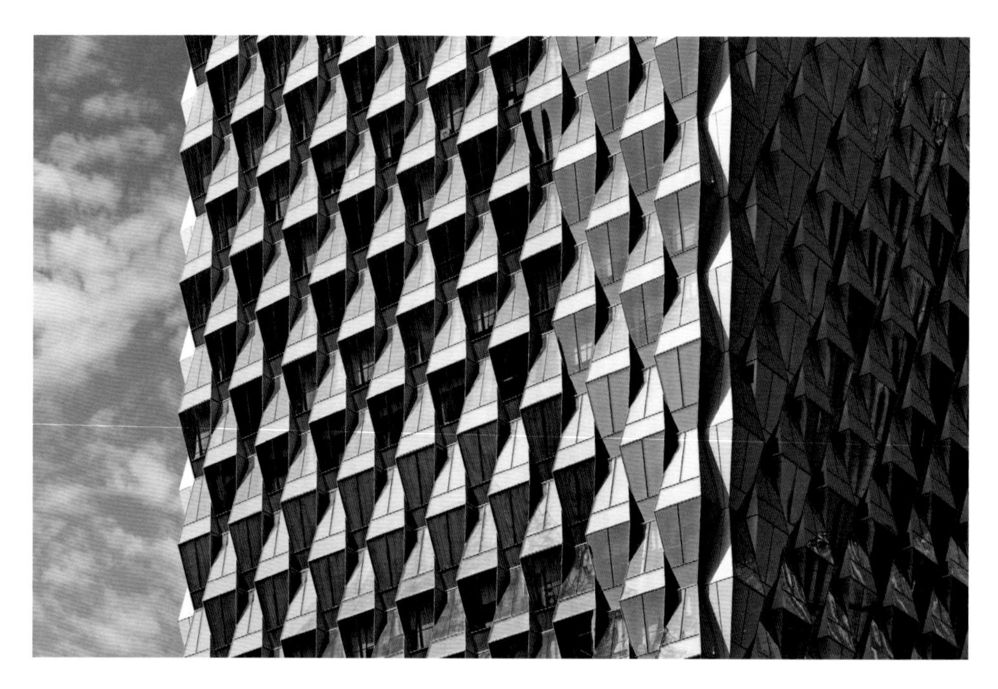

北京绿地中心地处大望京商务区中心位置，靠近机场高速路。设计以“中国锦”的形式为出发点，在外观上呈现出编织交错的肌理，特点突出，富有标志性。其塔楼和裙房分开布置在长条用地的两端。塔楼立面考虑到如何在白天和夜间创造出给人留下深刻印象的特征。模数化的玻璃幕墙单元在塔楼外立面上营造出波浪状此起彼伏的纹理，以不同的角度映衬周边的环境，为塔楼提供一种动态且不断变化的形象。在白天，精心设置的玻璃角度可以起到为室内遮挡阳光的作用，使之成为具备“自行遮阳”能力的超高层建筑。而到了晚上，这种微妙的变化又能折射出迷人的光芒，在夜幕下展现塔楼优雅的轮廓。商业裙房建筑外墙由竖向半透明玻璃面板组成，并结合大型标识和阳台创建了一个富有活力的零售环境。中庭将自然光引入购物中心室内，为举行活动和庆典提供了吸引人的场所。

Anchoring Beijing's burgeoning business district and nearing the airport expressway, Beijing Greenland Center is designed with the concept of "Chinese brocade", which endows the building with a special texture. The tower and the podium are located at opposite sides of the site. The tower is clad in an intricate glass curtain that creates a woven texture of light and shadow. The undulate glass modules provide self-shading that improves the building's performance in the day. At night, the intricate changes on the façade reflect lights from the street, showcasing the elegant profile of the tower. A retail building flanks the tower, featuring a central atrium and large, rectilinear, translucent-glass panels that fill the space with natural light.

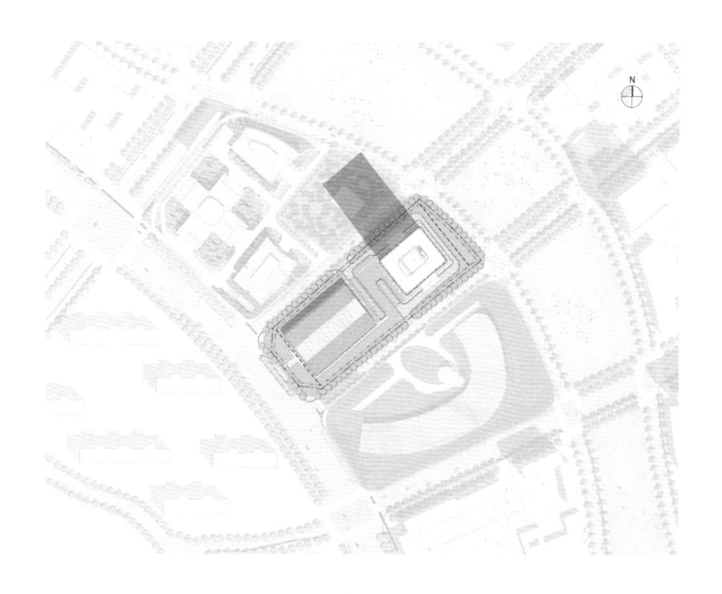

总平面图

绿地中心

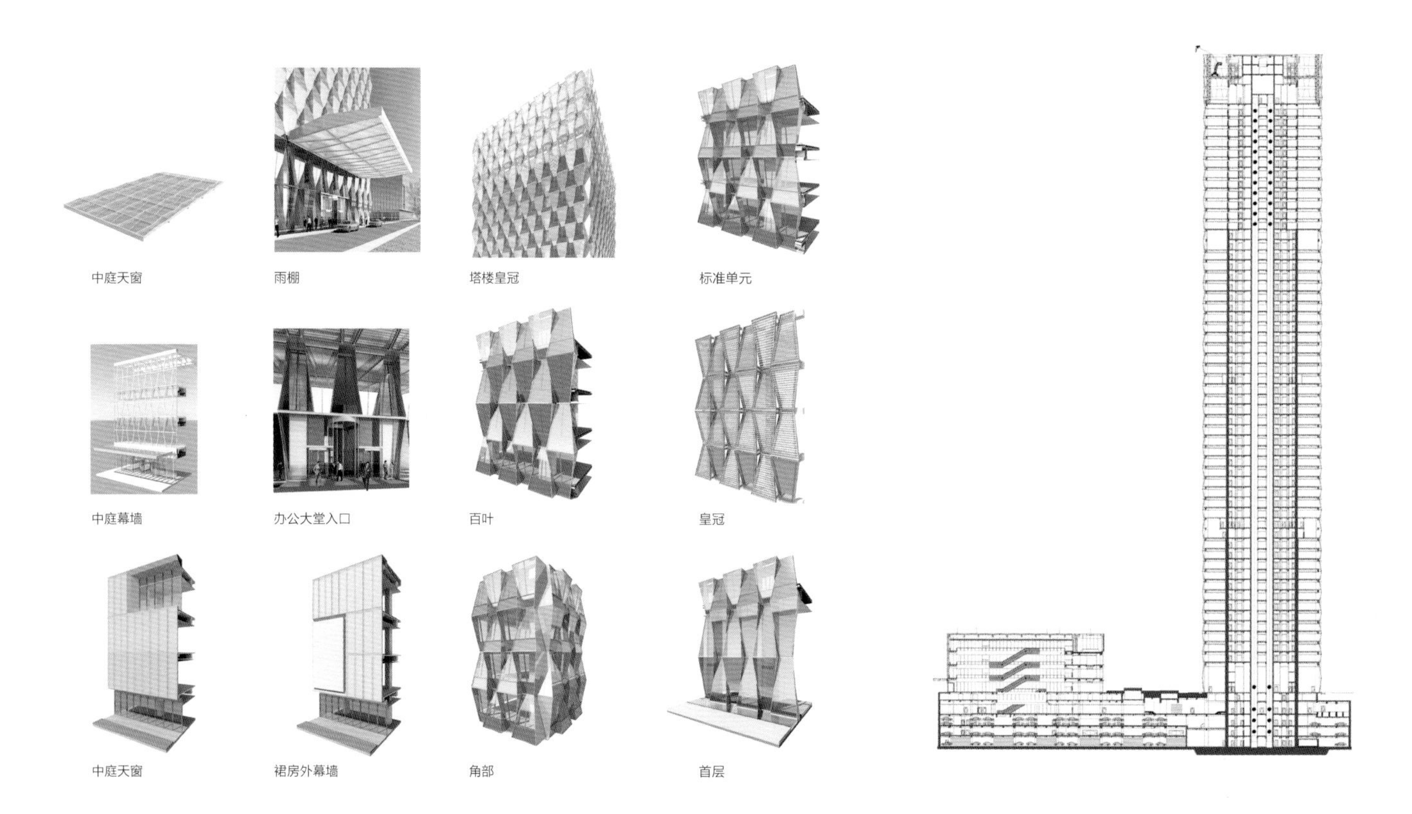

幕墙节点效果图

剖面图

招商银行深圳分行大厦 Shenzhen Branch of China Merchants Bank

地点 广东省深圳市 / 用地面积 7,594m^2 / 建筑面积 108,000m^2 / 高度 165m / 设计时间 2007年 / 建成时间 2017年

方案设计 崔　愷、于海为、杨益华
　　　　 周　宇、刘晏晏、韩　聪
设计主持 崔　愷、于海为、杨益华

建　　筑 刘晏晏、周　宇
结　　构 施　泓、史　杰
　　　　 鲍晨泳、马晓雷
给 排 水 黎　松、邢燕丽
设　　备 宋孝春、韦　航
电　　气 陈　琪、甄　毅
总　　图 白红卫
室　　内 刘　嵘
景　　观 赵文斌、刘　环
　　　　 颜玉璞、杨　陈
照　　明 北京宁之境照明设计有限公司

招商银行深圳分行大厦位于深圳市的金融中心，东邻深圳证券交易所（深交所）、西邻广电大厦。由于用地极为紧张且地面功能复杂，建筑紧贴红线布置，将底层敞开，形成十字形公共空间轴，让各功能体块相对独立。连接办公主塔楼和营业大厅的是正对深交所的城市客厅，空间通透明亮，并将室外景观延续至室内。办公主塔楼东、南两侧结合底层 M 形结构柱，遵循深圳城市设计导则，营造出骑楼空间，改善了建筑与城市街道间的关系，增加了公共性和亲和度。作为企业总部级办公大厦，建筑底部及中部的 M 形结构转换柱，分别在人行尺度及城市尺度层面与招商银行行徽呼应，在城市核心区营造标志性的建筑形象。建筑的照明设计同样以主体幕墙单元为基础展开，以单元网格均布灯光点阵，并以中部大型“M”标志为核心，渐远渐疏。而在“M”标志中，幕墙穿孔板透出的 LED 灯光则通过多级幻彩调控，向城市动态展示企业品牌。

As the site is compact and the functions are highly complex at the ground level, the building is designed along the red line, with its ground floor open to the outside to form an X-shaped axis, separating various functional sectors. A city lounge, going from west to east, introduces the landscape into the interior of the building. The east and south sides of the main tower, together with the M-shaped columns at the bottom of the building, form an arcade space connecting the building and the street. The M-shaped structural columns at the bottom and middle of the building resemble the logo of the bank, creating an iconic image in the central area of the city. The illumination design of the building is also based on the units of the curtain wall. Lighting dots have been aligned along the grids, with varied density determined by the dots' distance from the M-shaped logo.

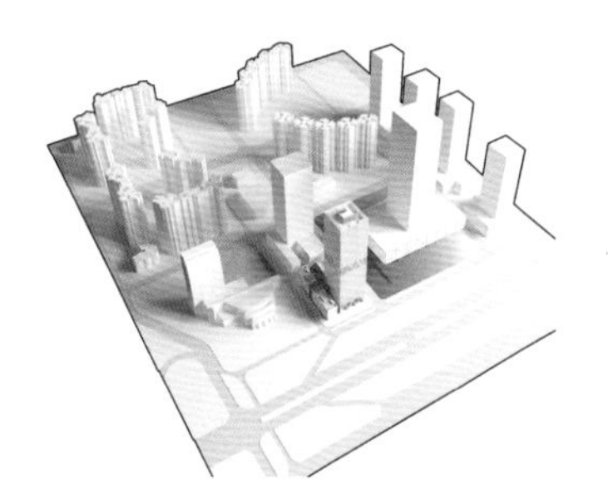

十字轴延续城市空间

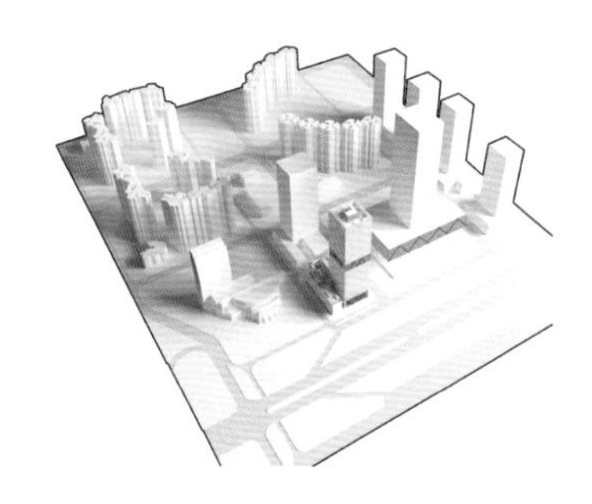

M形结构转换柱的标志性和与深交所的呼应关系

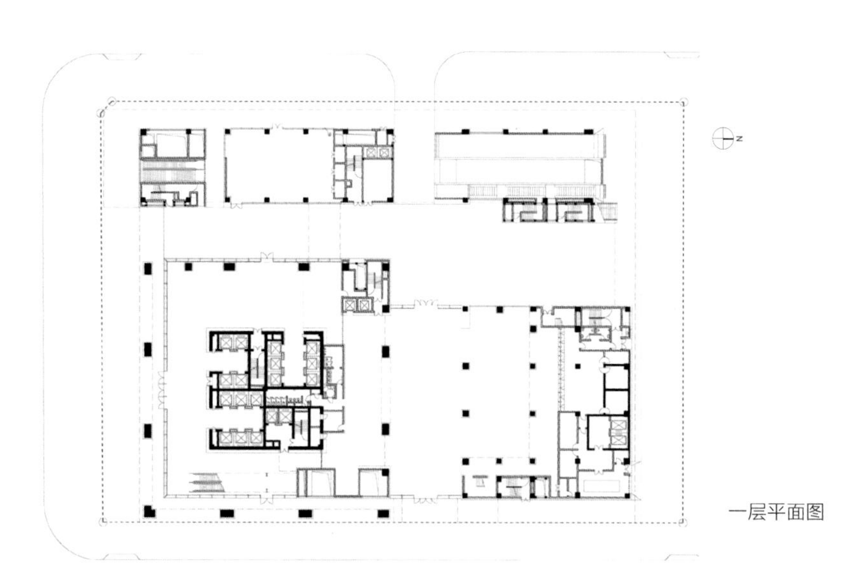

一层平面图

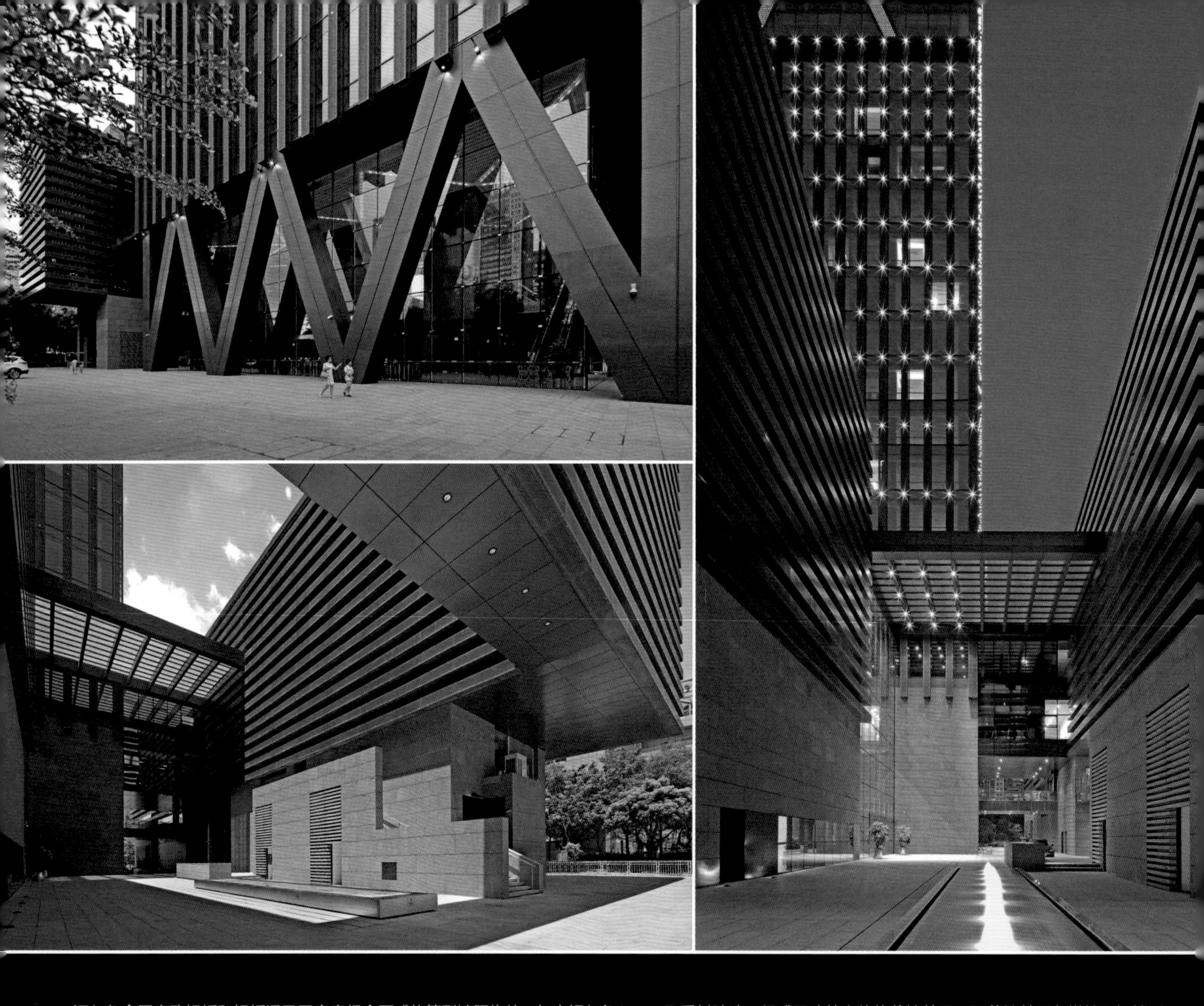

深灰色金属穿孔铝板和铝板通风开启扇组合而成的筒型遮阳构件，加之银灰色 Low-E 反射玻璃，组成了建筑主体的幕墙单元。而幕墙单元在塔楼和裙楼形成的垂直和水平金属线条，又简洁、明细地强化了建筑整体的体量关系。

Tube-shaped sunshade parts, together with silver Low-E reflective glass, form the curtain wall unit of the main part of the building.

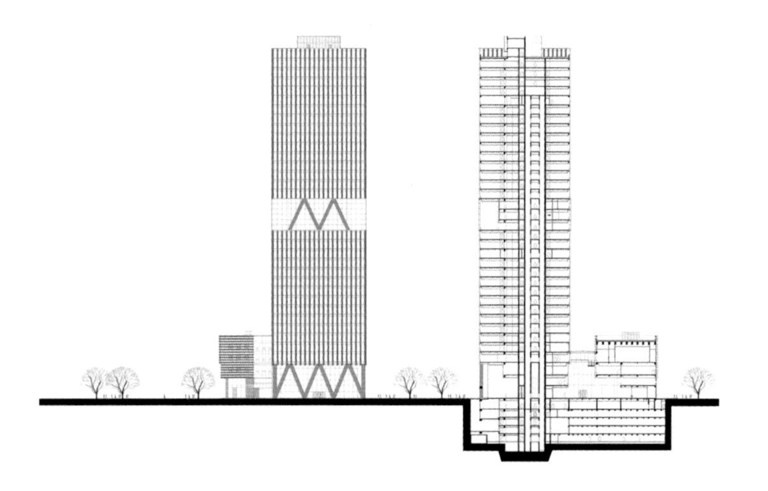

立面剖面图

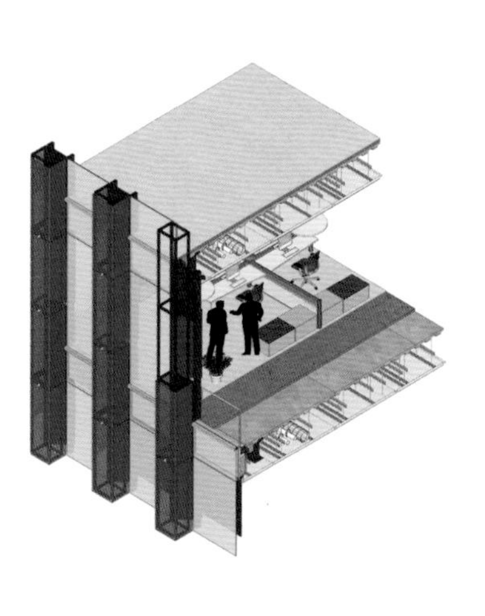

幕墙与室内管线的设计

招商銀行
CHINA MERCHANTS BANK

北京华都中心 Beijing Huadu Center

地点 北京市朝阳区 / 用地面积 27,000m² / 建筑面积 办公楼及酒店229,000m²、美术馆9,100m² / 高度 100m / 设计时间 2011年 / 建成时间 2018年

办公楼及酒店
方案设计 美国 KPF 事务所
设计主持 张 燕、肖晓丽
沈晓雷、胡水菁

建 筑 肖晓丽、沈晓雷、胡水菁
结 构 王 载、陈 巍、王文宇
给 排 水 王耀堂、王则慧
宋国清、刘永婵
设 备 李 莹、张亚立
电 气 李俊民、何 静
总 图 连 荔

美术馆
方案设计 安藤忠雄建筑事务所
设计主持 张 燕、沈晓雷

建 筑 杨承涵
结 构 王 载、王文宇、叶 垚
给 排 水 宋国清、刘永婵
设 备 张亚立、王 肃
电 气 李俊民、李建波
总 图 连 荔

合作设计 Thornton Tomasetti PB Enea

华都中心位于北京东三环的亮马河畔，是由两座商业办公楼、一座城市度假酒店（宝格丽酒店）和一处美术馆组成的城市综合体。项目所在用地面积紧张，且北侧为大面积居住区，设计从尊重城市的角度入手，避免对居住区的日照造成影响，并有效提升了该区域的城市空间品质，改善了滨河环境景观。

办公东西塔楼形体近似但略有不同，外轮廓形体均由多段弧线平滑相接构成，呈切削后的斜锥形，如同扬起的风帆，拥抱天空。东西塔楼之间为双曲面复杂空间体的中庭空间，其屋顶采用网壳结构，并将南北幕墙悬挂于网壳之上。位于双塔东侧的宝格丽酒店，平面近似于梯形，屋顶亦由北向南倾斜，以框架结构和石材立面为特征，与办公塔楼相互联系，又有所区别。三座塔楼的施工图阶段设计通过对机电转换、结构扭转、消防疏散、排雨水和融雪以及幕墙擦窗维护等方面问题的针对性处理，有效解决了塔楼屋顶的形体切削为结构和机电设计带来的种种技术难点。

美术馆入口设于北侧，通过长廊将人流导入位于河畔的建筑主体。其南立面采用双曲面内外双层清水混凝土，在均质的墙面上分布着不同尺度的锐角三角窗，成为建筑标志性的几何母题。室内空间延续了三角形的母题，若干三角形的中庭和展厅及过厅中三角锥形的空间，利用灰色的混凝土墙体与白色的方形展墙形成材质对比和透视错觉，使各个空间交融、流动起来。清水混凝土低调、内敛的特性，为藏品创造出一个纯净的空间舞台，使观者可以心无旁骛地漫步其中，充分感受藏品的艺术魅力。

Located along Liangma River by the 3rd Eastern Ring Road of Beijing, Huadu Center is an urban complex consisting of 2 commercial office buildings, a Bvlgari Hotel, and an art gallery. Built on a site with limited area bordering a large-scale community, the design has shown great respect for the city, effectively improving the quality of both urban space and landscape in the region.

With slight differences in the form, the eastern and western office towers are both composed of a series of curves, resembling sails facing toward the sky. A hyperboloid atrium space between the two towers is covered by reticulated shell structures with curtain walls hung on both sides. Bulgari Hotel to the east of the towers has a trapezoidal plan and a roof sloping from north to south, which is coherent with the office towers.

In the construction drawing phase, various targeted design approaches, such as MEP transformation and curtain wall maintenance, have solved the problems in structure and MEP brought by the special shape of the towers' roofs.

The gallery, with its entrance at the north of the complex, provides access to the main building along the river. The south façade of hyperboloid fair-faced concrete are dotted with triangular windows of varied sizes. Furthermore, gray concrete walls and white show walls form contrast while serving as a pure background for the exhibits.

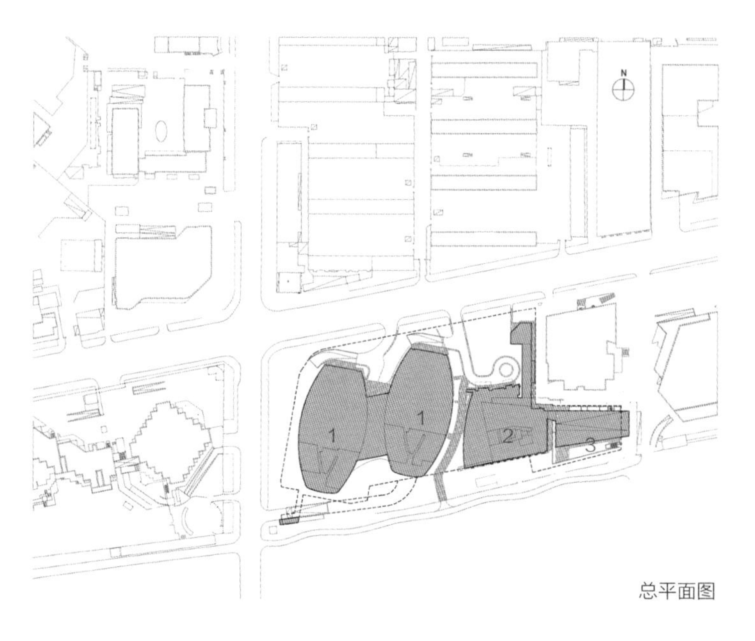

总平面图

1. 办公楼 2. 酒店 3. 美术馆

启皓
北京

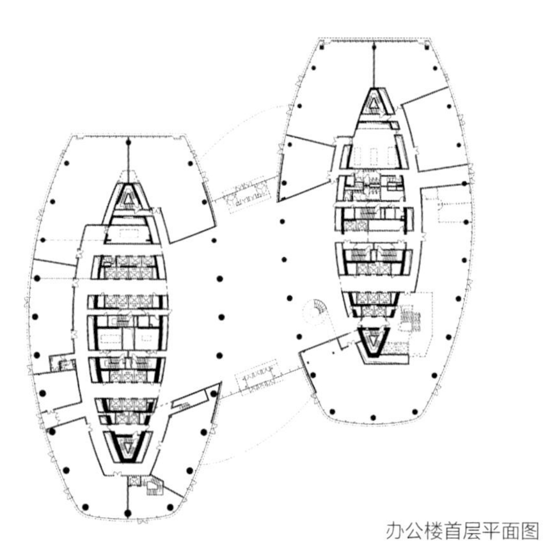

办公楼首层平面图

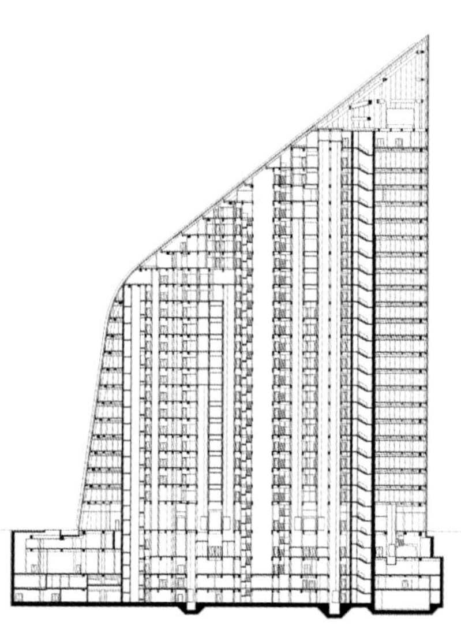

办公楼剖面图

BVLGARI
启皓
北京

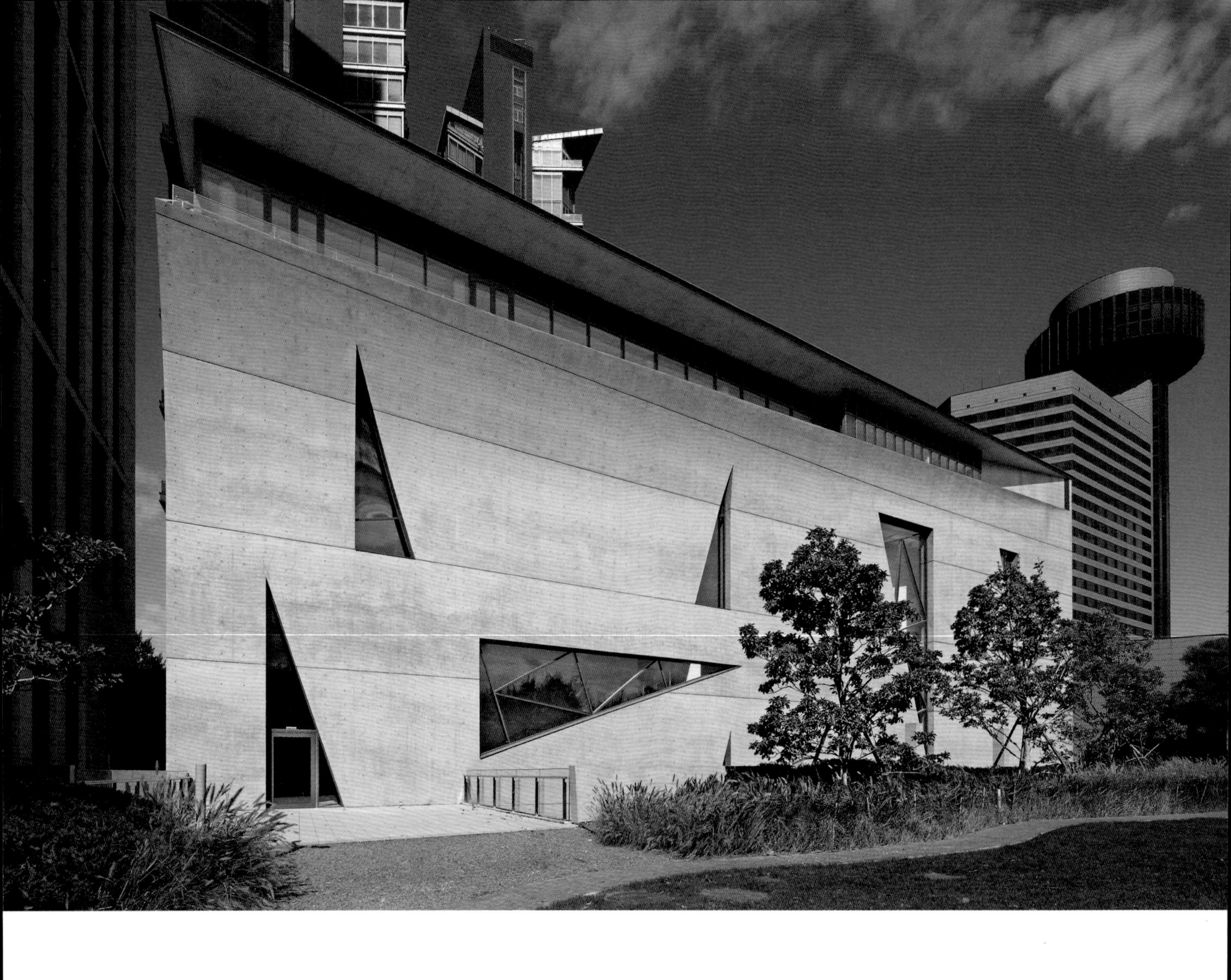

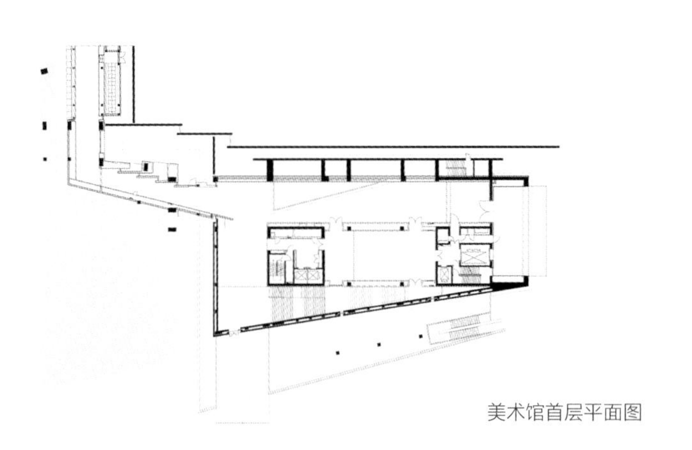

美术馆首层平面图

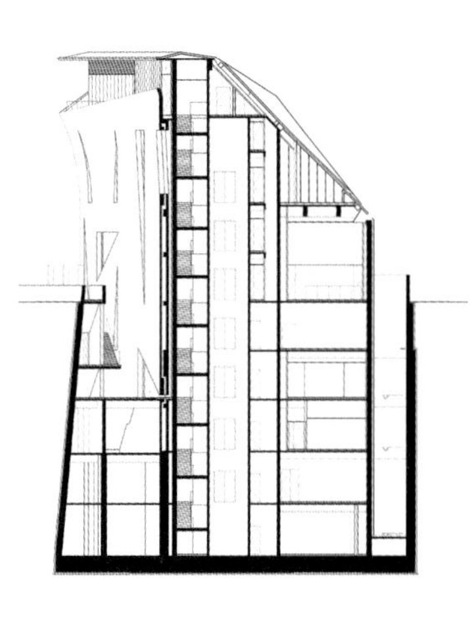

美术馆剖面图

北京通盈中心 Beijing Topwin Center

地点 北京市朝阳区 / 用地面积 11,584m² / 建筑面积 122,000m² / 高度 148m / 设计时间 2010年 / 建成时间 2015年

方案设计 美国 HOK 公司
设计主持 叶 铮、刘 勤

建　　筑 宋 焱、杨丽家、王 锁
结　　构 王 载、王文宇
给 排 水 赵 锂、陈 宁
设　　备 李雯筠、王 肃
电　　气 陈 琪、甄 毅
总　　图 吴耀懿
室　　内 CCD 香港郑中设计事务所

位于北京三里屯地区的通盈中心，是一处由五星级酒店、公寓及高端商业组成的超高层商业综合体建筑。其中的北京三里屯通盈洲际酒店是洲际酒店在北京的旗舰店，其时尚靓丽的外形已成为三里屯的新地标。项目用地为狭窄的长方形，设计在基地面积紧张的情况下，着重解决了酒店复杂的功能流线与周边城市肌理有机融合的问题，使建筑空间和城市空间有机结合，保持了空间的连续性。酒店塔楼外立面采用六角形铝合金型材及 Low-E 镀膜玻璃幕墙，形似蜂窝又模仿晶石的结构，结合不同颜色及透光率的玻璃，营造出质感丰富又不失光洁的立面效果，其极具雕塑感的体量从周边风格迥异的建筑中脱颖而出。

Beijing Topwin Center is a super high-rise complex with an InterContinental Hotel, luxury apartments and a boutique retail mall. The InterContinental Sanlitun distinguishes itself as a modern landmark with its stylish exterior. Challenged by a narrow site, the architects focused on the integration of the hotel's complex functions and the context of its neighborhood to achieve harmony between the building and the urban space. The hotel's façade features a hexagon-patterned steel structure integrated with LED lights that gives the tower a distinctive honeycombed texture.

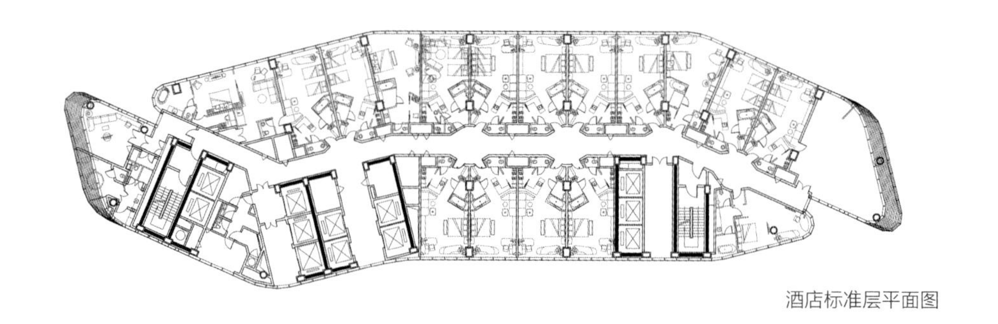

酒店标准层平面图

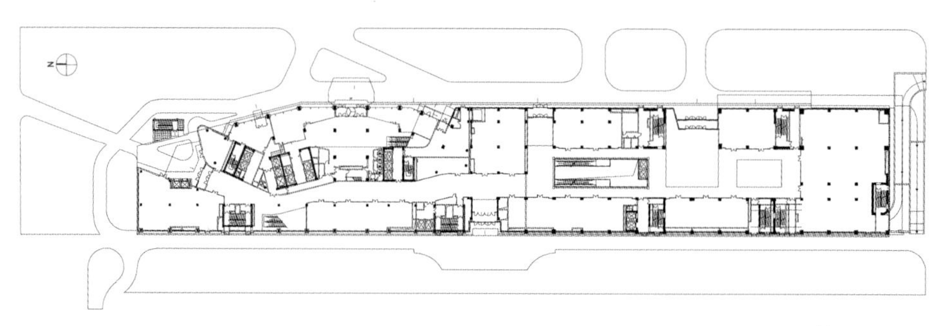

首层平面图

ROAD FC
3
ROAD FC
路德FC格斗健身酒吧 永利国际大厦B1
Fitness&spa

OPWIN CENTER 通盈中心
Gloria Jean's Coffees

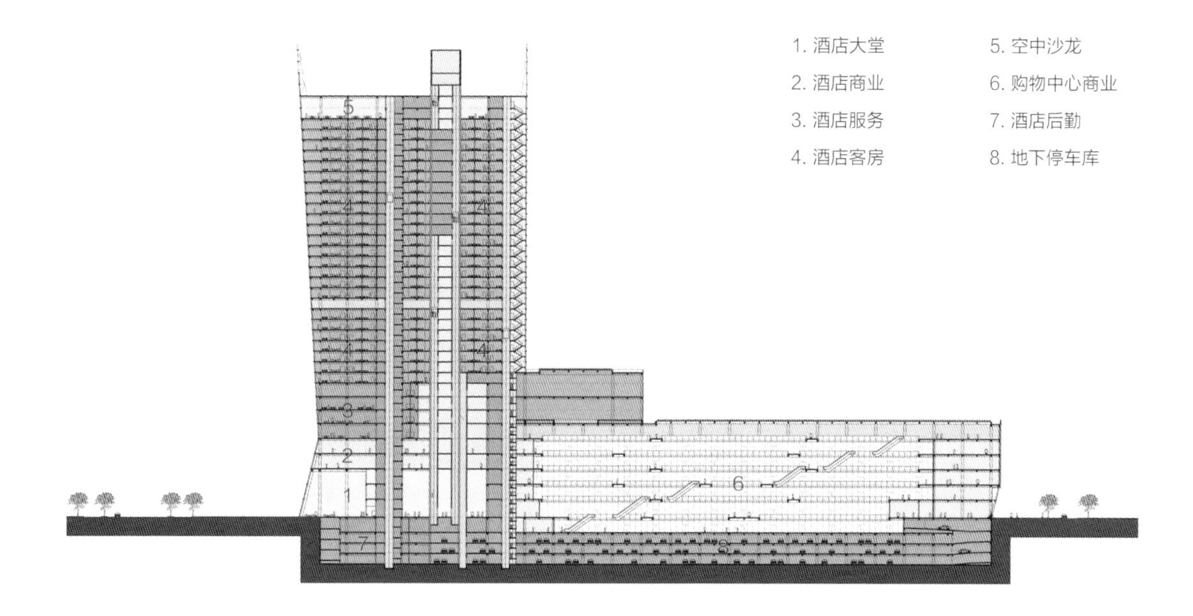
1. 酒店大堂
2. 酒店商业
3. 酒店服务
4. 酒店客房
5. 空中沙龙
6. 购物中心商业
7. 酒店后勤
8. 地下停车库

剖面图

北京浦项中心 Beijing POSCO Center

地点 北京市朝阳区 / 用地面积 20,021m² / 建筑面积 163,286m² / 高度 154m / 设计时间 2011年 / 建成时间 2015年

方案设计 POSCO A&C/Gansam
设计主持 逄国伟、陆 静、何 佳

建　　筑 何 佳
结　　构 张 猛、马玉虎、郭天焓
给 排 水 郭汝艳、石小飞
设　　备 孙淑萍、郑 坤
电　　气 王苏阳、李 磊、李天一
总　　图 高 治、高 伟

浦项中心以曲线形的裙房对城市空间形成半围合姿态，与周边建筑形成对话，并尽可能增加建筑的采光面。两座塔楼在高度和位置上产生一定的错落，丰富了城市天际线的节奏。裙房为三层高的采光中庭，宽阔的无柱空间内，造型独特的钢结构树成为中庭的焦点，既是玻璃顶棚的结构支撑，也是对浦项钢铁公司企业形象的展示。树木般流畅的造型同时将下沉广场、屋顶花园等不同高度上的公共空间联系起来。主楼的建筑表皮采用玻璃肋与弧形幕墙结合的形式，带有装饰感的钢构件进一步增加了表皮的层次，在严谨中透出细致微妙的光影变化。

Supported by a curved podium, the two towers of POSCO Center form a well-proportioned silhouette with varied heights and positions. In the column-free atrium with skylights, a unique steel structure becomes the focus of the space, which not only supports the glass roof, but also showcases the features of POSCO. The structure also serves as the connection among a series of public spaces. The glass ribs and steel components on the curved glass curtain wall have endowed the towers with a delicate atmosphere with the variations of light and shadow.

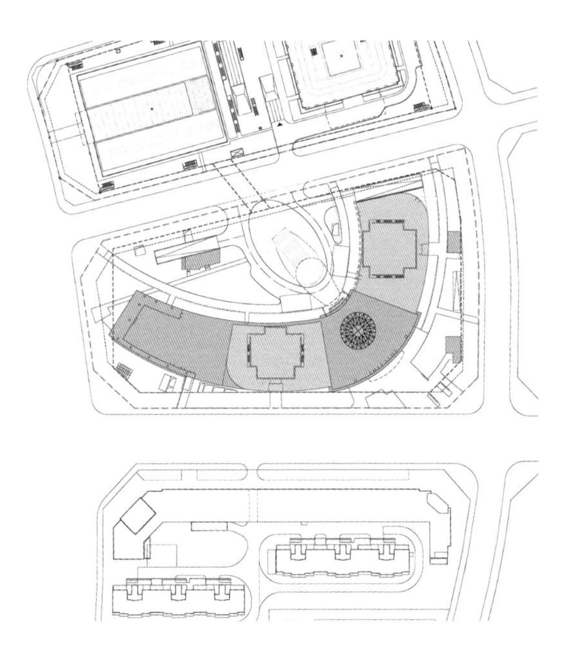

总平面图

posco
posco
posco

亦城财富中心 Yicheng Fortune Center

地点 北京市亦庄经济技术开发区 / 用地面积 33,712m^2 / 建筑面积 151,834m^2 / 高度 140m / 设计时间 2010年 / 建成时间 2016年

方案设计 汪 恒、曾宁燕
李 蕾、杨晓龙
设计主持 汪 恒、曾宁燕

建 筑 李 蕾、肖 婷
结 构 任庆英
给 排 水 夏树威
设 备 杨向红
电 气 贾京花
电 讯 凌 颉
总 图 连 荔
景 观 赵文斌

这一项目的布局从城市设计角度出发，以两栋标志性的商务办公塔楼为核心，与周边的其他建筑结合形成一组综合体建筑群，功能互补互利。低矮的裙房环绕塔楼布置，中间形成一个三角形平面的共享大厅。两座塔楼的造型具有强烈的虚实对比，顶部的屋顶花园采用半透明幕墙，加之造型上的上下收分，意图营造灯塔的意象。外立面材料为蓝绿、银灰亮色玻璃与咖啡色的石材，与周边建筑产生联系。三角形的共享大厅作为建筑群的枢纽，既是两座塔楼的入口大堂，沟通塔楼与裙房；也兼具新闻发布、展示、会客等多种功能，成为地区性的公共商务大厅。

The layout of the center focuses on two iconic commercial towers, forming a group of buildings complementing each other in functions. Low-heighted podium surrounds the tower, in the middle of which lies a lounge with a triangular plan. The two towers present a strong contrast between the solid and the transparent. The roof garden with translucent and battered curtain walls resembles the image of a lighthouse. Glasses with bluish green and silver colors, together with brown stones on the façade, are in accordance with the buildings in its neighborhood.

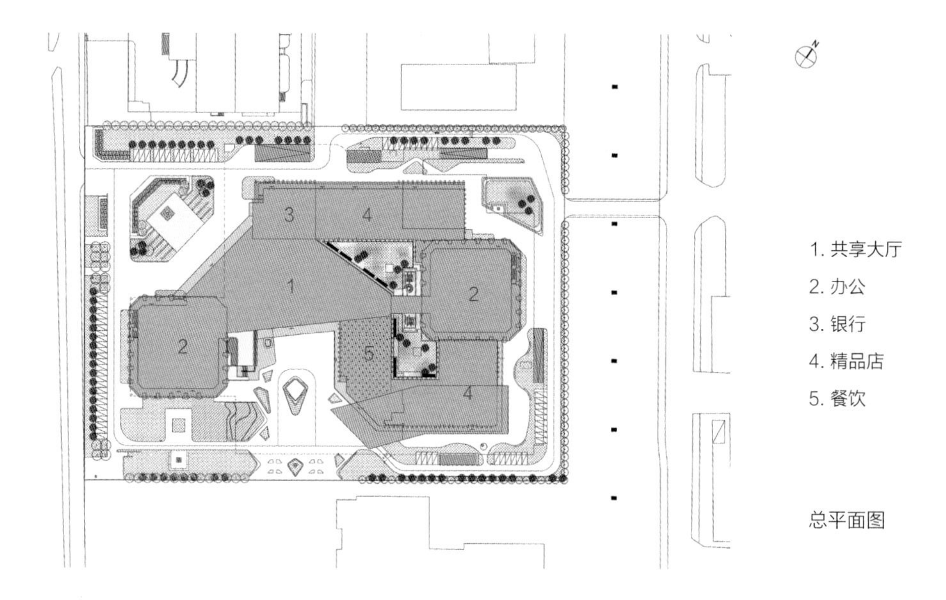

总平面图

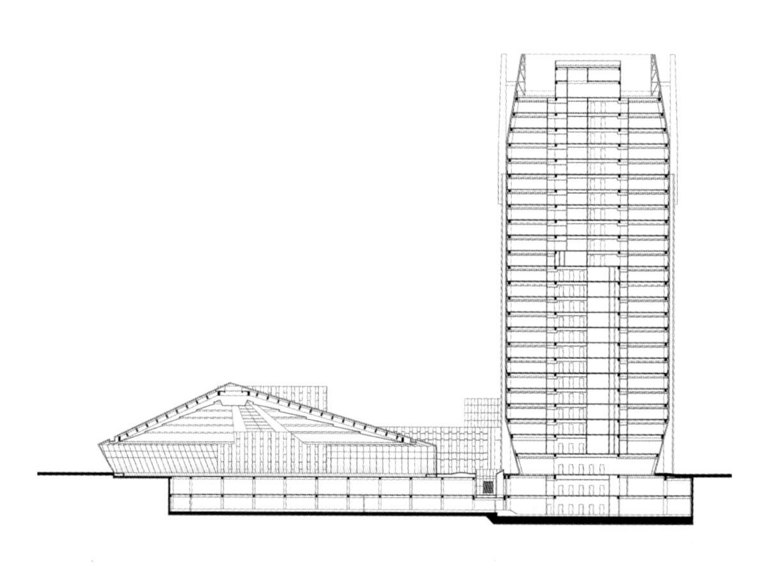

剖面图

亦城财富
亦城财富
互联网+创新园

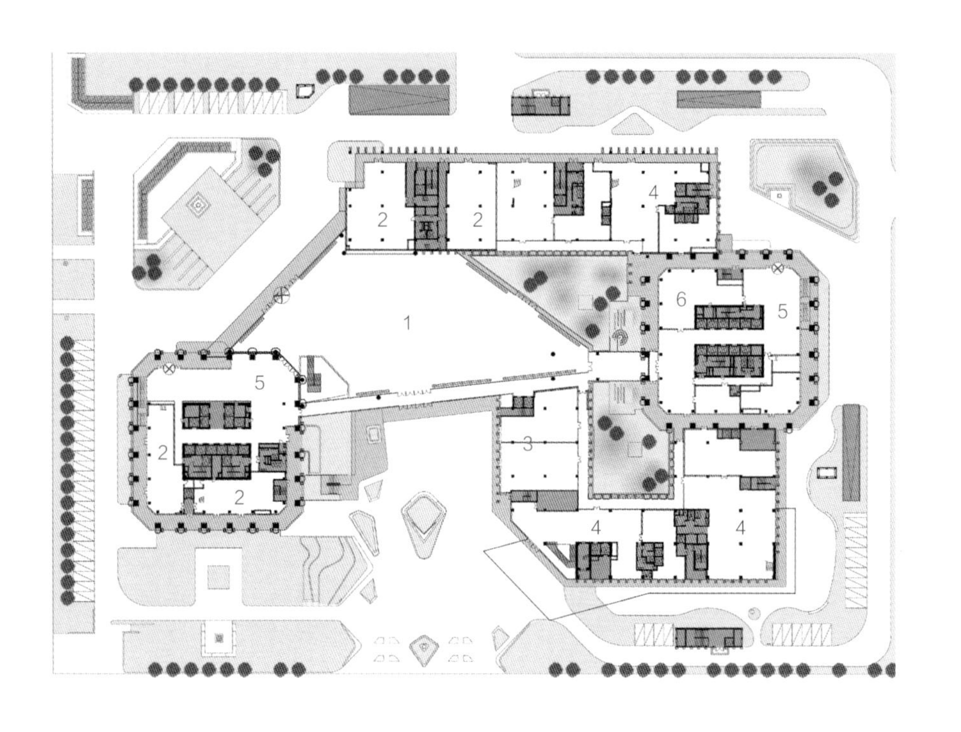

1. 共享大厅
2. 银行
3. 精品店
4. 餐厅
5. 办公楼大堂
6. 多功能厅

首层平面图

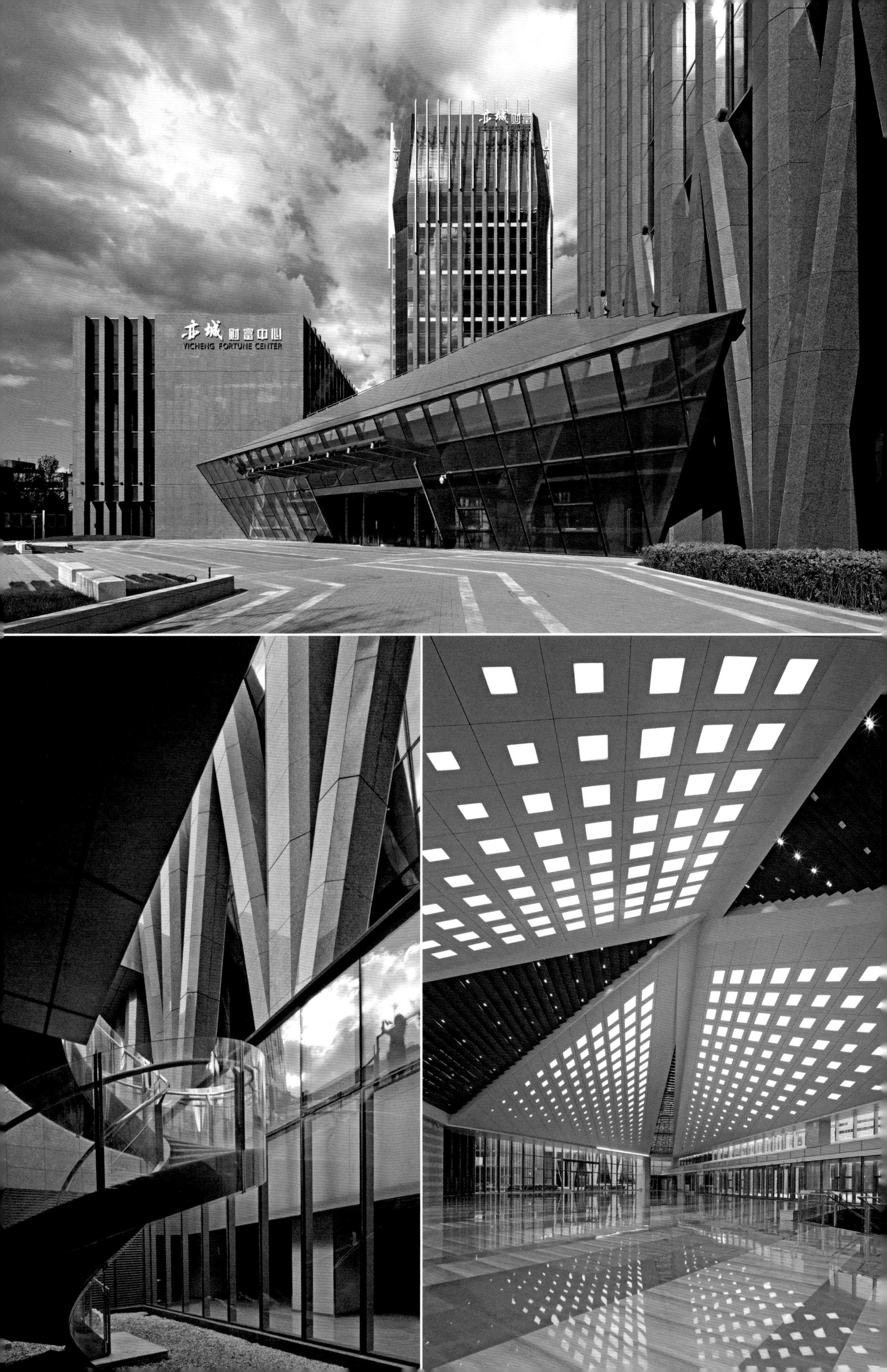
亦城财富中心
YICHENG FORTUNE CENTER

新华日报报业集团新闻传媒中心 News Media Center of Xinhua Daily Press Group

地点 江苏省南京市 / 建筑面积 136,382m² / 高度 156m / 设计时间 2009年 / 建成时间 2014年

方案设计 汪 恒、曾宁燕、李大丹
李 蕾、谭 瑶、陈 璐
杨晓龙
设计主持 汪 恒

建　　筑 曾宁燕、张 通
结　　构 尤天直
给 排 水 赵 锂、陈 宁
设　　备 孙淑萍
电　　气 贾京花
电　　讯 凌 颉
总　　图 连 荔

设计以园林化布局、水墨画格调的建筑群组来表达大江南的地域特征，体现"新华日报"集团的文化内涵。建筑群由三座呈品字形布局的塔楼及裙房组成，塔楼与中心广场和裙房屋顶的绿化组成的倒品字布局在空间上形成交叉，与错落的建筑形体和层次丰富的园林景观，共同营造出写意山水般的江南意境。主要入口面向二层的中心广场，广场同时也强化了主楼的中心地位。建筑色彩来自江南的白墙黛瓦，取意中国水墨画的意境，黑白灰的色调素朴而典雅，精致而简约。三栋主楼整体选用浅灰色 low-E 中空玻璃，外层为白色铝板遮阳翼，使建筑整体看起来有水墨画"润"的效果。两栋裙楼选用青黑色 low-E 中空玻璃、黑色磨光花岗石与黑色烧毛花岗石，整体色调深稳厚重。

The local features of Jiangnan region are revealed through the garden-styled layout and the ink-painting colors of the buildings, showcasing the cultural characteristics of Xinhua Daily Press Group. The towers and podium, together with the central square and the roof greening, create an atmosphere that reminds people of the natural beauty of South China. The main entrances of the buildings can be accessed from the central square on the 2nd floor, highlighting the dominating position of the main tower. Black, white and gray colors reveal a sense of elegance and simplicity. The three main towers are finished with light gray low-E insulating glass and white aluminum shading panels.

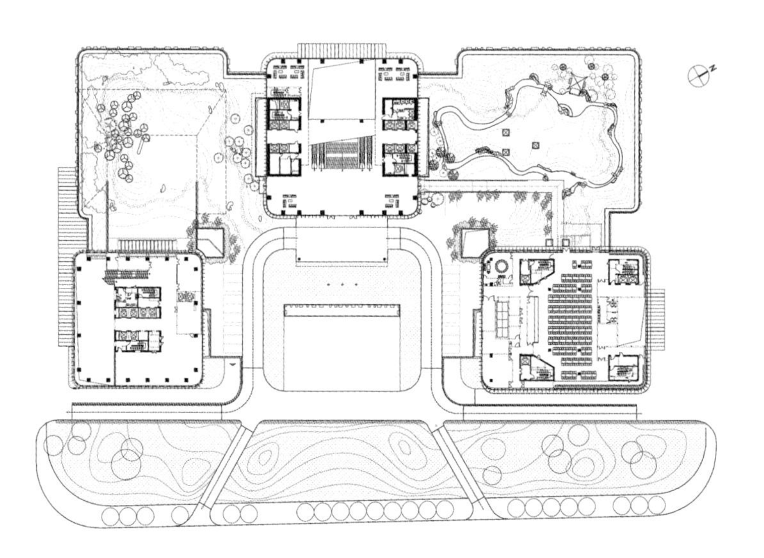

二层（入口层）平面图

建发大阅城一期 JF Metropolis, Phase I

地点 宁夏回族自治区银川市 / 用地面积 27,530m² / 建筑面积 313,476m² / 高度 99m / 设计时间 2013年 / 建成时间 2014年

方案设计 于海为、陈 宁、张玉明
韩 聪、季 欣、秦 筑
设计主持 陈 宁、张玉明

建　　筑 韩 聪、郭正同
结　　构 段永飞
给 排 水 匡 杰
设　　备 李雯筠
电　　气 贾京花
总　　图 邵守团

建发大阅城是银川阅海CBD周边的大型商业综合体，其一期包含综合商业、娱乐、演艺中心、办公、酒店、公寓等多种业态。其中办公塔楼及酒店板楼与商业裙房结合为一体，演艺中心为独立建筑，与二期的多层商业相夹形成基地内的商业街。室内外两条风格迥异的商业街首尾相连，结合室内外七处广场空间共同构成整个项目核心的流线构架。立面设计则借鉴了具有宁夏地方特色的黄河流水和贺兰山石意象。

As a large-scale commercial complex around the Yuehai CBD region in Yinchuan, the Phase I of the shopping plaza's development consists of various business types, including commerce, recreation, performance, offices, hotels and apartments. The office towers, the hotel and the commercial podium are connected, forming a commercial street together with an independent performance center.

1. 商业裙房 2. 办公楼 3. 酒店 4. 文化演艺中心 5. 底商停车综合楼 6. 汽车文化展厅 7. 创意办公 8. 公寓 9. 底商 10. 住宅

总平面图

HYATT PLACE
HYATT

凯悦嘉轩饭店
凯悦嘉寓饭店

室内商业街为其核心商街，自地下一层至地上六层整体贯通，水平联系酒店、办公的各个裙房及南端的商业出入口，沿街布置精品店和百货店，营造了开放、活跃的商业空间。室外商街则位于一期建筑和二期建筑之间，分别于出入口处形成景观广场和停车场，两侧设有沿街商铺。

The interior shopping street, lined with boutique stores and department stores, connects the hotel, the podium of office towers and the entrance of commercial spaces at the south, presenting a commercial space that is both open and vigorous.

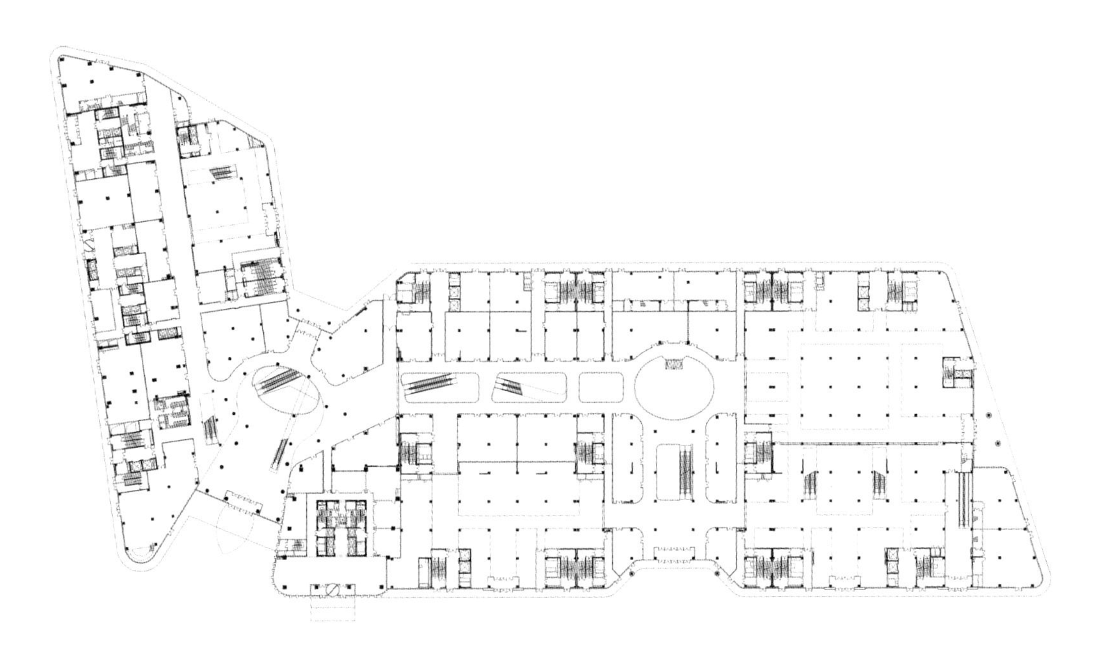

首层平面图

京西商务中心西区　West Plot of Jingxi Business Center

地点 北京市石景山区 / 用地面积 52,734m² / 建筑面积 368,740m² / 高度 115m / 设计时间 2014年 / 建成时间 2017年

方案设计 张　燕、李衣言、章　蔚
　　　　 杨承涵、龚子竹
设计主持 张　燕、李衣言

建　　筑 章　蔚、杨承涵
　　　　 龚子竹、付　婕
结　　构 王　载、王文宇
　　　　 叶　垚、陈　明
给 排 水 匡　杰、张源远
设　　备 李京沙、尹奎超、王　佳
电　　气 胡　桃、崔振辉、赵心亮
总　　图 连　荔
幕　　墙 上海尤安建筑设计公司

作为京西新首钢创意商务区的门户项目，这一复合型商业综合体由 9 栋高层塔楼、5 栋裙房组成，涵盖 5A 级商务总部、五星级酒店、办公、商业、餐饮等多种复杂业态功能。总体规划强化横向与纵深的空间序列组合，地块沿长安街展开的长度达 330m，通过均匀对称的形体获得总部办公楼整体稳重、严谨的形象。在纵向轴线上以多栋建筑围合形成回字形的商业空间，以下沉广场和连接通廊组成层次丰富的活力中心，向北辐射过渡为酒店和办公围合的安静的绿化庭院，形成富有张力和内涵的整体空间形象。作为全国首个获得绿色三星设计标识的大型办公、商业、酒店综合体，项目的设计和建设中应用了大量绿色建筑技术，包括太阳能光伏发电系统、屋顶绿化、地下室光导管采光、高热工性能围护结构、循环洗车台及水处理机房，以及其他节水、节能、节电技术措施。

As a major project of Shougang Innovation & Commerce District in West Beijing, the complex consists of 9 high-rise towers and 5 podiums, incorporating a 5A-level commercial headquarter, a five-star hotel, office buildings, commercial spaces and catering services. The site stretches 330 meters along Chang'an Street, and a symmetrical image of the office building presents prudent and rigorous feature. Rectangular commercial spaces are enclosed by various buildings, where a sunken square and corridors vigor and diversity for the center. As the first large-scale complex labeled as 3-star green building incorporating offices, commercial spaces and hotels, the project has adopted a variety of green technologies, such as photovoltaic power, roof greening, light pipe and other energy-saving methods.

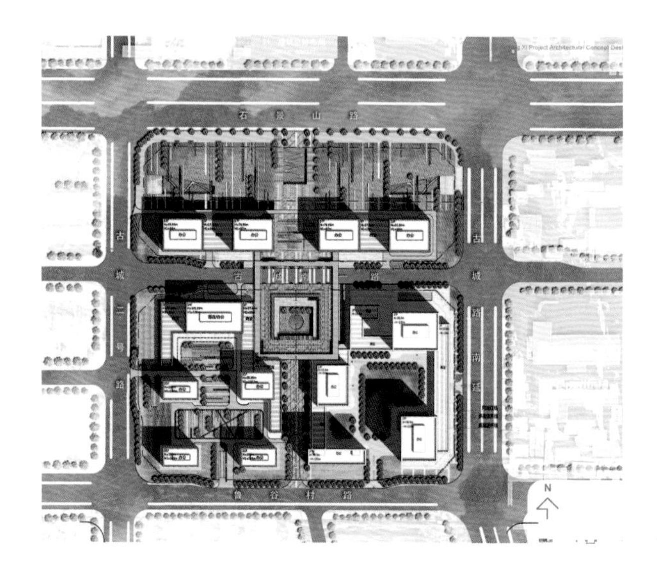

总平面图

北京诚盈中心 Beijing CCT Center

地点 北京市朝阳区 / 用地面积 27,300m² / 建筑面积 141,374m² / 高度 80m / 设计时间 2013年 / 建成时间 2016年

方案设计 柴培根、于海为
牛 涛、戴天行
设计主持 柴培根、于海为、周志红

建　　筑 牛 涛、李 颖
李柯纬、季 欣
结　　构 孙洪波、罗敏杰、刘会军
给 排 水 钱江峰、史建华
设　　备 刘 濒
电　　气 林 佳
总　　图 吴耀懿

北京诚盈中心坐落于一处南北向城市绿地上。设计概念即由绿意入手，尽可能将其引入办公区，形成坐落于“公园上的办公楼”。布局通过对消防车道的组织，将地块拆解为景观带和景观点，在重要节点布置绿地、下沉庭院、景观商业，7 栋办公楼则穿插于景观之上；围合的中心庭院成为公园绿地的延伸，且有利于形成建筑群的生态微气候。首层部分架空，既实现了空间上的通透，也形成了步行尺度上宜人的风环境。三栋高层办公楼沿主要城市道路布置，突出其标志性。外立面玻璃幕墙的细部设计是体现建筑品质的关键，地面两层采用明框做法，在人的尺度上形成丰富的层次；二层以上则为隐框做法以强化完整的几何体量。

Through introducing the green landscape into the office area of CCT Center, the design aims to create “office buildings on the park”. The carefully planned fire lanes have divided the site into several landscape stripes and nodes, where greenbelts, sunken courtyards and outdoor retailing spaces are situated. Seven office buildings are scattered on the site, enclosing a central courtyard as the extension of the park’s greening, facilitating the formation of the building cluster’s micro climate. The stilted first floor has created an open space with appropriate scale and easy access. Three high-rise office buildings line the main roads, standing out as landmarks.

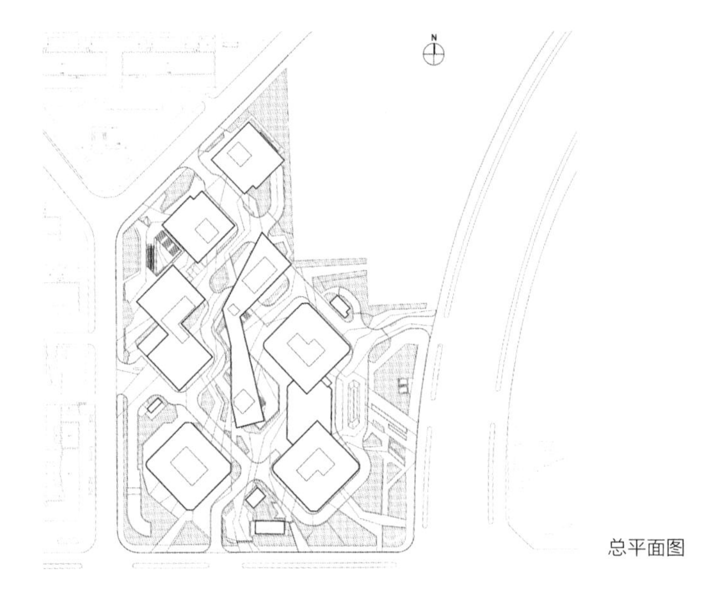

总平面图

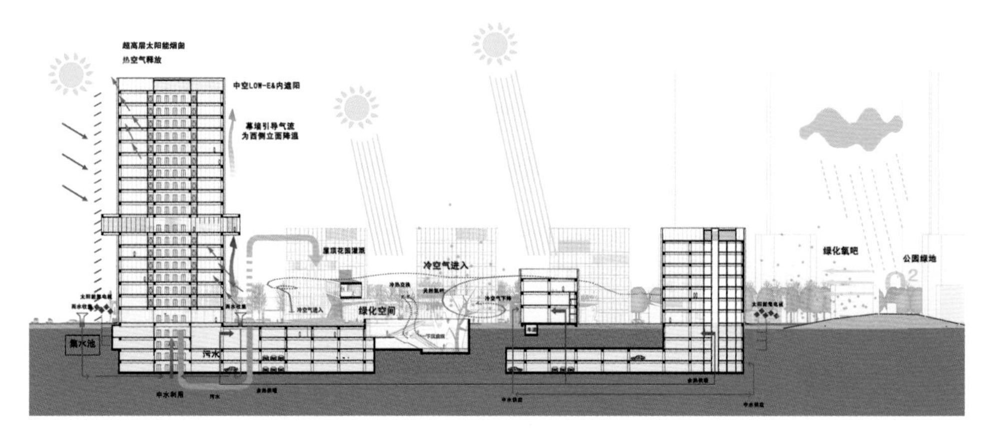

生态分析图

BURGER KING
富强 民主 文明 和谐 自由 平等
公正 法治
诚信 友善

北京冠捷大厦 Beijing Grand Yvic Building

地点 北京市朝阳区 / 用地面积 16,300m^2 / 建筑面积 64,100m^2 / 高度 80m / 设计时间 2011年 / 建成时间 2014年

方案设计 谷德庆、杨 华
设计主持 曾 雁、谷德庆

建　　筑 杨 华
结　　构 常林润、曾金盛
给 排 水 关 维
设　　备 张 昕
电　　气 王京生

冠捷大厦比邻北三环商务圈而建，建筑的体量简化到了极致。标准层最外侧柱间距为 9.3m，并向外悬挑 2.7m，通过合理的柱网排布，提高了室内空间的利用率和结构经济性。同时，核心筒布局采用完整墙体形式，将风井设在核心筒内，管井设置在交通核外，使得结构和功能更加合理。设计中融入被动式节能理念，降低全生命周期维护费。地下一层采用自然坡度的下沉庭院设计，将室外的阳光和空气直接引入地下餐厅，屏蔽了道路的喧闹和视线干扰。建筑幕墙采用有秩序感的微体型变化，尽可能实现丰富的效果。“鱼鳞”状的幕墙会因阳光的照射而在不同角度产生不同的效果，与近旁坝河的粼粼波光相得益彰。

Grand Yvic Building borders the business circle of the North 3rd Ring Road, and its volume has been simplified to the full extent. The standard floor plan has a column span of 9.3m with 2.7m of overhanging span, which has enhanced the economical and operational efficiency. The design of the service core has adopted a layout of “full walls”, where the ventilation shaft is located in the core while the tube shaft is outside the core. Passive energy-saving approaches have been applied in the aim of minimizing operational cost in the building’s whole life cycle. On the curtain wall, slight but orderly variations add to the diversity of the building’s image.

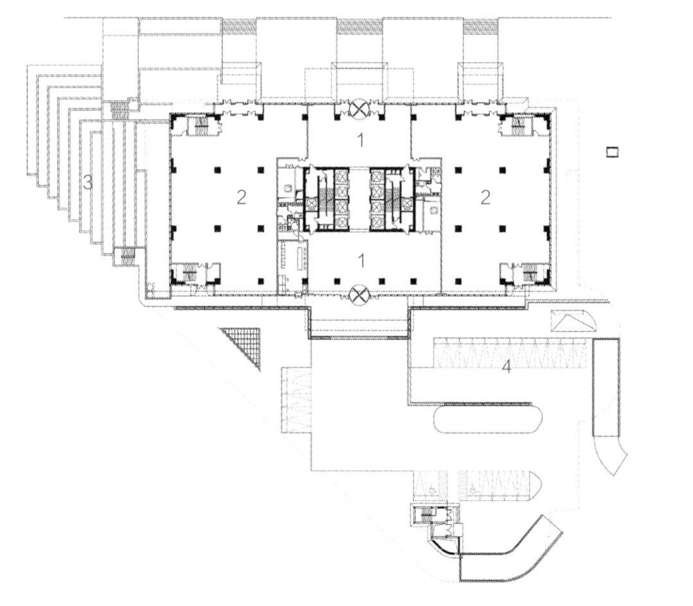

1. 入口大堂
2. 商业
3. 阶梯绿化
4. 停车位

首层平面图

冠捷大厦

北京专利大厦 Beijing Patent Office Building

地点 北京市丰台区 / 用地面积 37,076m² / 建筑面积 140,262m² / 高度 45m / 设计时间 2013年 / 建成时间 2017年

方案设计 崔海东、李东哲
徐 超、文 亮
设计指导 崔 愷
设计主持 崔海东、李东哲

建　　筑 徐 超、文 亮、李 曼
结　　构 张 猛、马玉虎、郭天焓
给 排 水 黎 松、唐致文、董新淼
设　　备 郑 坤、韩武松、徐俊杰
电　　气 李俊民、陈双燕、姜海鹏
总　　图 吴耀懿

作为专利技术研发和专利审查业务用房，并顺应规划条件，建筑呈外围规整、内部丰富的合院形式。通过丰富的空间变化，为以严谨为主的专利审核人员提供多样的环境。设计由此引入了“鲁班锁”的概念，这一传统木构榫卯的原型，以严密的逻辑，通过构件的穿插创造出多元形态。建筑的设计演绎恰如专利的发明创造，充满机巧和逻辑，繁衍出无限意趣。内院设置多处下沉庭院，为各类公共功能提供了舒适的环境，同时楼内设置的空中庭院则提供了短暂休憩的平台，共同组成了立体的园林体系。8.4m 的柱网模数转而生成 1.4m 的立面基础模数，所有部件均在模数体系下展开。外圈为竖向石材格栅组成的实体，实体之间是玻璃幕墙的“虚体”，虚实相间，形成立面上的榫卯。

Built for patent technology R&D and patent examination, the building adopts the form of a courtyard with regular outline and diversified interior spaces. The concept of “Luban lock” has been introduced into the design, where the prototype of traditional mortise and tenon work generates various shapes through combinations of components. The interior courtyard has several sunken parts, serving for various functions. Together with terraces for short stay, they have formed a multi-layered garden system. On the façade, the solid part composed of vertical stone grids and the transparent part of glazed curtain walls complement each other, forming a type of “mortise and tenon”.

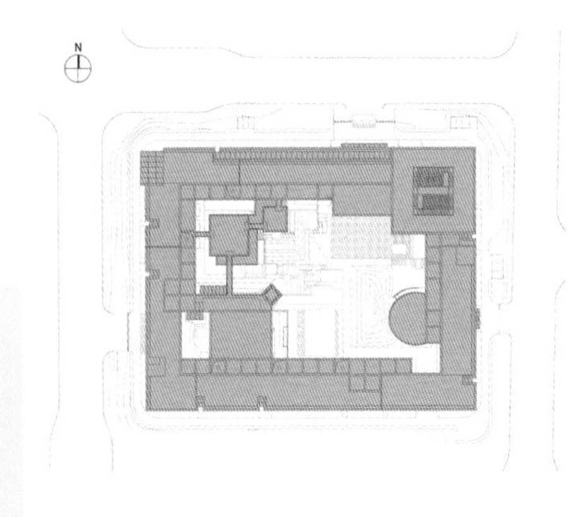

总平面图

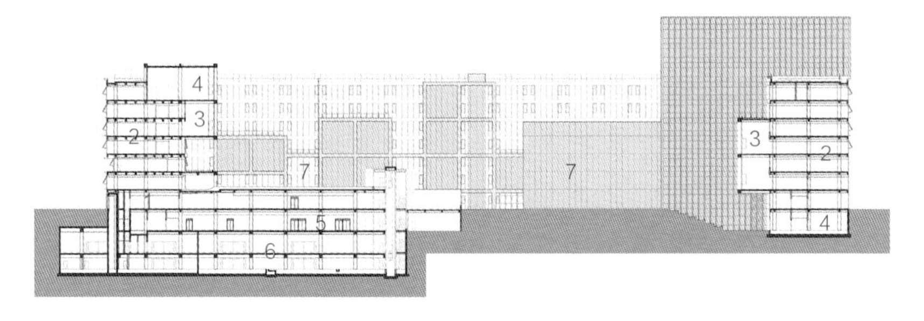

1. 入口门厅 2. 业务用房 3. 共享空间 4. 会议室 5. 食堂前厅 6. 地下车库 7. 室外庭院

剖面图

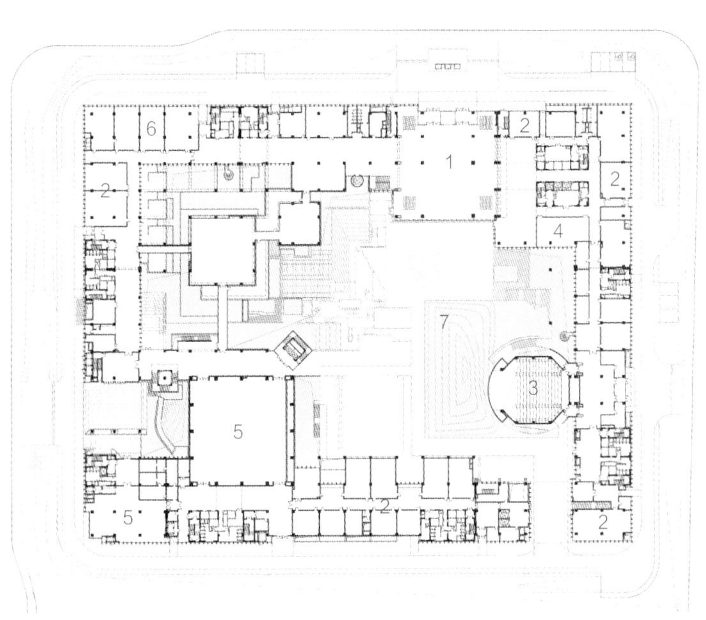

1. 入口门厅
2. 业务用房
3. 大报告厅
4. 会议室
5. 多功能厅
6. 培训教室
7. 室外庭院

首层平面图

北京中关村第三小学 Beijing Zhongguancun No.3 Primary School

地点 北京市海淀区 / 用地面积 23,500m^2 / 建筑面积 45,728m^2 / 高度 22m / 设计时间 2012年 / 建成时间 2015年

方案设计 美国 Bridge3 建筑事务所
设计主持 刘燕辉、王敬先

建　　筑 徐　超、覃　沁
结　　构 余　蕾、罗敏杰
给 排 水 匡　杰、陈　静
设　　备 马　豫
电　　气 孙海龙、史　敏
电　　讯 陈玲玲
总　　图 王　炜

中关村第三小学的新校区设计，以取自传统土楼的形式和高度复合化的功能组织，解决了用地紧张、需求众多的难题。饱满的弧形教学楼将操场包围在中央，其外围墙面厚实，确保安全；内侧玻璃通透，确保视线通达，以利师生交流。建筑的主要开口朝向南侧，为主楼争取了更多的采光面，也阻挡了冬季的北风，并利用遮阳、通风等策略增加对自然气候的适应。在位于地下室的下沉庭院周边布置了排练厅等需要采光的教室，周边安排游泳池、餐厅、教师餐厅、厨房等功能。

With its semi-enclosed layout derived from Tulou, a type of Chinese vernacular building, the project responds to complicated requirements and creates a nurturing and protective environment. A dynamic courtyard with a gymnasium can accommodate recreation and gatherings. The innovative "school within a school" design divides the building into smaller units to give a greater sense of identity and pride of ownership among students, faculty, parents and the community.

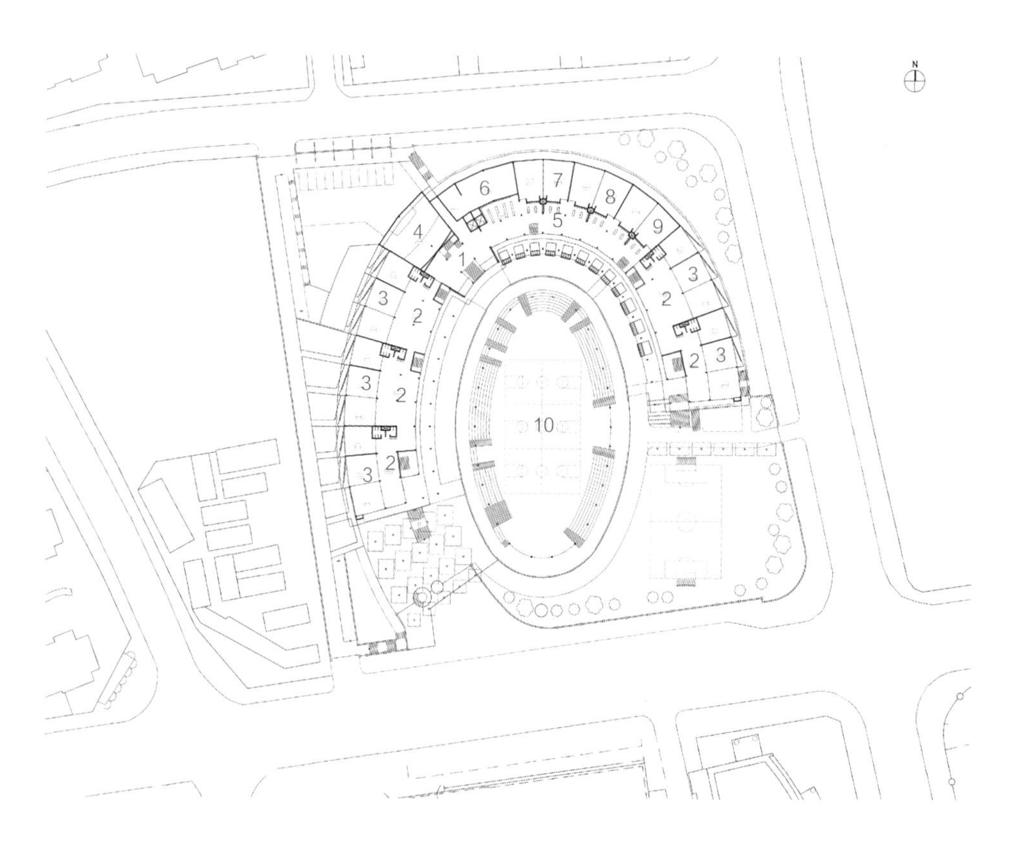

1. 门厅 2. 开放教室 3. 教室 4. 会堂 5. 读书廊 6. 教师办公室 7. 网络机房 8. 美术教室 9. 音乐教室 10. 风雨操场

首层平面图

室外操场位于风雨操场屋顶，因此在屋面钢结构架跨中设置了阻尼器，以满足在屋顶活动时的安全和舒适性需要。二到四层的外廊采用外挑楼板，为阳台绿化种植预留了条件。

An outdoor playground is planned on the roof, with dampers placed in the steel structure of the roof. Overhanging floor slabs are applied to the exterior corridors of 2-4 floors.

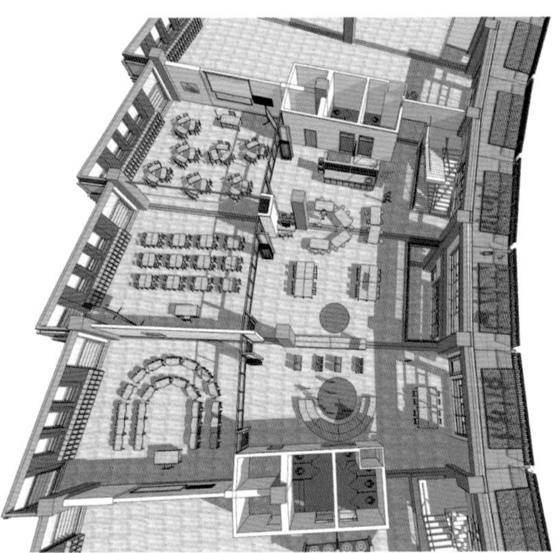

教室布局采用组团形式，每三个教室共用一个开放教室，形成班组群，通过灵活隔断实现教学空间的多样变化，可满足教学指导、集体讨论、自习、活动等多种功能需求。

A "learning pod" consists of three classrooms sharing an open space. Flexible partitions have offered various solutions to modern teaching activities. The arrangement creates opportunities to balance teacher-directed, whole-group instruction with flexible, learner-centered work and study spaces.

教学空间组合示意图

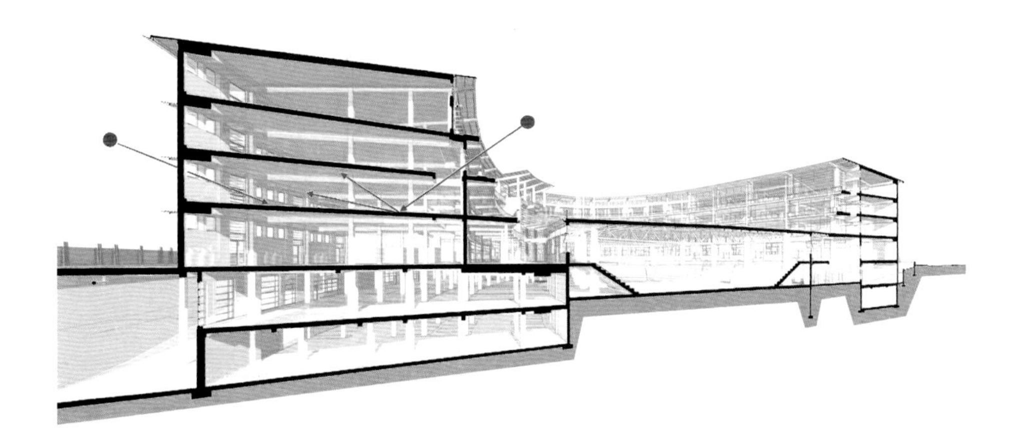

剖透视图

北京法国国际学校 French International School of Beijing

地点 北京市朝阳区 / 用地面积 38,000m² / 建筑面积 20,000m² / 高度 14m / 设计时间 2009年 / 建成时间 2016年

方案设计 法国 Jacques Ferrier Architecture
设计主持 马 琴、宋 焱

建　　筑 杨丽家
结　　构 王文宇、陈 明
给 排 水 王耀堂、王则慧
设　　备 李 莹、向 波
电　　气 曹 磊、刘征峥
总　　图 高 治、高 伟
室　　内 魏 黎

法国国际学校的校址曾经是一片果园，春天桃红柳绿，秋天硕果累累。设计希望新建的学校仍然能够承载场所的记忆，并将建筑和景观融合在一起。教学空间集中在场地东侧，在规则的矩形轮廓内创造了三个相对独立又有联系的三角形庭院，每个庭院都有一条边向校园敞开，形成了各自入口。整个建筑的外围被连续的、有疏密变化的木格栅包裹，形成了宁静且识别性很强的外观，兼顾采光和遮阳，也为室内活动提供了私密保护。

The design aims to carry forward the memory of the site and integrate both the buildings and the landscape. Classrooms are mainly located in the east of the site, forming three triangular courtyards. Each of the three courtyards, both independent from and interconnected to each other, has one side open to the campus as its entrance. Wooden gratings on the façade has formed a quiet and iconic look while satisfying both daylighting and shading requirements.

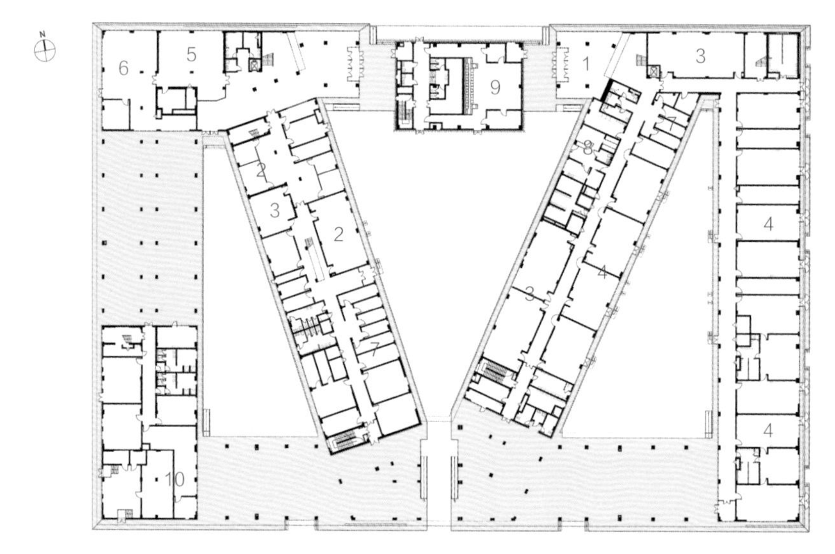

1. 门厅
2. 学习室
3. 活动室
4. 学前班教室
5. 资料室
6. 多媒体中心
7. 教师办公室
8. 医务室
9. 多功能厅
10. 设备用房

教学楼首层平面图

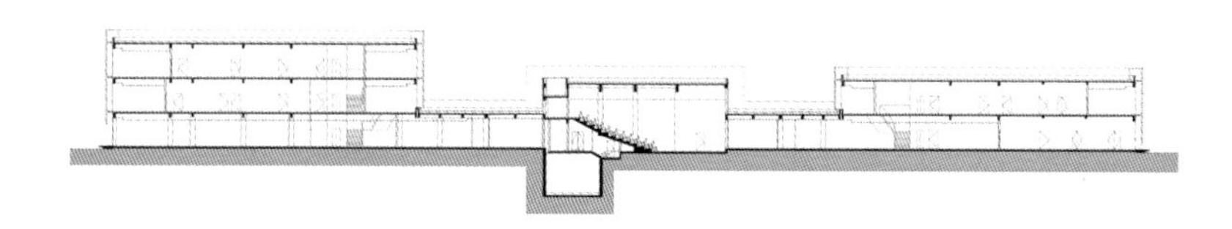

教学楼剖面图

文化共享

木格栅以全新的现代构造方式阐释传统中国花格窗的意象。选择天然耐腐抗虫的户外木材红雪松制成标准尺寸木砖，以钢管穿过其中加以固定，梯形截面的木砖有利于解决清洁和排水问题。

The wooden grids' design has been inspired by windows of traditional Chinese buildings and implemented in a modern way, where standardized wooden bricks are made of insect and corrosion-resistant red cedar.

北京三十五中学高中部 Senior Department of Beijing No.35 Middle School

地点 北京市西城区 / 用地面积 46,177m² / 建筑面积 61,772m² / 高度 16m / 设计时间 2008年 / 建成时间 2015年

方案设计 崔 愷、邓 烨、罗 荃、黄 琳、关 飞、黄文韬、梁洲瑞

设计主持 崔 愷、邓 烨
建 筑 邓 烨、罗 荃
结 构 郝国龙、程颖杰、李 谦
给排水 商 诚、董新淼
设 备 郭 然、李冬冬、郭丝雨
电 气 王苏阳、甄 璐、裴元杰
总 图 齐海娟
室 内 郭晓明、张栋栋、曹 阳、郭 林

合作设计 北京市古建研究所、北京建工建筑设计研究院、北京创新景观园林设计公司

校园处于城市新老过渡的区域，设计将原场地中保留下来的鲁迅家族旧居、前公用胡同古建院落、八道湾胡同，以及复建的志成楼与整体校园格局融合在一起，成为校园的文化核心与历史脉络。建筑主要采用现代中式风格，外界面方正平直，内部空间开放活跃，多层院落式的布局延续了北京旧城的城市肌理。胡同、长廊、庭院为学校和师生提供了丰富的展示和交流空间，形成了园林式的教学环境。校园内设置了不同类型、规模的下沉庭院和采光中庭，为学生使用的区域提供了良好的自然采光和通风。配合学校“五制”教育教学改革和培养高素质、创新型人才，学校建设了金帆音乐厅、中科院高端实验室等创新型内容。尊重历史传承、挖掘场地价值、提升教学理念，造就了校园的多样性和丰富性，体现了学校的人文精神。

The campus is located in a transitional region between the new and old areas of the city. The design has integrated the whole campus with existing structures on site, endowing the campus with unique historical context. The design of the campus creates a rigid enclosure and a lively interior space. Furthermore, the multi-layered courtyard style for the layout has inherited the traditional context of the old city of Beijing, with alleys, corridors and courtyards as diversified spaces for exhibition and communication. In conclusion, the efforts in the design, featuring respect for the history, discovery of the site's value and the improvement of teaching theories have endowed the campus with diversity and the spirit of humanism.

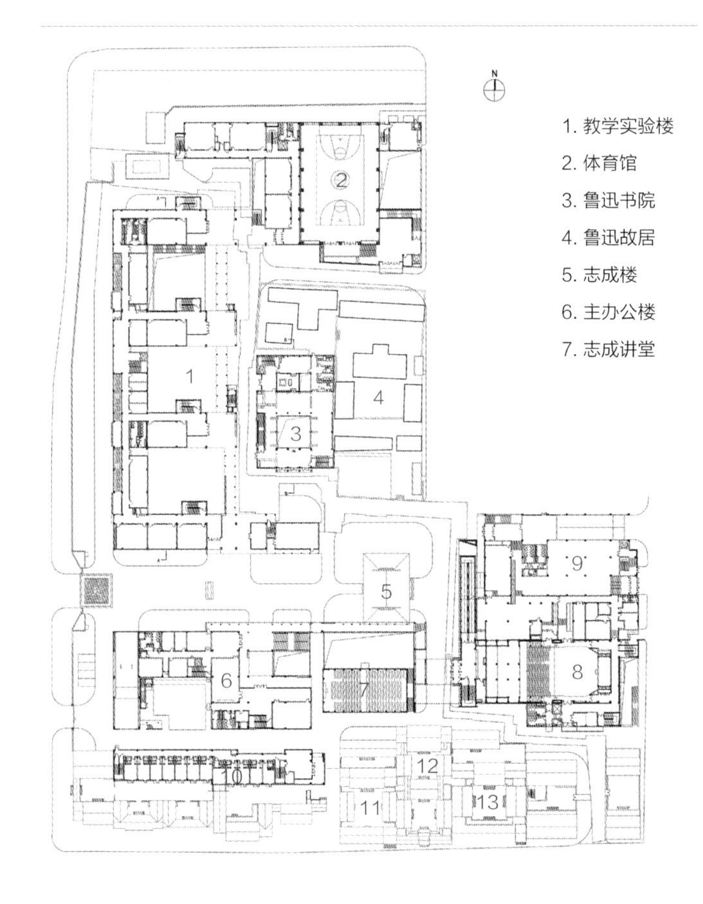

首层平面图

魯迅書院

教学设施具有相当的前瞻性，设有 800 座音乐厅，引入 7 个中科院实验室，并利用地下空间设置了篮球馆、游泳馆、学生食堂、阶梯教室、图书馆、自行车库及地下车库等场所。浅灰色的面砖、深色的坡屋顶、局部红色的装饰，以及修复的古建院落，使校园与周边的胡同风貌充分融合在一起。

The school is equipped with forward-looking facilities, such as a 800-seat concert and 7 laboratories of Chinese Academy of Sciences. Underground spaces are used for basketball hall, natatorium, canteen, etc. Light gray bricks, dark-colored sloped roofs, partially red decorations and the renovated courtyards are all elements that achieve harmony with the neighborhood.

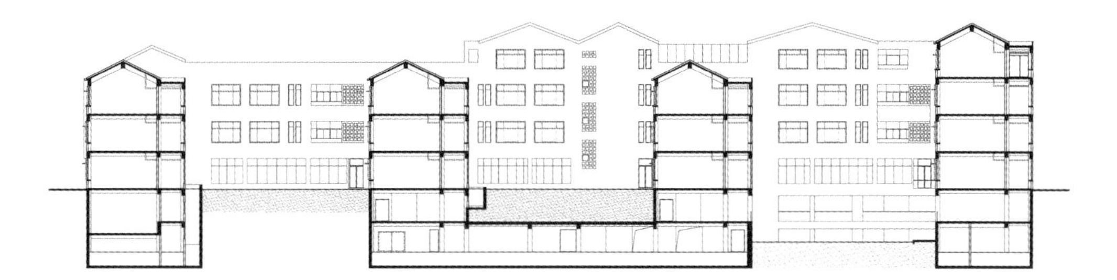

教学实验楼剖面图

西宁市第五高级中学 Xining No.5 High School

地点 青海省西宁市 / 用地面积 171,200m² / 建筑面积 105,940m² / 高度 24m / 设计时间 2011年 / 建成时间 2015年

方案设计 崔海东、王敬先、金海平
文 亮、华 翔、徐 超
杜 江、靳树春、林 莹
设计主持 崔海东

建 筑 王敬先、金海平
结 构 张根俞
给排水 高 峰
设 备 向 波
电 气 丁宗臣
总 图 刘晓琳

合作设计 青海省建筑勘察设计研究院

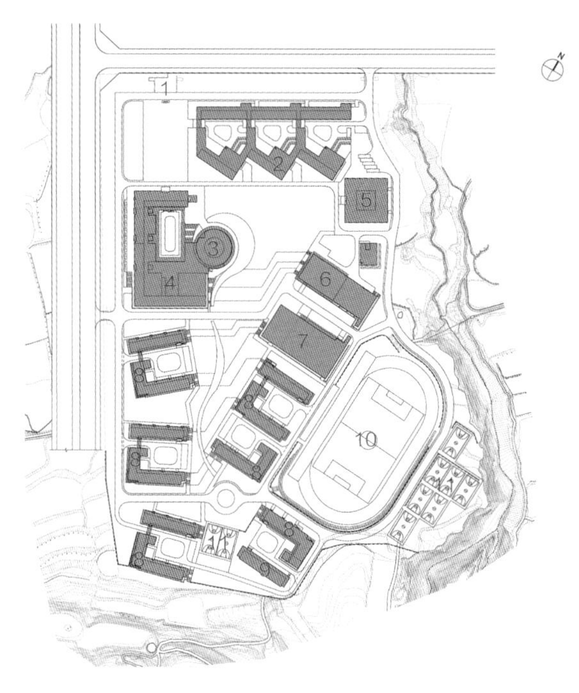

1. 校门
2. 教学楼
3. 图书馆
4. 综合楼
5. 实验楼
6. 食堂
7. 体育馆
8. 学生宿舍
9. 教工宿舍
10. 体育场
11. 篮球场

总平面图

西宁市第五高级中学位于西宁城南新区的一片郊野山地，是容纳了60班3000人的寄宿制高中，分为教学、综合、食堂、宿舍、体育五个区。布局因地制宜，将各部分建筑整合为方正的院落，并顺山势偏转，灵活布置。让图书馆等共享的建筑与景观相融合，形成连续的缓坡台地，嵌山、跌水、砌台、围院，营造多层级交流场所，增加师生的空间归属感。建筑外观汲取青海传统庄廓特色和河湟文化内涵，并注意适应西北地区的气候，整体朴实淳厚，提炼石灰白、土黄、古铜的主题色彩，与青山、绿树、蓝天相映。

Located on a hilly site, Xining No.5 High School is a boarding high school with five zones. In alignment with the terrain, some parts of the buildings form square courtyards with flexible adjustments in alignment with the site conditions. Together with terraces, slopes and waterfalls built on site, the public spaces have merged into the landscape. Inspired by traditional local culture, the exterior design presents a modest look with adaption to the local climate.

天津大学新校区主楼、综合实验楼 Main Building & Laboratory Building of Tianjin University New Camp

地点 天津市津南区 / 建筑面积 主楼85,928m² 综合实验楼38,761m² / 高度 31m / 设计时间 2011年 / 建成时间 2015年

方案设计 崔　愷、任祖华、梁　丰
　　　　 朱　巍、曹　洋
设计主持 崔　愷、任祖华

建　　筑 叶水清、朱　巍、梁　丰
　　　　 彭　彦、李　欣
结　　构 孙海林、段永飞、陆　颖
　　　　 高彦良、吴先坤
给 排 水 匡　杰、黎　松
　　　　 潘国庆、唐致文
设　　备 王　加
电　　气 史　敏、刘　畅
总　　图 吴耀懿
室　　内 邓雪映、张全全
景　　观 冯　君

天津大学新校区主楼位于该校区东西主轴线的东端，是从主入口进入学校时见到的第一组建筑。由于校园东西主轴线、入口广场轴线、行政办公轴线交叉于此，设计采用圆形广场对应轴线交点处，自然地实现了空间的转折。轴线贯穿建筑，形成了对景和连续的开放空间。主楼因此分为中央广场、南侧、北侧三组体量。综合实验楼位于校园主轴线上，是其序列空间的重要节点。轴线两侧的建筑入口因此一圆一方，一高一矮，既对称又有所差异。各个实验楼设置在轴线南北两侧，形成校园的主轴空间。计算机中心位于西南侧的方形用地，便于安排较大的机房。设计延续天津大学原有校园特色，将页岩砖作为主要饰面材料。建筑开窗与镂空砌筑相结合，在统一中蕴含变化。通过庭院、单廊布局，尽量减小建筑进深，获得较好的自然通风效果。

Located on the eastern end of the main axis of the campus, the main building is the first structure to be seen from the entrance of the campus. Since the building is situated at an intersection of three axes of the campus, a circular square, with its center at the point of intersection is planned. The main axis goes through the buildings to form diverse open spaces, and the main building is divided into three parts. As the comprehensive laboratory building serves as an important node on the main axis, the two lobbies on both sides of the axis are located symmetrically. Carrying forward the existing features of the university, the façades are dominated by shale bricks. The positions for hollowed-out parts and window openings are arranged in a unified style with slight variations.

1. 主楼
2. 综合实验楼
3. 综合体育馆

新校区总平面图

主楼结合建筑功能需求，将北洋会堂和理学院围绕中心广场设置，材料学院办公教学楼单独设置于北侧，文法学院、马克思学院、职教学院和第四教学楼等统一设于南侧。统一的外观形式让各组团有所联系，并根据各自特点而有所差异，提供了一系列开放、多层次的学生交流空间。

The buildings has an exterior style that is largely unified and slightly different. A series of open and multi-tiered communication spaces for the students are also planned in the buildings.

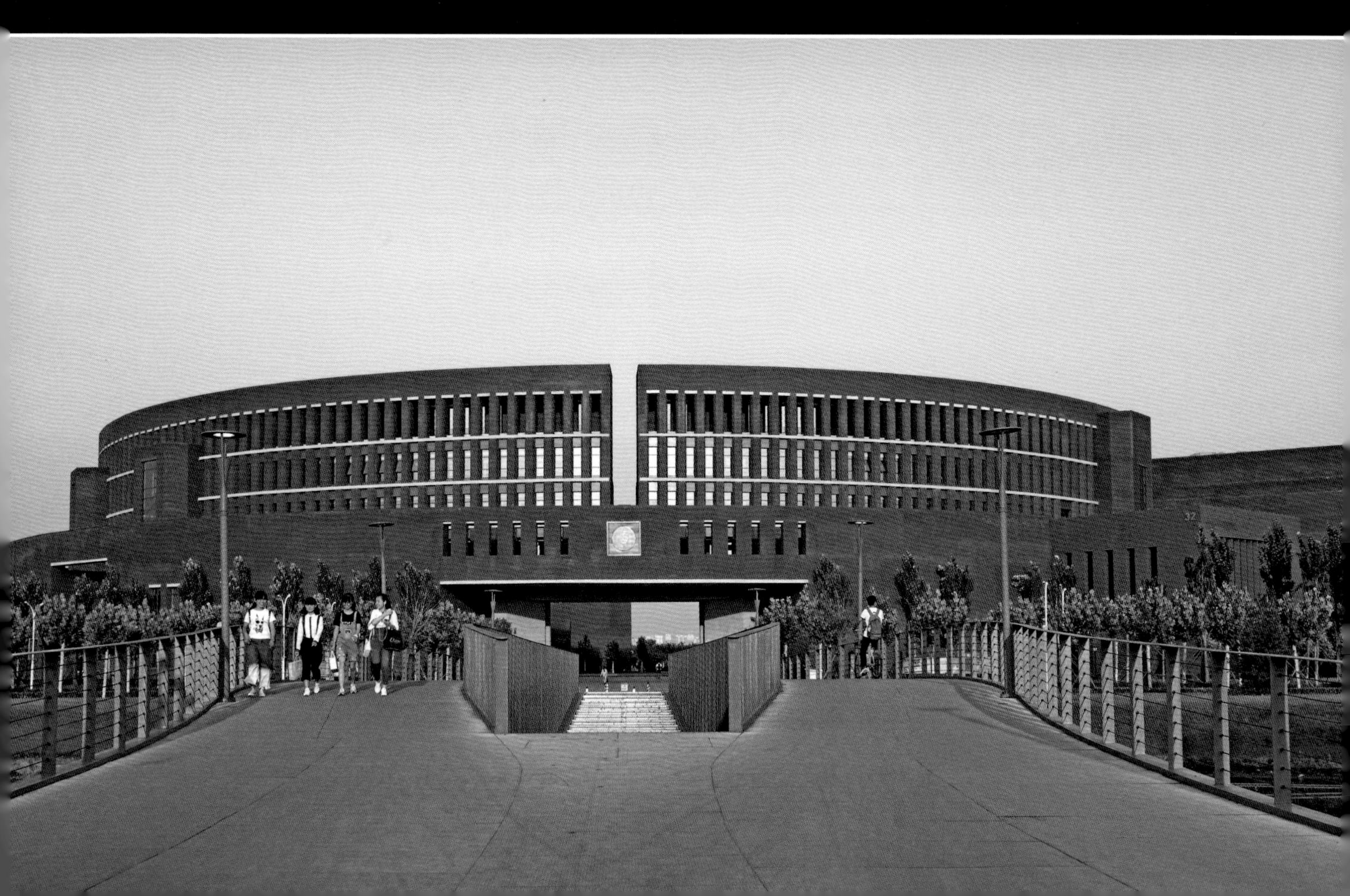

1895-1951
實事求是

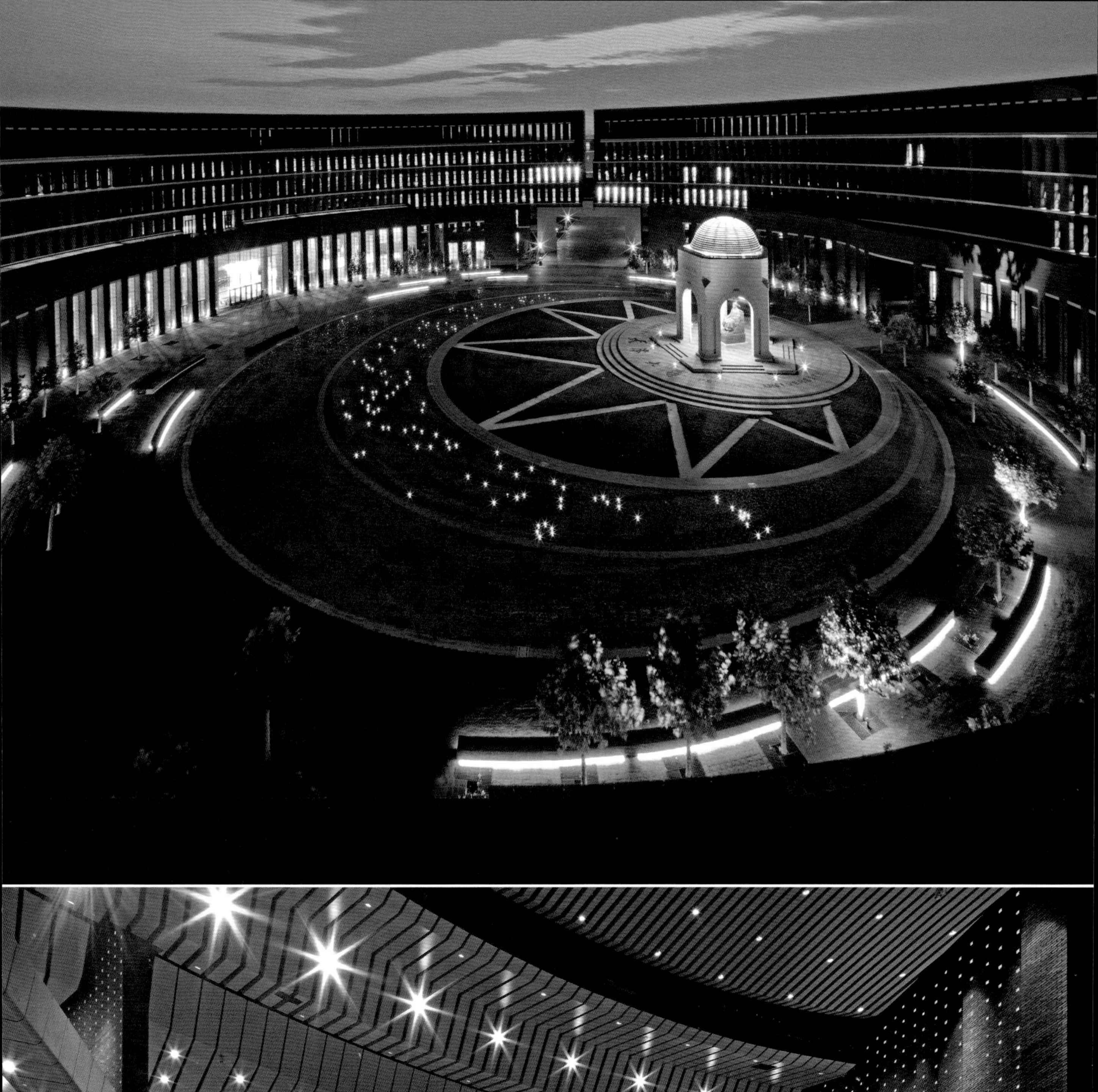

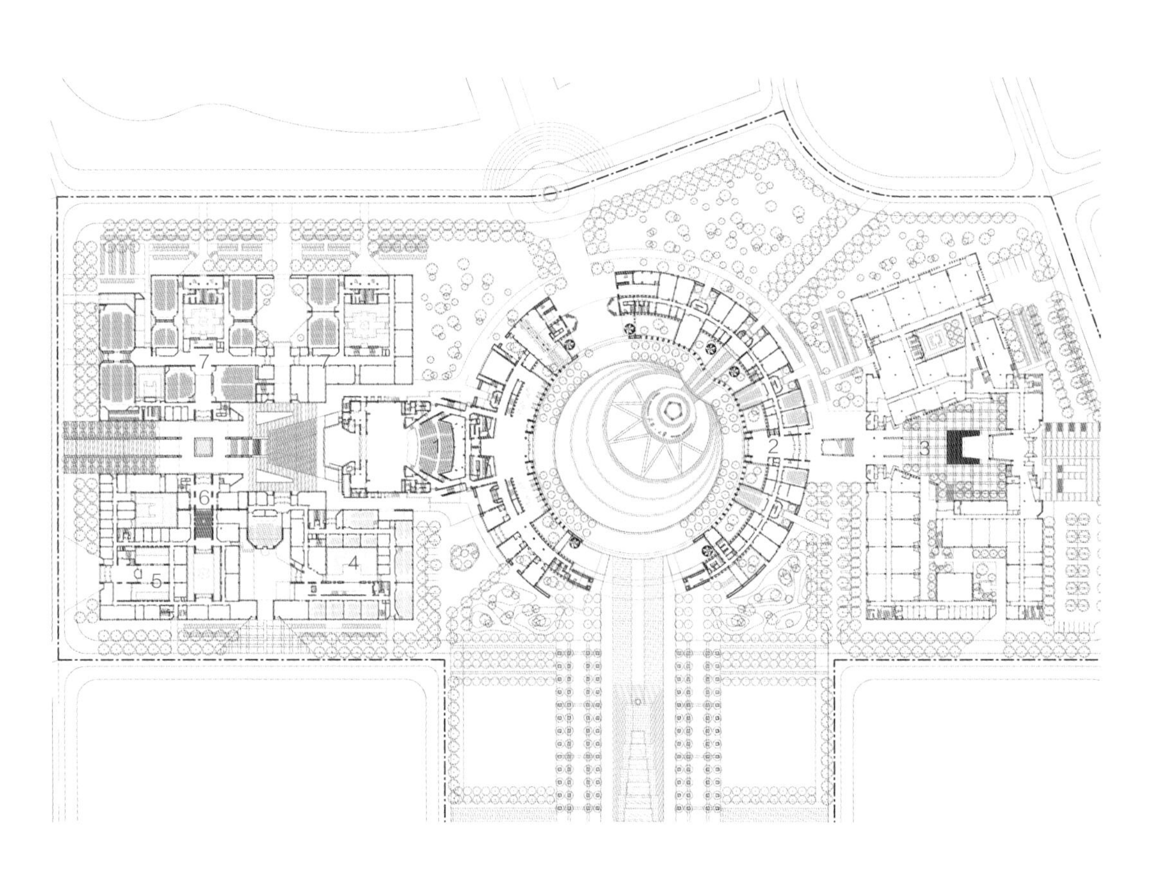

1. 北洋会堂 2. 理学院 3. 材料学院 4. 文法学院 5. 职教学院 6. 马克思学院 7. 第四教学楼

主楼首层平面图

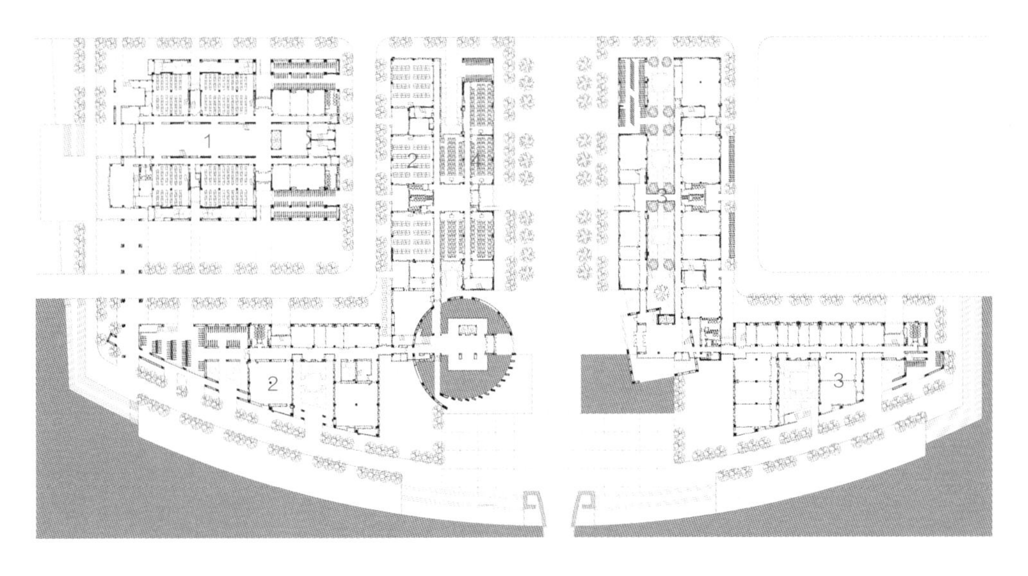

1. 计算机实验中心
2. 电子电气实验中心
3. 物理教学实验中心
4. 语音教室

综合实验楼首层平面图

天津大学新校区综合体育馆 Gymnasium of Tianjin University New Campus

地点 天津市津南区 / 用地面积 33,950m² / 建筑面积 18,798m² / 高度 24m / 设计时间 2012年 / 建成时间 2015年

方案设计 李兴钢、张音玄、闫 昱、易灵洁、梁 旭
设计主持 李兴钢、张音玄

建　　筑 闫 昱、梁 旭
结　　构 任庆英、张付奎
给 排 水 赵 昕
设　　备 王微微、唐艳滨
电　　气 林佳旸
总　　图 刘晓林

图　　片 李兴钢工作室提供

天津大学新校区综合体育馆位于校前区北侧，包含室内体育活动中心和游泳馆两部分，一条跨街的大型缓拱形廊桥将两者的公共空间连接起来，形成环抱式入口广场。各类室内运动场地依其对平面尺寸、净空高度及使用方式的不同要求紧凑排列，并以线性公共空间叠加、串联为一个整体，不仅增强了整个室内空间的开放性和运动氛围，而且天然造就了错落多样的建筑檐口高度。公共大厅屋面采用了波浪形渐变的直纹曲面形屋面（空心密肋屋盖结构），其东侧为长达140m的室内跑道。运动场地空间的屋顶和外墙，使用了一系列直纹曲面、筒拱及锥形曲面的钢筋混凝土结构，带来大跨度空间和高侧窗采光，在内明露木模混凝土筑造肌理，在外形成沉静而多变的建筑轮廓，达到结构、空间与形式的完美统一。外部材料主要采用清水混凝土结合具有天津大学老校区特色的深棕红色页岩砖饰面。

Located on the north side of the campus's front area, the gymnasium of Tianjin University consists of a sports center and a natatorium, the public spaces of which are connected by a large-scale covered bridge over the road. Sports fields with various demands and sizes are arranged in a compact way and connected by a linear space, forming an undulating eave for the gymnasium. The public lounge has a wavy roof of hollow ribbed structure, to the east of which is a 140-meter indoor track. The roofs and exterior walls are built with reinforced concrete structures with ruled surfaces, barrel vaults or conical surfaces, forming a large-span space with clerestories. Together with the interior walls that expose the textures left by concrete casting, they have achieved the harmony of structures, spaces and forms.

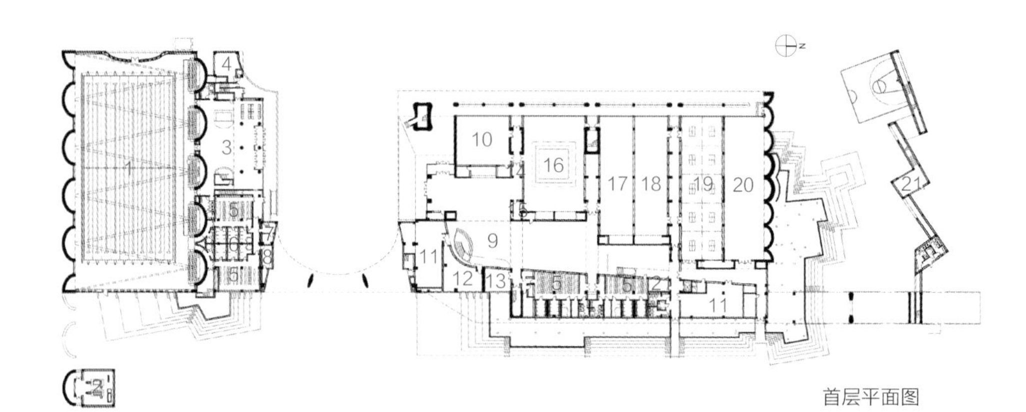

首层平面图

1. 游泳馆 2. 锅炉房 3. 游泳馆门厅 4. 消防控制室 5. 更衣室 6. 淋浴室 7. 值班室 8. 急救室 9. 门厅 10. 多功能厅 11. 新风机房 12. 体质检测中心 13. 群体竞赛中心 14. 服务间 15. 器材库 16. 跆拳道馆 17. 体操馆 18. 健身馆 19. 乒乓球馆 20. 武道馆 21. 管理室

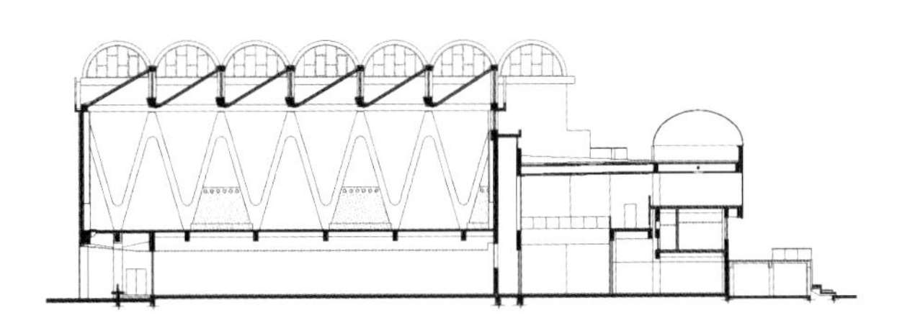

健身馆剖面图

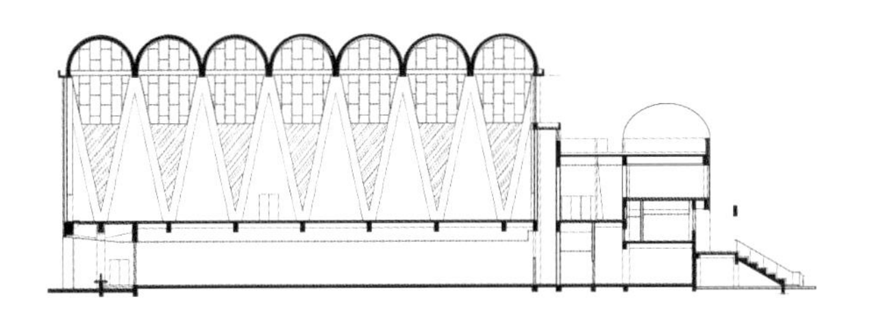

武道馆剖面图

设计强调在几何逻辑控制下对建筑基本单元形式和结构的探寻，重复运用和组合这些单元结构，以生成特定功能、光线及氛围的建筑空间。

The form and structure of fundamental units have been explored under the geometric logic of the building.

北京工业大学第四教学楼组团 No.4 Teaching Complex of Beijing University of Technology

地点 北京市朝阳区 / 用地面积 41,337m² / 建筑面积 77,504m² / 高度 50m / 设计时间 2010年 / 建成时间 2012年

方案设计 崔　愷、柴培根、于海为
　　　　 谢　悦、田海鸥、潘天佑
设计主持 柴培根、于海为
　　　　 谢　悦、张　东

建　　筑 田海鸥、潘天佑、李　楠
结　　构 任庆英、曾金盛、刘新国
给 排 水 黎　松
设　　备 韦　航
电　　气 胡　桃
总　　图 王　玮
室　　内 韩文文
景　　观 刘　环

第四教学楼组团是已基本成型的北京工业大学校园中最后一个大规模建设项目。相较由主校门、行政主楼、图书馆构成的传统校园轴线，我们希望这个新的教学组团能和体育馆遥相呼应，构成一个展现活力的开放性的场所。设计一方面在庞大的教学体量中提供更多的功能空间，另一方面也希望在高密度的前提下建立起街道、平台、广场等一套公共空间的体系，融入校园的整体环境中，并通过不同学科的集中，形成各专业的相互交流和碰撞，其高效复合的状态可以被称为"校园综合体"。设于首层的大型实验室屋顶通过连桥成为连续的平台，面向校园主要空间的大台阶则让平台更具公共性，艺术教学工作室如聚落般置于其上，以区别于其他教学体量的材料和色彩，吸引工科学生前来感受艺术的氛围。结合室外疏散楼梯增设的平台和绿化墙，则在竖向上为建筑增添了一抹绿意。

The No.4 Teaching Complex is the last large-scale construction project in Beijing University of Technology. Besides providing more functional spaces in the huge volume of the teaching building, the design aims to integrate the streets, platforms and squares into a system of public space which blends into the campus while facilitating the communication among students of various majors. The roof of a large-scale laboratory at the first floor forms continuous platforms, and the grand stairs facing the main space of the campus has added openness to the platform, on which workshops with distinguished colors and materials are scattered. Green walls and platforms around outdoor emergency staircases have added a tint of green to the buildings.

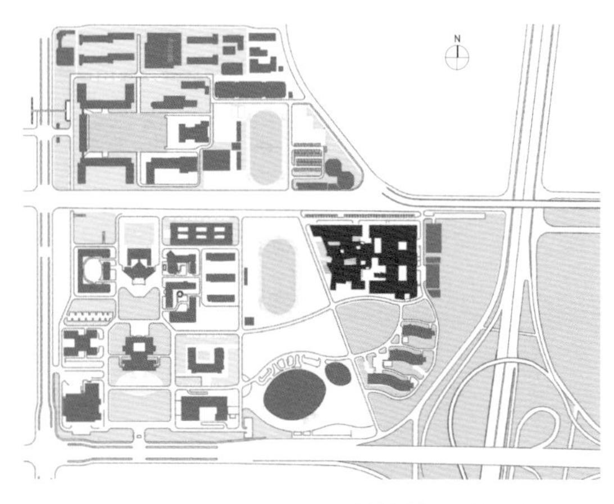

建筑在校园和城市中的位置

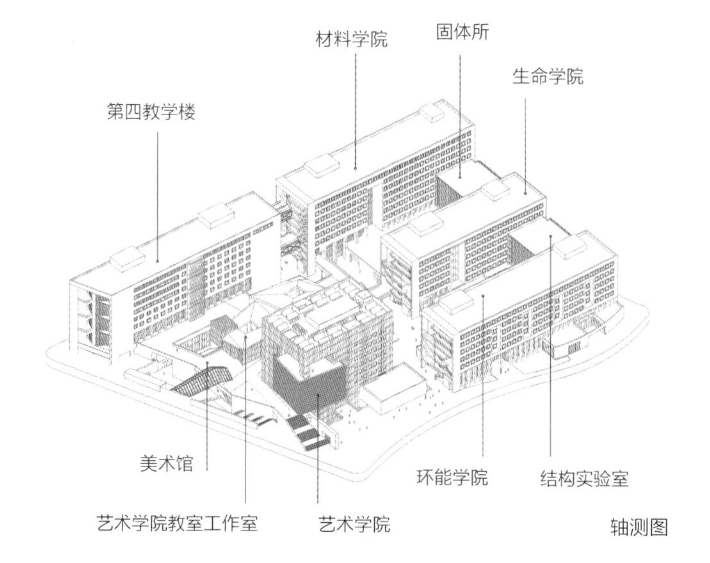

轴测图

设置于平台上的艺术教学工作室聚落，外部均采用黑色压花钢板，内侧的彩色涂料跳跃、闪动，在黑色的映衬下，增添了整组建筑的活跃度。

The workshops on the platform are covered by black embossing steel panels, which intensify the palette of bright-coloured inner walls and activate the open spaces of the whole complex.

北京邮电大学教学综合楼、图书馆 Teaching Building & Library of BUPT

地点 北京市昌平区 / 用地面积 14,950m² / 建筑面积 35,500m² / 高度 24m / 设计时间 2012年 / 建成时间 2014年

方案设计 崔　恺、邢　野、张军英
　　　　 高　凡、曹　洋
设计主持 崔　恺

建　　筑 邢　野、胡水菁、李　喆
结　　构 刘松华
给 排 水 王耀堂、陈　宁
设　　备 宋孝春、李超英
电　　气 胡　桃
总　　图 刘晓琳、郝雯雯

北京邮电大学沙河校区一期校园规划的调整以尊重高教园区的街区尺度和路网体系为出发点，将一横一纵的市政道路作为校园的基本骨架和形象轴线，分为东南西北四个区域，分别作为宿舍、教学、运动和景观之用，各类公共综合建筑则居于中部。单体建筑以合院式布局为主，不同尺度的院落空间连缀起来形成丰富的校园空间。

The planning of the campus has been carried out with respect for the scale and road system of the region. The campus are divided into four parts, which are used for living, teaching, sports and landscape respectively. Public buildings are located in the middle of the site, and courtyards with varied scales are planned.

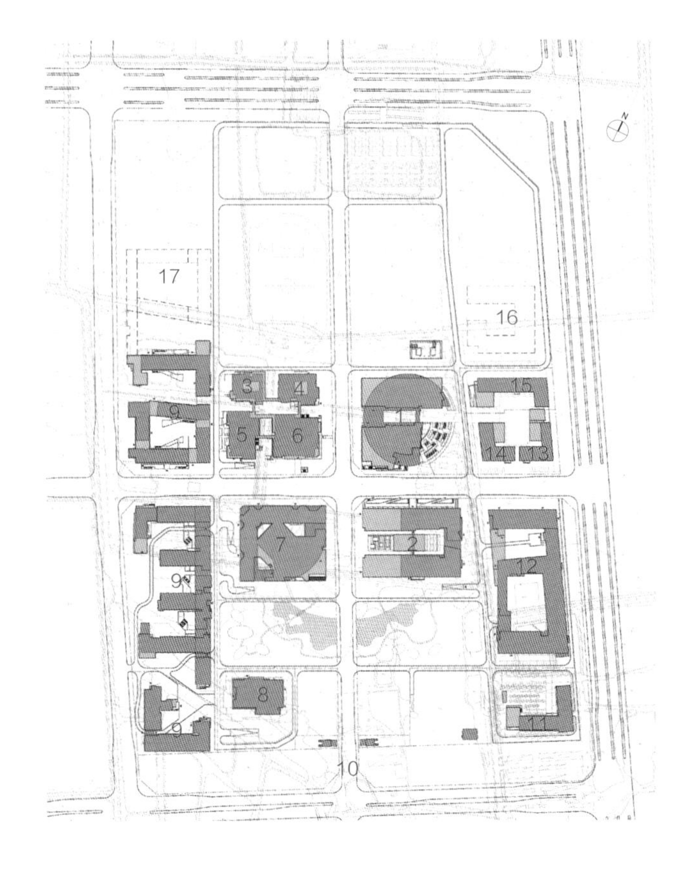

1. 图书馆
2. 教学综合楼
3. 教工食堂
4. 联合办公楼
5. 学生食堂
6. 活动中心
7. 公共教学楼
8. 南区食堂
9. 学生宿舍
10. 南校门
11. 学院楼
12. 实验楼
13. 邮政学院
14. 数字媒体与设计艺术学院
15. 网络安全学院
16. 预留教学楼用地
17. 预留宿舍用地

校区总平面图

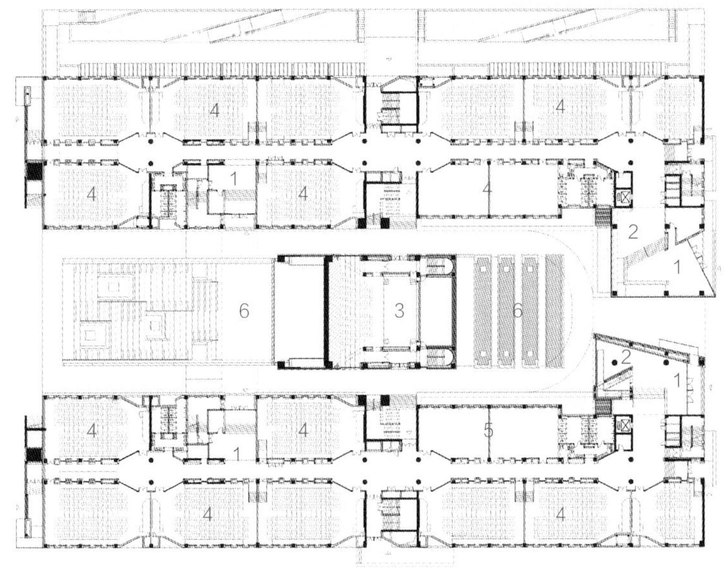

1. 门厅 2. 休息厅 3. 报告厅 4. 多媒体教室 5. 素描教室 6. 室外剧场

综合楼首层平面图

为塑造从校门看向图书馆的中轴视线，综合楼体量由东向西跌落，形成丰富的屋顶绿化平台。平台之间以大楼梯相连，既为师生提供景观优美的室外休闲、交流场所，又满足课间人流疏散的需要。内廊两侧密布各类教室单元，走廊的夹壁墙内容纳教室的拔风井、排烟井、玻璃砖竖窗等功能设施。为突出学术报告厅的重要性，设计让相邻建筑界面与之对话，由此派生出多种公共空间。

The comprehensive teaching building descends from the east to the west, forming green terraces connected by grand stairs, which can serve as both a communicating place and the access to classrooms. Classrooms line along the interior corridors, where smoke exhausting wells and vertical windows of glass tiles are located in double walls.

图书馆位于校园主轴线交叉点的东北角，场地中存有一条东西向的林道和一片高大的杨树林，保留林道和树林同时成为设计的难点和机会。既有条件将场地划分为四个象限，建筑底部以方形体量占据树林之外的三个象限，容纳报告厅、办公区、咖啡厅等功能空间，上部再插入以共享大厅和藏阅空间为主的圆形体量。方形体量以红砖和深窗洞产生厚重感，窗洞间的壁龛则提供小尺度的阅览空间。圆形体量以玻璃和铝格栅体现轻盈感，通透的界面透射出阅读和交流的场景。

The library located at the northeastern corner of the intersection of two main axes, made great efforts in preserving the poplar woods and a shady path on the site. The lower part of the library takes on a square form and consists of lecture halls, offices and a cafeteria, while the circular-shaped upper part is composed of a lounge and book storage spaces.

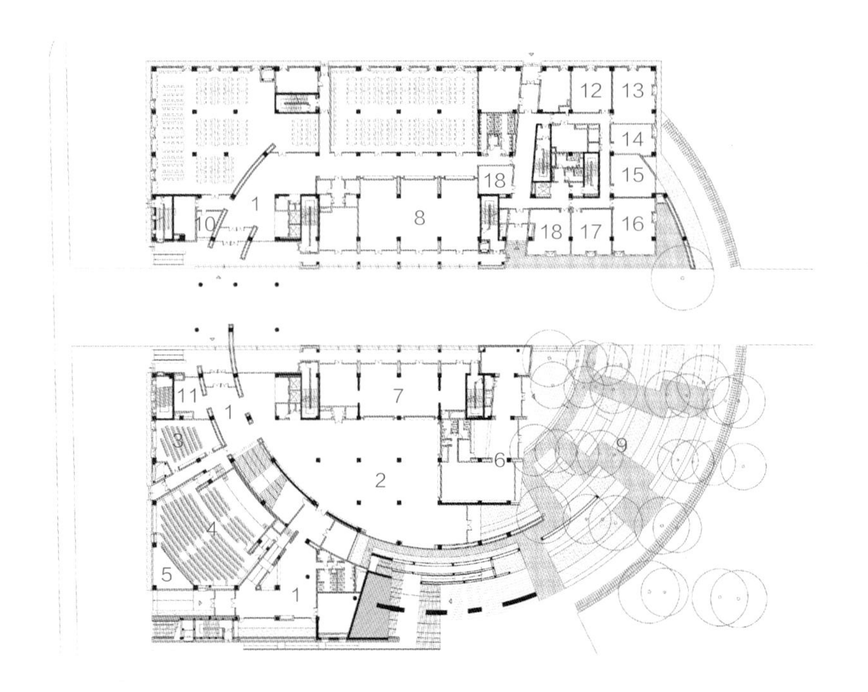

1. 门厅
2. 展示
3. 放映厅
4. 报告厅
5. 传译厅
6. 咖啡厅
7. 书店
8. 超市
9. 林下空间
10. 自助还书
11. 存包处
12. 采访室
13. 流转室
14. 典藏室
15. 加工室
16. 剔旧室
17. 报刊收登室
18. 编目室

图书馆首层平面图

北京大学学生中心　Student's Center of Peking University

地点 北京市海淀区 / 用地面积 40,299m² / 建筑面积 19,266m² / 高度 18m / 设计时间 2011年 / 建成时间 2014年

方案设计 张　祺、王　媛、胡　斯
设计主持 张　祺、刘明军

建　　筑 王　媛、胡　斯
结　　构 郝　清、袁　琨
给 排 水 王泽惠、陈　静
设　　备 杨向红、路　娜
电　　气 许冬梅、熊小俊、王　峥
总　　图 齐海娟

北京大学南门教学区位于北京大学主校区南校门至百年讲堂之间，原为学生宿舍，改建后则以教学和科研功能为主，形成南部校区的核心风貌。整组建筑由六栋单体建筑组成，其中的学生活动中心位于西北角，容纳了阳光大厅、服务大厅、社团工作室、多功能厅、小剧场、琴房等多种多样的学生活动场所。建筑借鉴了地块上原有建筑及静园等三合院形制，采用半围合布局，并以四坡屋顶、灰白相间的清水砖墙等形式，呼应了北京大学的百年历史文脉和建筑风格。

The Student's Center is located at the northwestern corner of a building cluster at the university's south gate. A variety of activity spaces are planned in the center, including the lounge, the service hall, the community's workshop, the multi-functional hall, the small theater and piano rooms. The design has adopted a semi-enclosed layout consistent with those of the existing buildings on site, so that it conforms to the overall style and the historical context of the university.

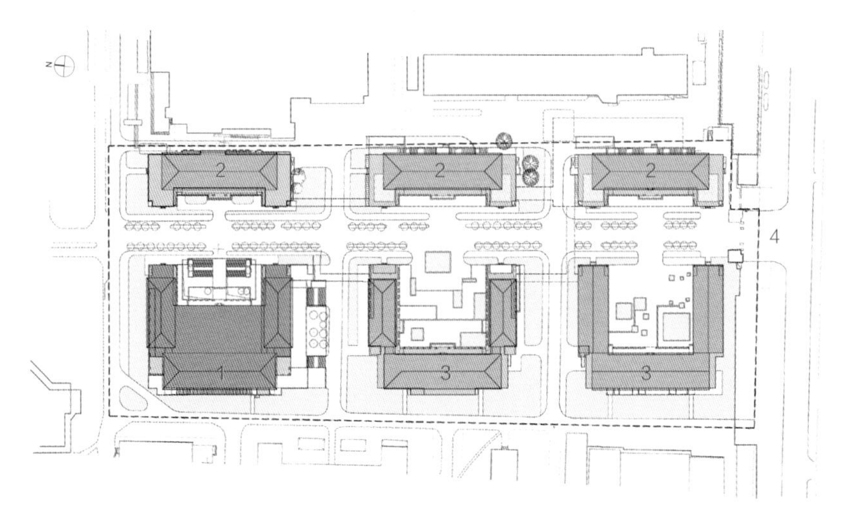

总平面图

1. 大学生中心 2. 新建学院楼 3. 原有建筑 4. 南门

首都师范大学南校区行政楼、教学楼 Administration & Teaching Building of CNU South Campus

地点 北京市海淀区 / 用地面积 190,300m² / 建筑面积 51,087m² / 高度 45m / 设计时间 2012年 / 建成时间 2016年

方案设计 崔 愷、吴 斌、郑 虎
张新星、范国杰、高 翔
设计主持 崔 愷、吴 斌

建　　筑 辛 钰、汪嘉绍
结　　构 徐 杉、张凌云
李黎明、张 彪
给 排 水 黎 松、董 立、翟瑞娟
设　　备 刘燕军、马 媛、刘筱屏
电　　气 王苏阳、陈 瑛
总　　图 王雅萍、高 治

作为原行政楼、教学楼的改造重建工程，建筑基本布置在原有建筑的旧址上，比较完整地保留了原广场的绿地和树木。行政楼正对学校东门，与南北两侧的教学楼依然围合成U字形，整个建筑群采用了沿水平方向分段的处理方式，下面三层形成“基座”，与原有校园尺寸吻合，并通过内院的柱廊强化基座的整体感。教学楼基座内布置公共教室，办公楼的底部则尽量安排与学生联系较为密切的部门。基座以上部分采用了较为简洁的处理方法，行政楼朝西的主立面以竖向格栅起到遮阳作用。内部大厅以“十年树木，百年树人”的寓意，将自由布置的圆柱呈现“树”的意象。

The project is renovation & reconstruction of the original administrative and teaching buildings. Largely built on the original location of those buildings, the project has preserved the original greeneries and trees on the square. The building cluster have all adopted a horizontal division of façade, so that the three floors at the bottom serve as the foundation of the building, conforming to the scale of the campus. Public classrooms and departments with close contact with students are located in those floors for the students' easy access.

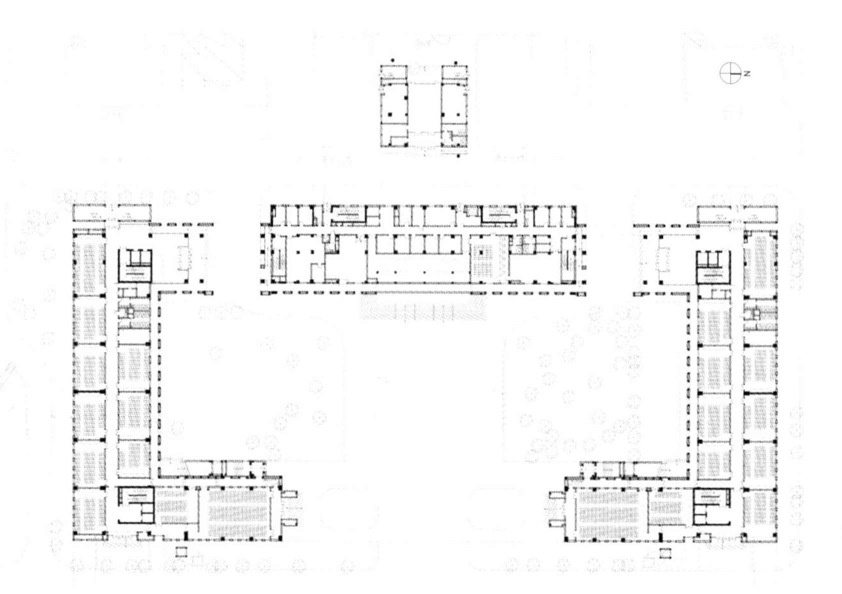

首层平面图

中央财经大学沙河校区图书馆 Library of Shahe Campus of Central University of Finance and Econo

地点 北京市昌平区 / 用地面积 12,652m² / 建筑面积 30,501m² / 高度 24m / 设计时间 2012年 / 建成时间 2016年

方案设计 崔 愷、崔海东
李东哲、张婷婷
设计主持 崔海东、李东哲

建　　筑 张婷婷
结　　构 余 蕾
给 排 水 马 明
设　　备 李 莹
电　　气 何 静、陈沛仁
电　　讯 任亚武
总　　图 吴耀懿
室　　内 饶 劢

图书馆位于中央财经大学沙河校区的核心位置，也是一期校园建设的收官之作。建筑设计从校园有机生长的总体规划和周边建筑对位出发，确定秩序成长框架。四组纵列阅览藏书单元以三组服务单元分隔，服务单元内整合楼电梯、卫生间及共享中庭于一体。七组单元空间以包裹三处共享空间的“书架墙”划分并形成序列。共享空间的屋顶选用锯齿形天窗，引入天然采光，让阅览空间围绕共享空间形成带有多重空间序列的“室内书院”。在外部形象上，四组阅览单元成为厚重的基座，并呼应周边街坊式布局作了适当的扭转。上部建筑外观突显混凝土质感，简朴而富文化意味。

The design for the library has been based on the overall planning and the positions of surrounding buildings. Four book storage & reading units are separated by three service units, where staircases, elevators, washrooms and a lounge are located. Jagged skylights on the roof introduce daylight into the public spaces. Seen from outside, the four storage & reading units are solid bases, which are rotated to conform to the directions of the streets. The upper part of the building reveals a sense of simplicity with its pure concrete texture.

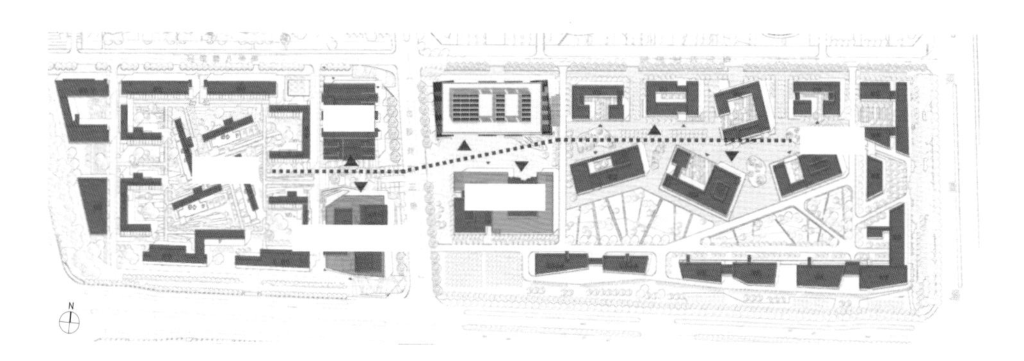

建筑在校园中的位置

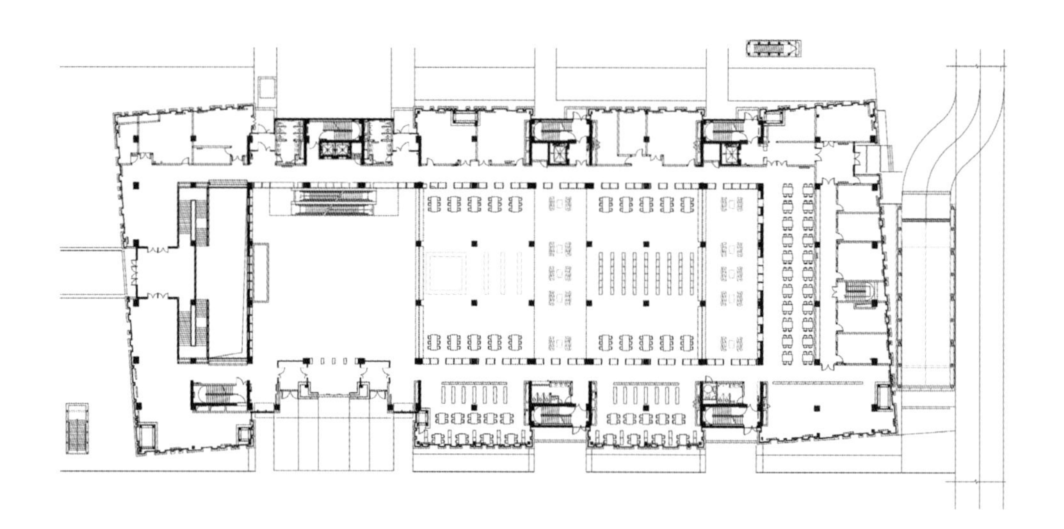

首层平面图

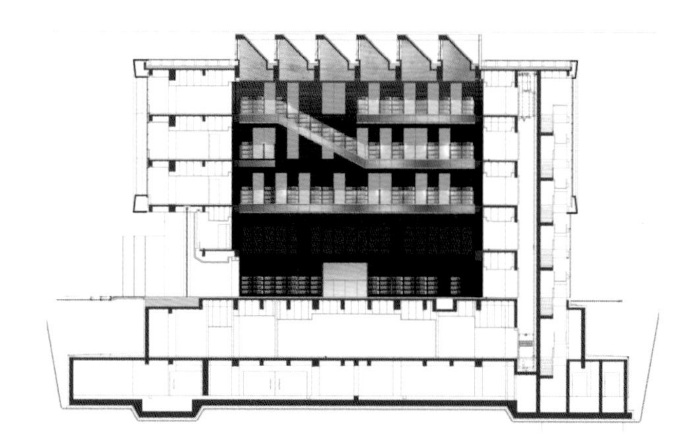

横向剖面图

中 2 区
Central Area 2

南京理工大学图书馆 Library of Nanjing University of Science and Technology

地点 江苏省南京市 / 用地面积 84,896m^2 / 建筑面积 45,796m^2 / 高度 47m / 设计时间 2012年 / 建成时间 2015年

方案设计 杨金鹏
设计主持 杨金鹏

建　　筑 钱玉斋、解思茹、王霖硕
结　　构 李　谦、石　雷
给 排 水 周　博
设　　备 李京沙、姜　红
电　　气 李战赠、何　穆
总　　图 郑爱龙

新建图书馆位于校园中轴线尽端，南侧是中轴线上重要的视觉焦点——逸夫楼。图书馆以环抱之姿呼应圆形的逸夫楼，形成半围合的校园主入口广场，在图书馆北侧形成与教学主楼共同限定的大型活动广场，强化了校园主轴线的秩序感。外立面的遮阳格栅和竖向开窗突出了竖向线条，为图书馆阅览空间提供优质的光环境。阅览空间围绕宽敞的中庭布置，提供优化的光环境，空间开放，结合书的展示和放置，形成了富有书卷气的“书箱”。

The newly-built library is located at the end of the middle axis of the campus. It has an encircling posture in response to the circular form of Yifu Building to its south, forming a semi-enclosed square at the main entrance of the campus while defining the boundaries of a large-scale square together with the main teaching building. Vertical windows and sun-shading gratings on the façade have guaranteed a pleasant luminous environment for the open reading space.

方形图书馆与
圆形逸夫楼

图书馆与逸夫楼
结合、相切

建筑形体
划分层次

前后层次
交错演变

形态生成分析图

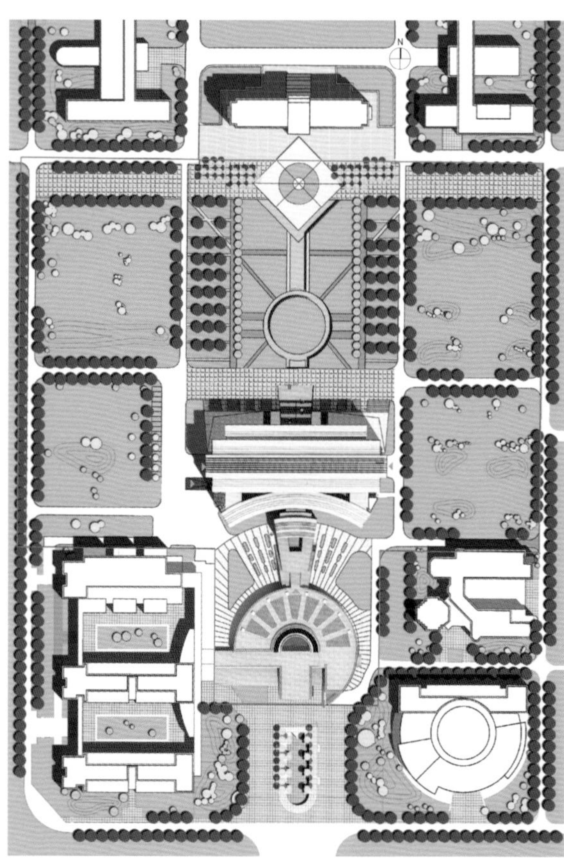

总平面图

南京理工大学校园中轴线和北侧中山陵的中轴线重合，新建图书馆位于校园中轴线尽端。

The newly-built library is located at the end of the middle axis of the campus, which coincides with the axis of Sun Yat-sen Mausoleum to its north.

北京外国语大学综合教学楼 Comprehensive Teaching Building of Beijing Foreign Studies University

地点 北京市海淀区 / 建筑面积 60,515m^2 / 高度 45m / 设计时间 2009年 / 建成时间 2016年

方案设计 崔 愷、王 祎、辛 钰、马喻强
设计主持 崔 愷、张念越、王 祎

建　　筑 张念越、王 祎、辛 钰
结　　构 喻远鹏、许忠琴、岳彦博
给 排 水 李 严
设　　备 郭晓南
电　　气 姜晓先、刘 涛
电　　讯 马 宁
总　　图 夏菡颖
室　　内 饶 劢、郭 林

综合教学楼位于北京外国语大学西校区，处于风格统一的外研社办公楼和西校区主校门之间。为了与二者取得呼应，该建筑同样以红色面砖作为主要材料，并采用类似的竖向元素和大尺度拱券，突出建筑的可识别性。楼内设有十余个院系的教学和办公功能。平面布局面向北侧校园广场呈环抱之势，增加了广场的围合感。为适应各院系对房间面积的多样化需求，建筑底部安排礼堂、报告厅、资料室、书店等公共功能，中部为各类教室，顶部为教学办公室。

To achieve harmony with its neighborhood, the façade of the teaching building is dominated with red bricks used on buildings and the campus gate to its adjacency, while large-scale arches and vertical lines highlight the identity of the building. Spaces for more than 10 departments are planned. Classrooms are located above the first floor, where an auditorium, a lecture hall, a bookstore and reference rooms are situated, while teachers' offices are located at the top floors.

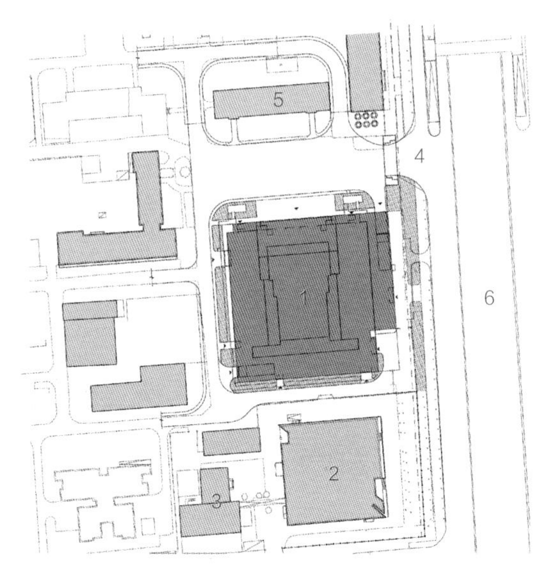

1. 综合教学楼
2. 外研社办公楼一期
3. 外研社办公楼二期
4. 主校门
5. 教学楼
6. 三环路

总平面图

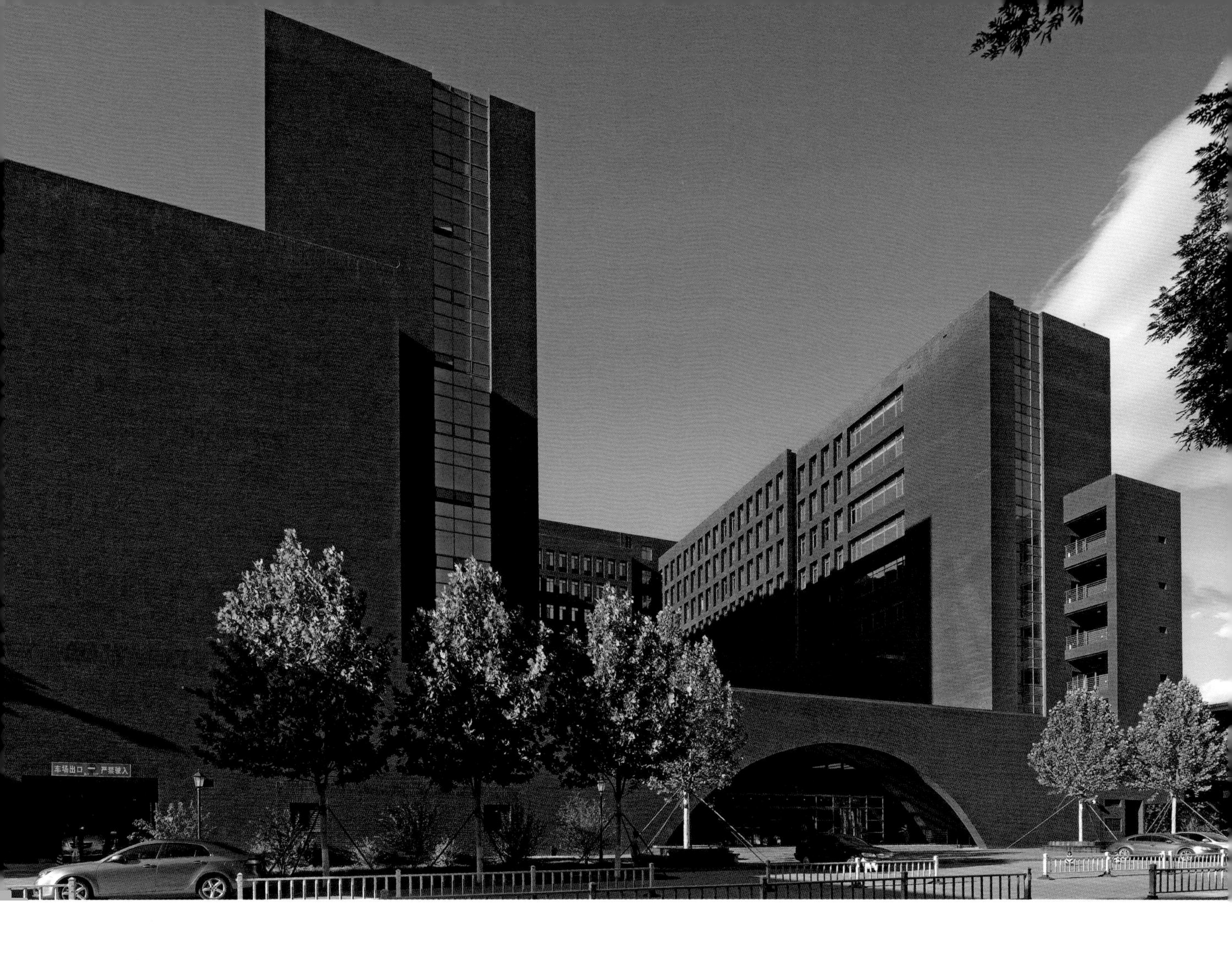

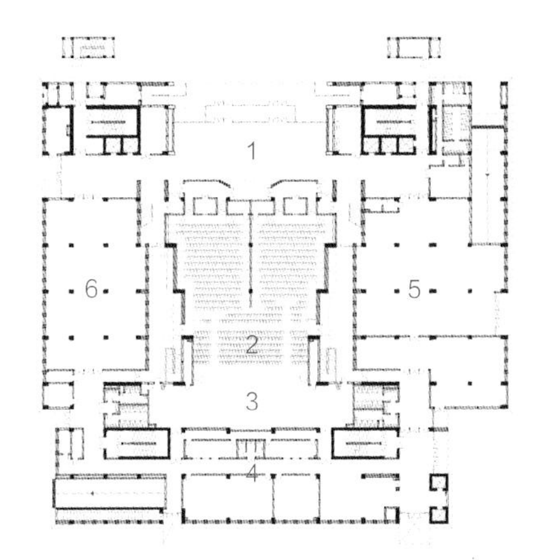

1. 门厅
2. 观众席
3. 舞台
4. 化妆、接待区
5. 书店
6. 资料室

首层平面图

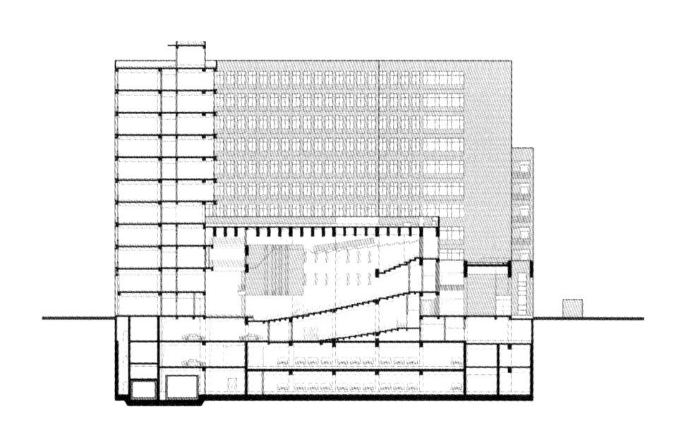

剖面图

江苏建筑职业技术学院图书馆 Library of Jiangsu Vocational Institute of Architectural Technology

地点 江苏省徐州市 / 用地面积 54,878m² / 建筑面积 27,896m² / 高度 18m / 设计时间 2009年 / 建成时间 2014年

方案设计 崔 愷、赵晓刚
周力坦、李 喆
设计主持 崔 愷

建 筑 赵晓刚、周力坦、李 喆
结 构 孙海林、石 雷、刘会军
给 排 水 杨东辉、董新淼
设 备 王 加
电 气 李俊民、何 静、陈沛仁
总 图 高 治
室 内 饶 劢、郭 林

江苏建筑职业技术学院坐落于徐州南郊。营造一个树下读书场所，是设计的主旨。井字梁与斜撑组成的清晰的混凝土结构，以及层层叠置的平台，在建筑形体上象征了树的寓意。图书馆平面采用8.4m×8.4m矩形柱网，所有的变化均在矩阵的控制下进行，而转折的外边界，则将更多的外部环境融入阅览区的景窗内。由于图书馆位于校门与教学楼之间的区域，在位置上担负着顺承周边地势并连接步行交通的任务。设计将底层部分架空，使之成为开放的交往空间，并将咖啡厅、书店、展厅、报告厅等公共功能安排于此，与近旁的水面结合，形成校园内吸引学子的文化广场。主要开架阅览空间则放置在二层到四层的开敞区域。五层为电子阅览室，并为研究室设置了屋顶花园，可供科研人员在休息间隙凭栏远眺。

Located in the south suburbs of Xuzhou, the institute has been designed to create an ambience of reading under the tree. The concrete structure with cross beams and diagonal bracings, together with the terraces, resembles trees with their forms. With an 8.4m×8.4m column grid, all the seemingly random variations are controlled with a matrix system. As the library is located between the gate and the teaching building, the library has been stilted to provide access as well as open spaces, with a cafeteria, a bookstore, an exhibition hall and a lecture hall around, forming a space appealing to the students. The open-shelf reading areas are located on the 2nd, 3rd and 4th floor, while electronic reading rooms are located on the 5th floor.

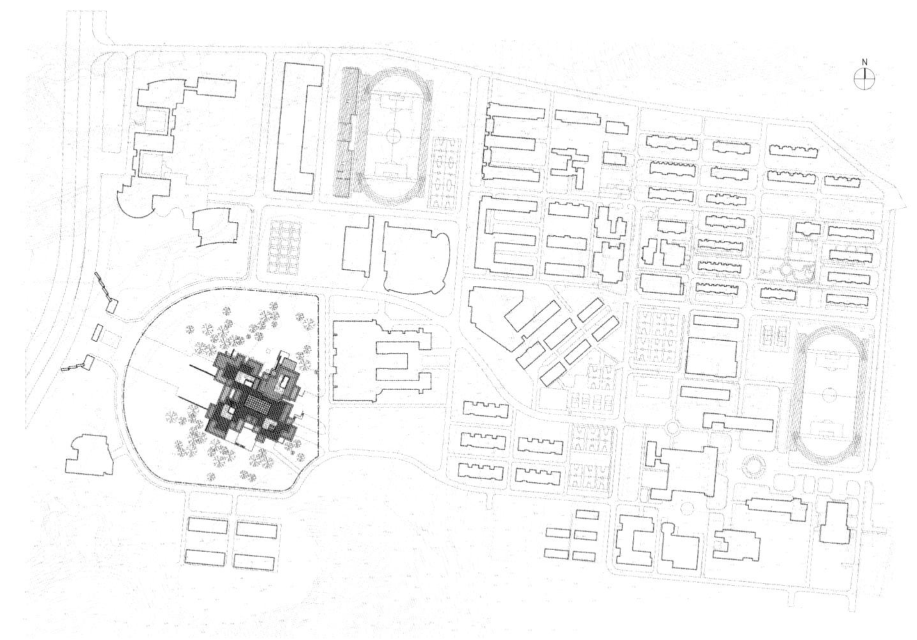

建筑在校园中的位置

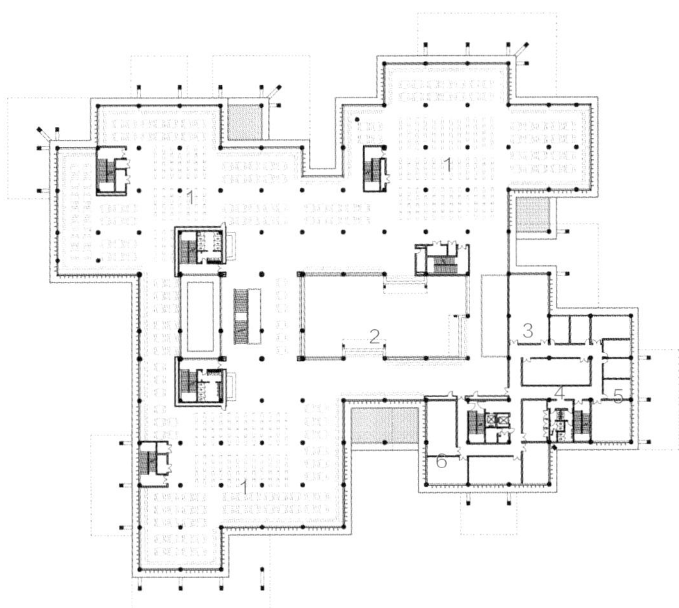

1. 文献开架借阅区
2. 中庭及上空
3. 数字化制作区
4. 会议室
5. 办公室
6. 信息技术区

三层平面图

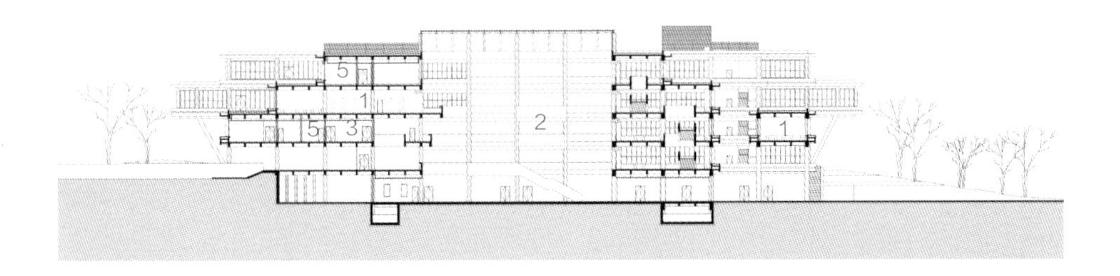

剖面图

为处理好角部斜撑与不同向度构件交接处的协调关系，斜撑的截面被处理为八角形。

The section of diagonal bracings at corners is made into octagon to coordinate with other building parts with different dimensions.

大连城堡酒店 The Castle Hotel, Dalian

地点 辽宁省大连市 / 用地面积 48,800m² / 建筑面积 117,000m² / 高度 99m / 设计时间 2008年 / 建成时间 2014年 / 客房数量 300间

方案设计 美国 WATG 设计公司
设计主持 崔海东、张 雅

建　　筑 杜 江、徐 超
　　　　 文 亮、王 飞
结　　构 张 猛、杨 婷、张根俞
给 排 水 李万华、董 超
设　　备 宋孝春、宋 玫
电　　气 王 琼
总　　图 白红卫

合作设计 HBA 室内设计公司

城堡酒店的前身，是原星海城堡博物馆，坐落于莲花山上，俯瞰星海湾。设计总结欧洲城堡的模式语言，充分运用山、城、楼、塔、尖等元素有机组合，实现了对城堡的当代诠释。设计在原有城堡形式下合理安排功能，提供了超大宴会厅、各类特色餐饮、四季屋顶、室内外泳池和300套酒店客房，形成丰富的酒店功能。所有客房均面向大海，依山就势展开。服务区则沿背面展开，通过地下窗井采光廊相连，实现良好的通风采光。背山面海的裙房屋顶花园已成为当地婚纱摄影和户外活动的优质场所。外立面主要选用砂岩石材与GFRC材料组合装饰，角塔、柱式、扶壁、檐口、窗花、栏杆、雕饰、线脚等装饰使酒店的外观丰富而不失细腻。

Formerly a castle museum, the Castle Hotel of Dalian is situated on Lianhua Mountain. The design started with the research on ancient castles in Europe, and presented a modern castle with the combination of mountains, cities, houses, towers and spires. A large ballroom, various restaurants, an all-weather roof terrace, outdoor swimming pools and 300 hotel rooms have endowed the hotel with a luxurious and diversified identity. Diversified decorations have endowed the hotel with a rich and delicate façade, which is dominated by sandstones and GFRC materials, forming contrast between the façade and the metal roof.

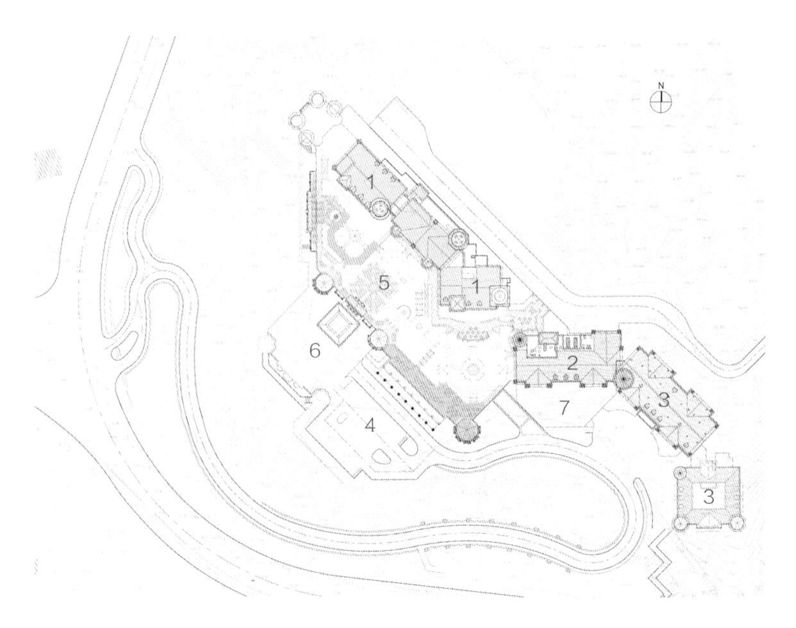

1. 酒店主塔
2. 服务式公寓主塔
3. 公寓主塔
4. 室外停车场
5. 酒店裙房屋面
6. 酒店入口观景台
7. 服务式公寓观景台

总平面图

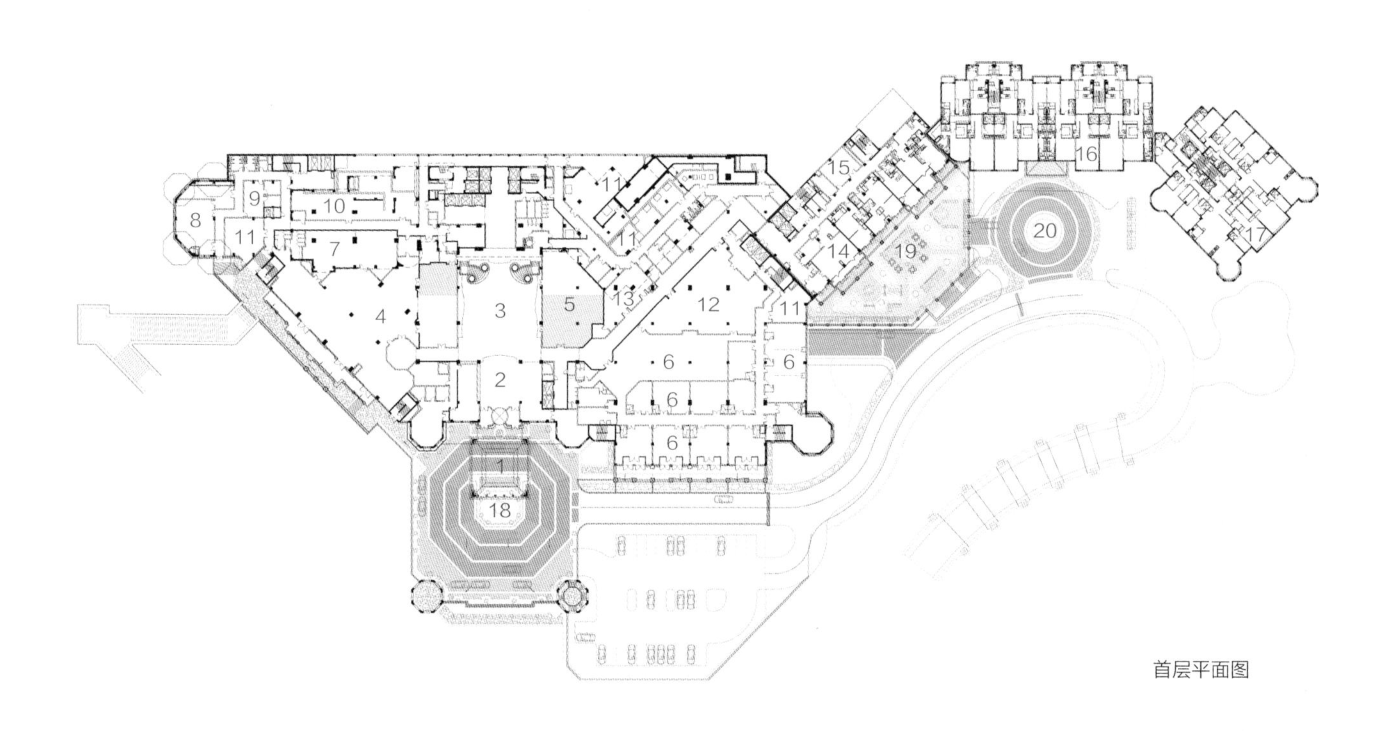

首层平面图

1. 酒店主入口门廊 2. 酒店大堂 3. 中庭 4. 全日餐餐厅 5. 酒吧 6. 中餐厅 7. 全日餐餐厅厨房
8. 员工食堂 9. 管家办公室 10. 洗衣房 11. 机房 12. 中餐厅厨房 13. 零售商店 14. 服务式公寓标间
15. 服务式公寓健身房 16. C1高级公寓 17. C2高级公寓 18. 酒店入口平台 19. 服务式公寓景观平台 20. 公寓入口平台

大连君悦酒店 Grand Hyatt Dalian

地点 辽宁省大连市 / 用地面积 15,057m² / 建筑面积 95,000m² / 高度 196m / 设计时间 2009年 / 建成时间 2014年

方案设计 美国 GP 公司
设计主持 叶 铮、刘 勤

建　　筑 魏 辰、余 洁
结　　构 施 泓、史 杰
给 排 水 赵 锂、钱江锋
设　　备 李雯筠、王 肃
动　　力 熊育铭、马 豫
电　　气 贾京花、史 敏
电　　讯 陈 琪
总　　图 余晓东

设计在用地起伏较大的情况下，着重解决了酒店复杂的功能流线与周边城市肌理有机融合的问题，保持了空间的连续性。建筑被尽量统一到完整的形体之中，塔楼采用三角形平面，端部均设倒角，减少冬季强风对细长楼体的影响，并保证所有客房都有南面朝向和最佳的海景视野。酒店的外墙主要采用玻璃、石材和金属，通过内部和外部的遮阳系统来减少阳光直射，并实现必要的客房私密性。从地面到顶棚的通长玻璃面，将客房视野最大化。而透明质感的表皮则将建筑对环境的视觉影响降低到了最小。

Located on Xinghai Square, the hotel with complex functional flows, has successfully achieved harmony with its neighborhood while standing out as a landmark. The tower has a triangular plan, which ensures that all rooms have both the sea view and the southern sunlight. The angles of the triangular plan are chamfered to reduce the negative impact of winter winds to the tower. Glasses, stones and metal dominate the façade, which is equipped with interior and exterior sun-shading systems.

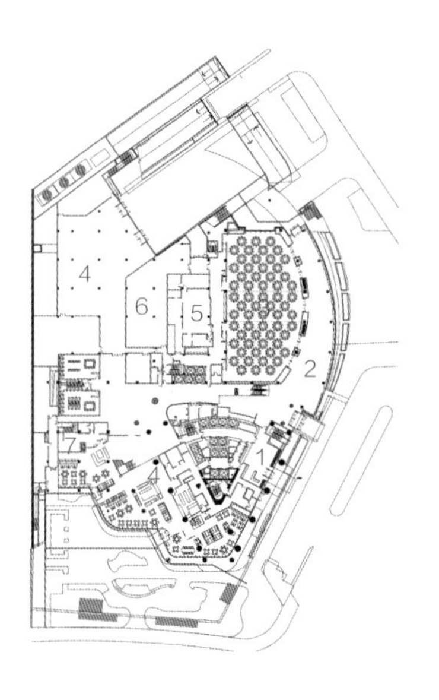

公寓入口层平面图

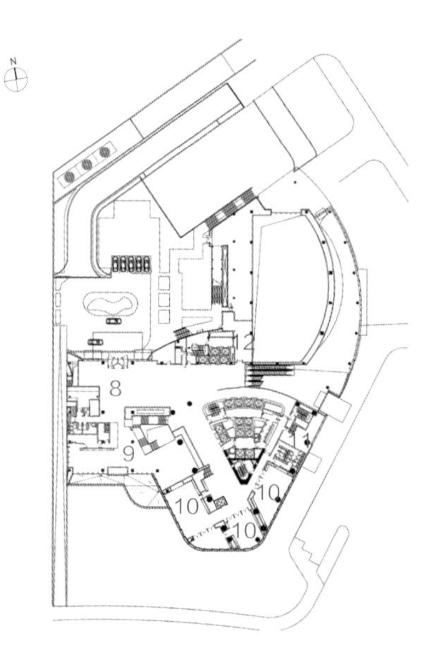

酒店入口层平面图

1. 公寓门厅
2. 多功能厅前厅
3. 多功能厅
4. 餐厅
5. 宴会厨房
6. 餐厅厨房
7. 煮食区
8. 酒店入口
9. 休闲区
10. 会议室
11. 备餐间
12. 商务中心

北京金茂万丽酒店 Renaissance Beijing Wangfujing Hotel

地点 北京市东城区 / 用地面积 9,857m² / 建筑面积 44,435m² / 高度 45m / 设计时间 2012年 / 建成时间 2014年

方案设计 美国 Gensler 建筑设计公司
设计主持 詹 红、曹晓昕

建　　筑 郑 菲、王 冠
结　　构 石 雷
给 排 水 王耀堂、周 博
设　　备 刘玉春、张志强
电　　气 李俊民、廖建军
总　　图 刘 文

金茂万丽酒店，原为建于 1987 年的王府井大饭店，酒店西侧可俯瞰故宫全景，但多年来已出现酒店形象落后、设施老化、安全隐患丛生等问题。改造保留了酒店有价值的部分，拆除并重建了裙房部分，使其拥有可分隔式宴会厅。同时拆除并重建停车场，在其上增设一处与城市周边关联的四合院会所。为扩大原本仅有 26m² 的狭小客房，采用两种方式：一种将沿街面的楼板向外出挑 1.8m，使房间面积达到 35m²；另一种将面向故宫，具有良好景观的客房三间并为两间，客房面积增至 41m²，以提升客房价值。在层高不足 3m 的情况下，通过细致的管线设计，确保了改造后较为舒适的房间及走廊净高。

Built in 1987,Wangfujing Grand Hotel, the predecessor of Renaissance Beijing Wangfujing Hotel, enjoys a panoramic view of the Forbidden City. Due to its outdated image, aging facilities and safety risks, the hotel is renovated with the preservation of valuable parts, as well as the demolition and reconstruction of the podium. The parking lot is reconstructed, where a courtyard club is built with respect for the urban context. The extension of floor slabs and the merging of hotel rooms have increased the area of a single hotel room for better experience. With a story height less than 3 meters, meticulous layout of pipelines has achieved appropriate net heights for both rooms and corridors.

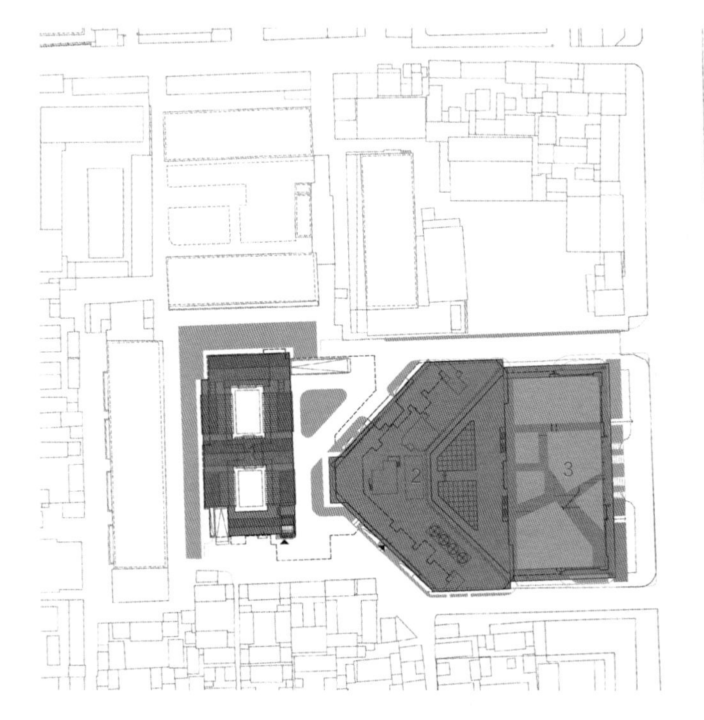

1.保留四合院
2. 塔楼部分
3. 拆除原有裙房后新建部分

总平面图

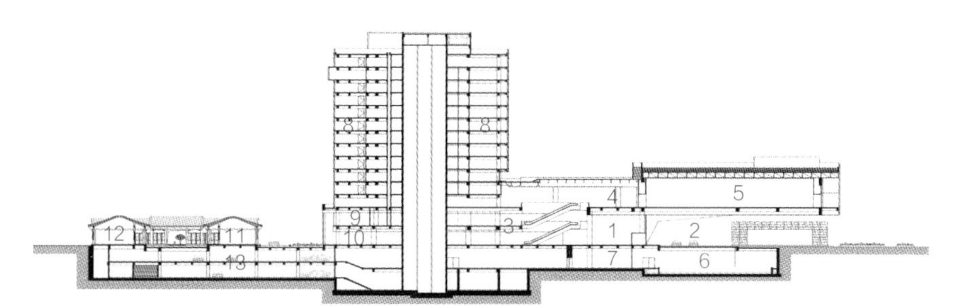

1. 大堂
2. 车道
3. 休息厅
4. 前厅
5. 宴会厅
6. 泳池
7. 水果吧
8. 客房
9. 中餐厅包房
10. 大堂吧
11. 会客室
12. SPA
13. 车库

剖面图

博鳌虎头岭1号酒店 No.1 of Tiger Head Ridge, Bo'ao

地点 海南省琼海市 / 用地面积 66,486m² / 建筑面积 23,000m² / 高度 15m / 设计时间 2012年 / 建成时间 2013年

方案设计 于海为、刘晏晏
靳哲夫、季 欣
设计主持 于海为、刘晏晏

建　　筑 靳哲夫
结　　构 杨 婷
给 排 水 董 超
设　　备 周 锐
电　　气 王 铮
总　　图 吴耀懿

虎头岭 1 号位于博鳌一处自然坡地之上，场地北侧环湖，植被丰富。规划布局以环境条件为出发点，尽量利用场地景观资源，减少对自然的破坏，将综合区主要功能沿湖一字排开，以获得最佳的景观视野。综合楼居中在交通和功能上联系各单体建筑。警卫指挥中心占据基地最高点，朝向博鳌论坛会址方向，可直观掌控会场情况。空间处理上借鉴了传统园林手法，配合依山的布局和循序渐进的空间节奏，将多层次的丰富景致穿插层叠，做到步移景异。金属结构坡屋面，用新结构与材料再现了亚热带地区传统民居特色。

Located on a natural sloping land, the No.1 of Tiger Head Ridge avails itself of the landscape resources to pose minimum interventions to the environment. The mixed-use area is located along the lake for best views, and the buildings of this area are connected by a multi-functional building. The command center for guards occupies the highest point in the site and faces the site of conference of Bo'ao Forum. Metal-structured sloping roofs are applied to all buildings, where the feature of folk houses in subtropical regions are presented.

1. 大堂
2. 会议中心
3. 指挥中心
4. 训练馆
5. 后勤楼
6. 宿舍楼

总平面图

三亚山海天酒店 Shanhaitian Resort Sanya, Autograph Collection

地点 海南省三亚市 / 用地面积 19,000m² / 建筑面积 51,600m² / 高度 81m / 设计时间 2013年 / 建成时间 2017年

方案设计 新加坡 WOW 设计公司
设计主持 逢国伟、李慧琴

建　　筑 苗铁默、隋倩婧、张　萌
结　　构 王　载、王文宇、叶　垚
给 排 水 朱跃云、关若曦
设　　备 胡建丽、苏晓峰
电　　气 曹　磊、刘征峥
总　　图 郝雯雯

摄　　影 李　季

酒店作为山海天片区最新建成的部分，引入万豪集团的最新品牌“傲途格精选”。该品牌通常以所属地区的独特生活作为设计主题，在这里突出的元素则是“水的灵动”。建筑的位置一方面让出整个片区的视觉通廊，一方面也注意让户户观海。新建部分与原有建筑共同围合出中央绿地，并保留了入口处标志性的古榕树。设计同时还达到了中国绿色建筑三星级标准，屋顶绿化、透水地面的设计使绿化率和透水率分别达到42% 和 90%。酒店客房设计为全海景朝向，客房设有内置可调节百叶的中空玻璃，保证遮阳并改善室内热环境。地下一层设置了部分导光管引入自然光。屋顶的太阳能集热器为酒店提供了生活热水。

As the latest completed part of Shanhaitian Resort, the hotel features the unique lifestyle of the city, where the “fluidity of water” is manifested in each element of this hotel in the coastal city of Sanya. The location and layout of the building have provided all the rooms with direct view of the sea, while enclosing a central greenery together with original buildings on site, and preserving an old banyan at the entrance area. The design has acquired a 3-star Green Building Label of China. Green roofs and permeable ground has increased the rate of greening and permeability to 42% and 90% respectively.

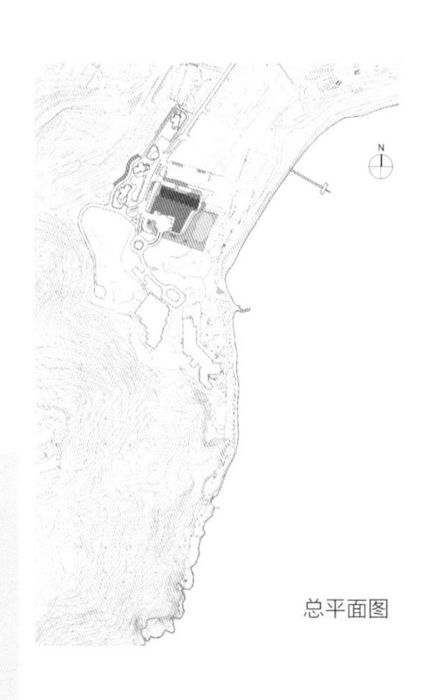

总平面图

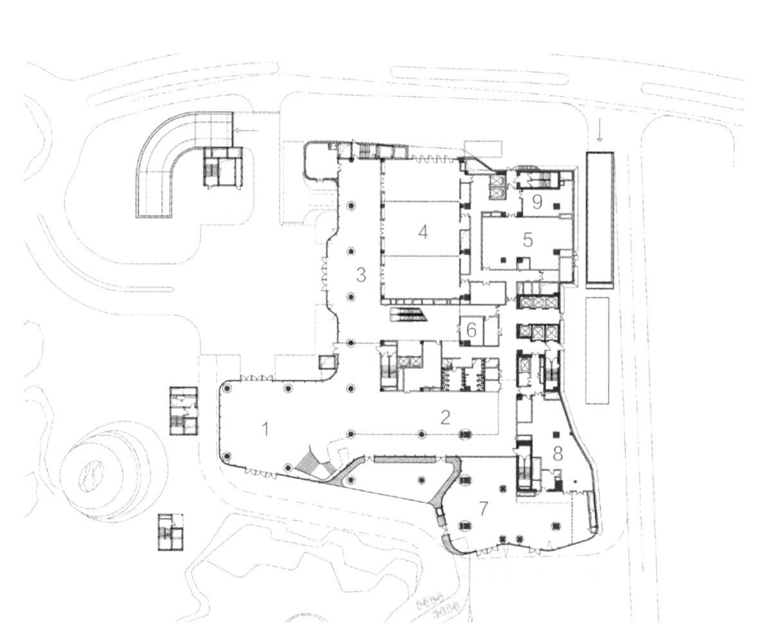

1. 大堂
2. 休息区
3. 宴会厅前厅
4. 宴会厅
5. 宴会厅厨房
6. 销售商店
7. 餐厅
8. 餐厅厨房
9. 消防控制室

首层平面图

曲阜鲁能JW万豪酒店 JW Marriott Hotel Qufu

地点 山东省曲阜市 / 用地面积 56,200m² / 建筑面积 54,736m² / 高度 9m / 设计时间 2012年 / 建成时间 2018年

方案设计 崔 愷、陆 静、何 佳
设计主持 崔 愷、陆 静、何 佳

建　　筑 成心宁
结　　构 徐 杉、高芳华、高彦良
给 排 水 王世豪、王 睿
设　　备 朱永智、刘权熠
电　　气 李维时
总　　图 王 炜

酒店的选址，与闻名遐迩的曲阜孔庙仅一街之隔，如何在尊重历史文物的前提下，将五星级酒店的需求融入院落式布局的建筑群中是设计的首要难点。布局采用当地建筑群南北向层层院落递进的形式，整体轴线与孔庙轴线平行。曲阜孔庙为宫殿式建筑群，孔府则偏向于住宅。酒店设计以公共区和客房区与二者分别对应，体现对传统的尊重。公共区屋顶举折明显，形成优美的屋面曲线，客房区则相应平直，采用无屋脊的卷棚顶，轻巧舒适，也衬托出东侧孔庙的雄浑气势。公共空间吸取山东园林的特点，以方亭作为空间核心；客房部分则以精巧的民居院落为原型。

The site of the hotel was just across the street from the renowned Confucius Temple, and the design was thus challenged by the requirements of respecting historic relics and accommodating the functions of a hotel into a courtyard-style layout. The axis of the whole project is parallel to that of the Temple of Confucius. Since the Temple consists of palatial buildings while the Mansions of Confucius are dominated by residences, to reflect such a contrast, the hotel's divided into a public area and a room area. In the public area, the raising of truss is adopted; while ridge-free roofs of the room area remain low-profile.

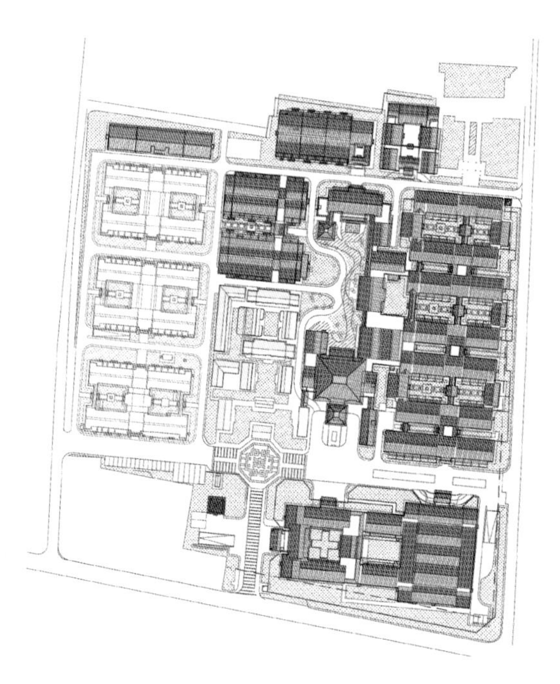

总平面图

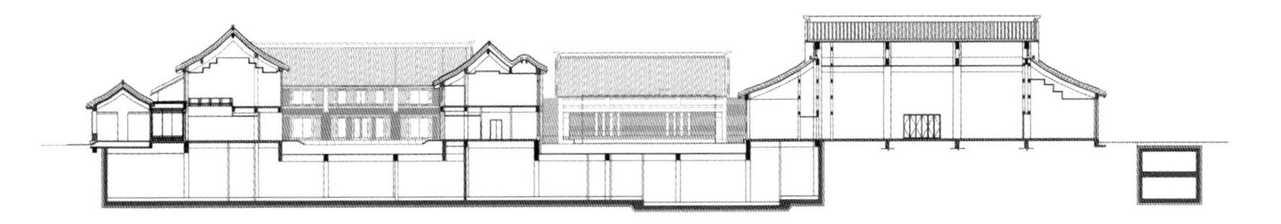

剖面图

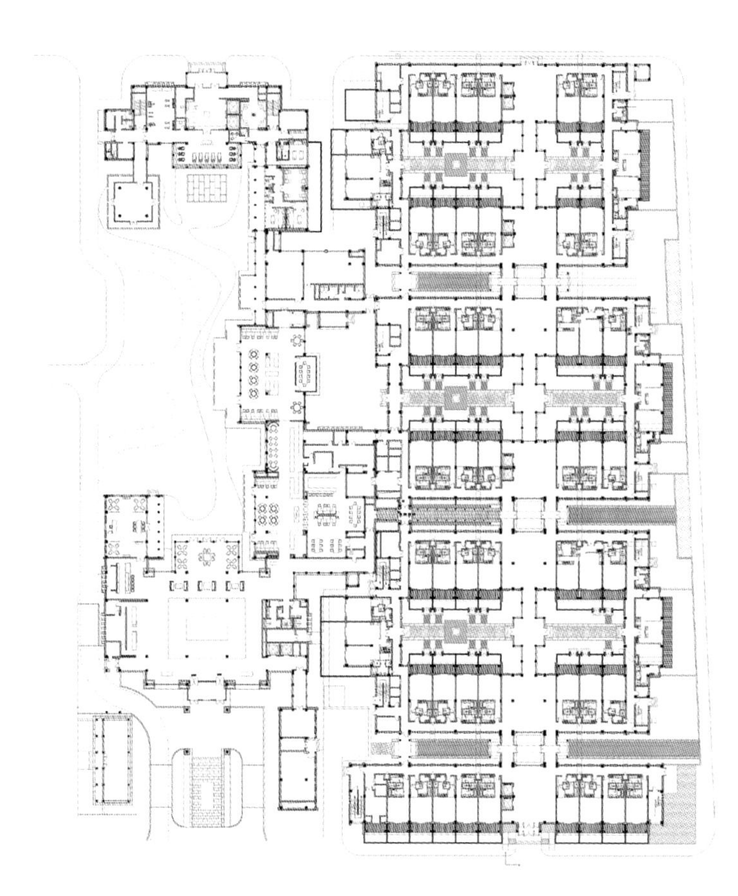

客房部分首层平面图

垂花门、照壁和孔庙叠落的五花山墙等传统元素，也都通过现代化的阐释应用于建筑之中。

Traditional building elements such as floral-pendant gates and spirit screens are applied in the buildings in a modern manner.

2018中国（南宁）国际园林博览会 2018 China (Nanning) International Garden Expo

地点 广西壮族自治区南宁市 / 园区面积 2,757,000m^2 / 建筑面积 57,937m^2 / 高度 15m / 设计时间 2013年 / 建成时间 2015年

规划设计总指导 崔 愷

总体规划及景观设计主持
李存东、史丽秀、赵文斌

建筑设计主持
崔 愷、景 泉、崔海东
黎 靓、金海平、杨 磊

项目经理
赵文斌、雷洪强、张军英

南宁园博园位于南宁市东南方向的顶蛳山，西临八尺江，其规划设计注重生态环境的恢复和可持续利用，注重地域文化的传承创新和游览体验。全园规划了九大功能区，形成了“三湖十八岭”的山水格局，结合现状水体并考虑了行洪蓄洪要求，形成了不同尺度的水系景观，并在保留现状十八座山体的基础上，通过生态手法和景观手法，形成三大山岭景观脉系。

Located at mountains in southeast Nanning, the Expo Garden highlights the restoration and sustainable utilization of the eco-system, as well as the local features and visitors' experience. Divided into 9 functional areas, the garden forms a layout of "3 lakes and 18 hills" with consideration of floodwater storage and drainage.

中国建筑设计研究院参与部门

环境艺术设计研究院 园区规划、景观设计、精装修设计

本土设计研究中心 建筑单体设计审定

第一建筑专业设计研究院 建筑单体设计、园区规划

第一工程设计研究院 机电设计

第二工程设计研究院 建筑结构、机电设计、小市政机电设计

第三工程设计研究院 建筑结构、机电设计

设计咨询管理中心 设计管理、项目建议书、可研、能评、项目概预算

第一总图市政设计研究所 道路、桥梁设计

第二总图市政设计研究所 场地平整及土石方工程

智能工程中心 智慧园区专项及建筑智能化

绿色设计研究中心 绿色建筑

交通规划研究中心 交通规划

建筑历史研究所 遗址保护及展示工程

同属中国建设科技集团机构

中国城市建设研究院有限公司 市政路设计审定

中旭建筑设计有限责任公司 建筑单体设计

外部合作机构

南宁市古今园林规划设计院 种植设计

北京多义景观规划设计事务所 矿坑花园设计

南宁市勘察测绘地理信息院 土方调配及地基处理

广西交通规划勘察设计研究院有限公司 车行桥结构设计

北京市建筑设计研究院有限公司 海绵城市设计

北京清华同衡规划设计研究院 健康花园设计

北京宁之境照明设计有限公司 照明设计

深圳市中世纵横设计有限公司 标识设计

深圳市大地幕墙科技有限公司 幕墙顾问

德润诚工程顾问（北京）有限公司 幕墙顾问

中通服广西设计院 智慧园区顾问

东盟馆横跨东盟湾而建，以广西特有的风雨廊桥为原型，架于东西两山坡之间。十国展厅均为六边形标准单元体，“手拉手，环环扣”，外廊则由互相搭接的防雨百叶覆盖。展厅室内悬挂帐篷状的膜结构，喷绘代表各国特色的花卉图案。通透的格栅掩映出膜帐的朦胧美。

标志性建筑清泉阁是整个园区的制高点。塔身通过扭转对应园区景观大道，并与市区龙象古塔遥相呼应，喻迎宾之意。纯粹的受力构件体现了结构之美，层层金属的外檐让人联想到少数民族传统的统密檐塔。扭转的形体如同“A”与“V”的结合，自然形成底部大空间与顶部观景大平台，也让建筑的不同侧面呈现出不一样的形态。

Built over the ASEAN bay, the ASEAN Pavilion is based on the prototype of the covered bridge of Guangxi. Exhibition halls of ten countries, composed of standard hexagon units, are arranged side by side, with the exterior corridor covered with overlapped rainproof louvers. Tent-like membrane structures in the exhibition halls are painted with patterns of each nation’s representative flowers.

Qingquan Pavilion, the garden’s landmark building with a commanding height, is rotated to coordinate with both the landscape boulevard and the ancient Longxiang Tower in the distance. Dense metal eaves, supported by steel structure, remind people of the traditional dense-eave tower while showcasing the beauty of its structure.

游客中心的设计考虑开园期间大量游客排队等候的需要，以及南宁炎热多雨的气候特点，以广西传统村落中成片屋顶的形式作为有效应对方式。其布局顺应南北山坡，260m 的连续弧形屋顶沿山坡覆盖 10 个单体建筑，并形成 3 个层次，采用曲线钢桁架上搭纵向直线钢檩条，形成有韵律的廊下空间。
宜居 · 城市馆利用原有的两座小山坡，将建筑一层嵌入山体；并将展览空间打散，形成小空间、聚落化的布局；充分利用园区建设中的“废料”，利用碎石、红土等原料形成由石笼、夯土、木色格栅、毛石组成地方材料系统。

Taking the queuing of tourists and the rainy weather into consideration, the design for the tourist center has adopted the form of “roof group” seen in traditional Guangxi villages. The layout of the center conforms to the terrain of the hills, with a 260-meter curved roof covering 10 individual buildings and forming 3 groups. The Livable City Pavilion has its first floor embedded into the existing hills on site. The exhibition spaces are scattered to form small clusters. Furthermore, the “waste materials” of the construction, such as gravels and rubble, are reused to form a system of local materials.

体验馆所在原有场地植被因采矿被破坏。设计从生态修复的角度出发，以广西干栏式民居为原型，整体架空。赛歌台受少数民族河边树下对歌的启发，屋顶为三角形几何分格，形成上下凹凸的锥体，如同榕树硕大的树冠，满足了演出、赛歌等功能需要。贝丘博物馆如同一片落在山野的贝壳，以圆润弧墙将人导引到展陈序列中，展示了园博园内顶蛳山上发掘出土的新石器时代遗址。

芦草叠塘、玲珑揽翠、花阁映日等八大景区，结合各自的地形地貌形成不同可供游览的小品，展现了人与自然和谐共生的生态理想。

The vegetation of the Living Experience Pavilion's site was damaged by previous mining activities. As a result, the design starts from ecological restoration. The singing stage has a cone shape with its roof divided into triangles. Beiqiu Museum, like a shell in the field, has curved walls that guide the visitors into the exhibition while showcasing the relics of Neolithic age.

Accessorial buildings built in the 8 scenic spots, varied in their forms, showcase the ideal of harmony between people and nature.

北京世界园艺博览会生活体验馆 Life Experience Pavilion at Beijing International Horticultural Expo

地点 北京市延庆区 / 建筑面积 36,000m² / 设计时间 2016年 / 建成时间 2019年

方案设计 郑世伟、罗 云、李 越
设计指导 崔 愷
设计主持 郑世伟

建　　筑 罗 云、史 倩
　　　　 李 越、李慧敏
结　　构 邵 筠、冯启磊、刘文阳
给 排 水 吴连荣、郭瑞雪
设　　备 杨向红、郭 超
电　　气 李维时、于天傲
电　　讯 唐 艺
总　　图 吴耀懿、王 炜
　　　　 路建旗、白红卫
室　　内 曹 阳、马萌雪、闫 宽
景　　观 朱燕辉、李 飒、戴 敏
　　　　 管婕娅、王 悦
绿　　建 王陈栋、林 波、王芳芳
照　　明 马 戈、刘冰洋、王博源

生活体验馆位于世界园艺博览会园区贴近延庆县城一侧。作为二者之间的纽带，生活体验馆旨在成为一个开放的“社区活动中心”，因而营造为温暖亲和的北方村落形态。16 个 27 米见方的“盒子”由横纵交织的街巷联系，适合多种使用方式。4 个盒子上部挖空，形成空中院落，并分别以“果、药、菜、茶”为主题，提供别致的园艺体验。以低技方式将土、木、砖、石等传统材料作为外围护材料，强化了建筑北方地域的厚重感。被恢复的原始农业景观，则将播种、灌溉、收获等过程直观地呈现给游客，传达了绿色生态的生活观。

The Life Experience Pavilion at Beijing International Horticultural Expo is positioned as an open “community center”, with a welcoming courtyard-styled layout. 16 “boxes” with sides of 27 meters are connected with crossed streets and alleys, forming 7 clusters for various purposes. 4 of the 6 boxes are hollow at the top, forming upper courtyards with various themes that offer unique experiences. Materials such as soil, wood, brick and stone, are applied in low-tech manners to present the solidness of buildings in Northern China.

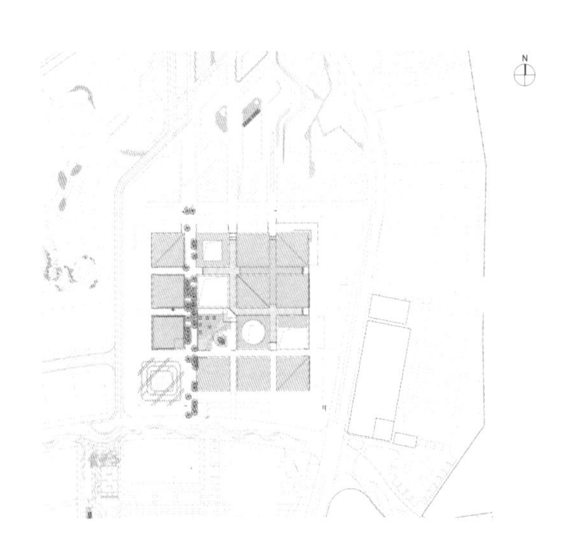

总平面图

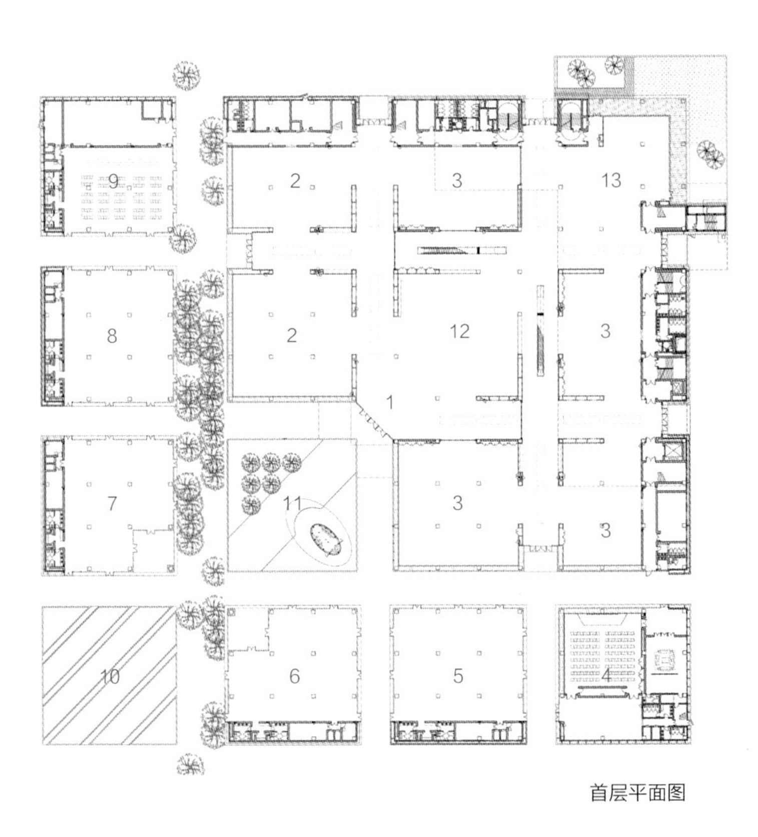

首层平面图

1. 主展厅 2. 自然园艺花园 3. 生活园艺社区 4. 多功能厅 5. 艺创生活馆 6. 众筹生活馆 7. 茗传四海馆 8. 众创时尚馆 9. 餐厅 10. 主入口前广场（麦田） 11. 集散区（种子） 12. 景观坡道（灌溉） 13. 生活园艺社区（果实）

李季 摄

建筑外界面能够反映延庆地区特征的材料，包括由妫河石组成的石笼墙、木格栅墙、青砖墙，还有就地取土筑成的夯土墙。

Materials with local features are applied to exterior walls, which are made of wood gratings, black bricks, rammed earth or stones from Guirui River.

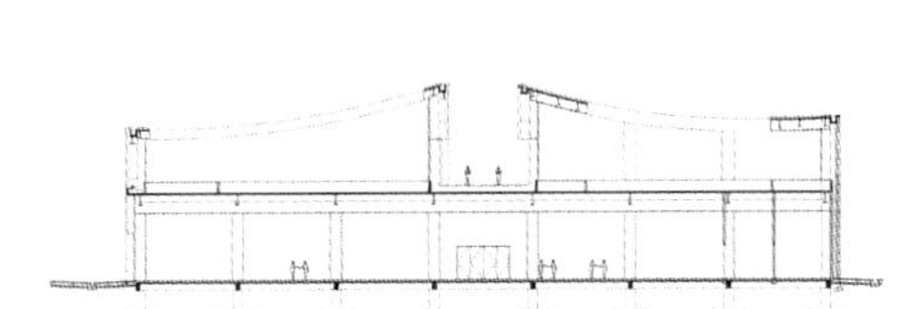

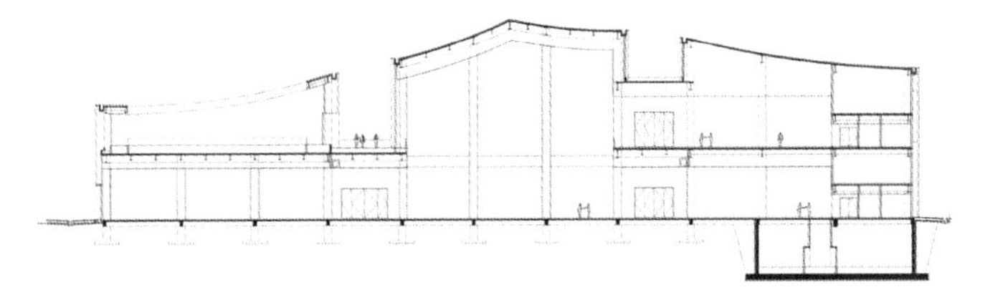

剖面图

北京世界园艺博览会妫汭剧场 Guirui Theater at Beijing International Horticultural Expo

地点 北京市延庆区 / 建筑面积 6,335m^2 / 高度 20m / 设计时间 2016年 / 建成时间 2019年

方案设计 汪 恒、李 蕾、徐 超
陈 璐、赵一霖、张思雨
陈 桐

设计主持 汪 恒

建　　筑 徐 超、李 蕾、张思雨
李 城、陈 桐、菅 睿
罗 颖

结　　构 张 路、朱禹风
崔小连、郭 强

给 排 水 杨东辉

设　　备 孙淑萍

电　　气 王苏阳、姜海鹏
沈 晋、张 龙

总　　图 吴耀懿

室　　内 王 强、纪 岩、赵 阳
刘子贺、李 甲、曹 诚

景　　观 史丽秀、刘 环、孙 昊
齐石茗月、曹 雷、魏 华

摄　　影 李 季

妫汭剧场位于北京世园会核心景观区，是开、闭幕式及重要国家日的主要舞台。建筑外形如一只展翅欲飞的彩蝶，其主体钢结构屋面翼展双向跨度120m × 115m，轻盈灵动的建筑造型融于山水格局之中。可用、可观、可游、可赏的半室外剧场建筑呼应了世园会“绿色生活，美丽家园”这个主题。剧场主体为半室外空间，辅助用房为覆土建筑，均尽量减少能耗需求。主体钢结构由26榀钢桁架构成，结构自平衡并通过钢索加强备份，悬挑最大达到47m，通过高低错落模拟出蝴蝶轻盈、飘逸的双翅。

Guirui Theater is the main stage for the expo's opening ceremony, closing ceremony and national-day activities. The form of the building resembles a fluttering butterfly. The spans of its steel roof are 120m x 115m in two dimensions, presenting a light and vigorous form that achieves harmony with the mountains and waters. The main steel structure consists of 26 steel trusses, with a self-balancing structure reinforced with steel cables, reaching a maximum overhanging length of 47 meters, revealing the lightness of a butterfly's wings.

1. 室外集散平台
2. 观众入场景观台阶
3. 室外观众座席
4. 大剧场升降舞台
5. 小舞台
6. 湖面

看台层平面图

1. 彩色EFTE膜
2. 钢桁架
3. 高开孔率铝合金丝勾花网
4. 低开孔率铝合金丝勾花网
5. 大剧场看台
6. 大剧场用房
7. 大剧场舞台
8. 小剧场看台
9. 小剧场用房
10. 小剧场舞台
11. 景观绿坡

建筑空间生成

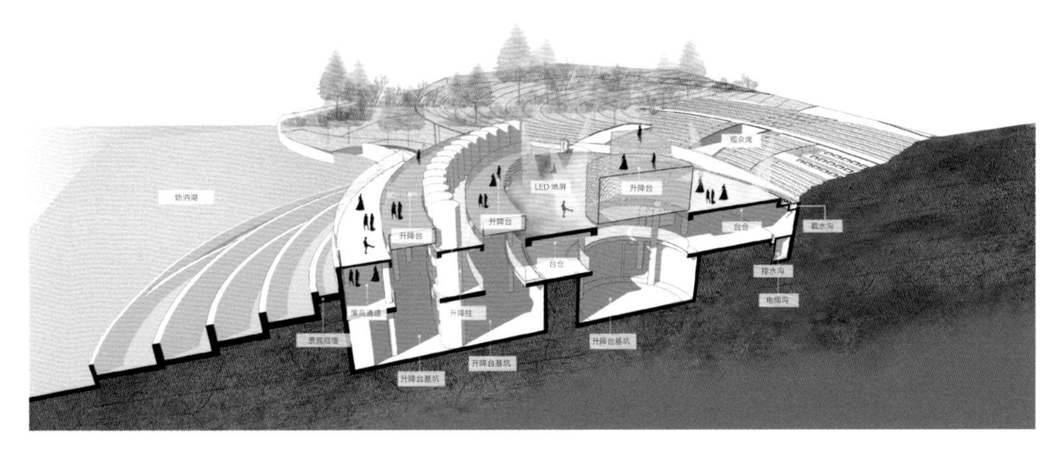
妫汭湖
观众席
LED 地屏
升降台
升降台
升降台
台仓
台仓
截水沟
排水沟
电缆沟
演员通道
升降柱
景观挡墙
升降台基坑
升降台基坑
升降台基坑

舞台剖透视图

海口市民游客中心 Haikou Citizen & Tourist Center

地点 海南省海口市 / 用地面积 39,191m² / 建筑面积 28,976m² / 高度 21m / 设计时间 2017年 / 建成时间 2018年

方案设计 崔 愷、康 凯、张一楠
王庆国、马 欣、朱 巍
设计主持 崔 愷、康 凯

建 筑 朱 巍、张一楠
结 构 史 杰、郑红卫
给 排 水 杨东辉、董新淼
设 备 郑 坤
电 气 王苏阳
电 讯 张月珍、白雪涛
总 图 齐海娟
室 内 郭晓明、魏 黎
景 观 关午军

海口市民游客中心位于海口滨海公园内，靠山邻水，容纳城市服务及旅游服务等政府功能，同时设有城市形象展示场所。建筑整合城市及沿湖的空间，将当地传统的骑楼街巷的图底关系融入建筑中，在一层和地下一层分别形成内街和带型下沉广场，将建筑自然划分为东西两条。东侧靠山的部分与山体相结合，错落有致地处理为几个盒状体量，插入山中；西侧与现有的青少年活动中心和港湾小学形成连续界面，并遮挡附近变电站对沿湖视觉环境的不良影响；内街南端以一个伸入公园内湖的体量形成对景。

Haikou Citizen & Tourist Center provides services to both the citizens and tourists while showcasing the image of the city. The "figure-background" relationship of traditional arcades and alleys was applied in the design with an interior street on the 1st floor and a belt-shaped sunken square dividing the building into two parts. The eastern part has several cubical volumes embedded into the mountain; the western part forms a continuous façade together with the surrounding buildings while weakening the negative visual impact of a power grid company's building on the lakeside.

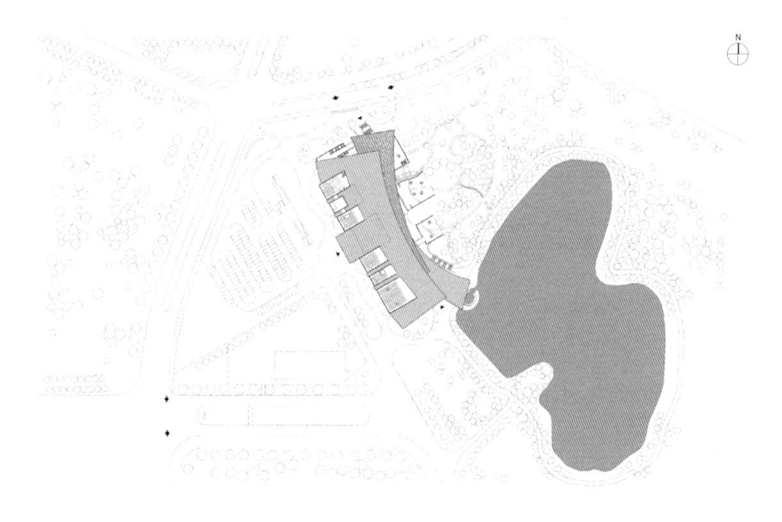

总平面图

建筑屋顶形态取意海口独有的民居、海洋、自然形态，三片木结构屋面由若干 V 形钢柱依次跌落撑起。起伏的形态不仅形成了丰富的立面视觉效果，更是营造出遮阳挡雨的半室外开放街巷空间。

The form of the roofs symbolizes the images of folk houses and oceans, with three pieces of timber roofs supported by several V-shaped steel columns. The undulating form of the building has not only enriched the features of the façade, but also created a shelter in the alley.

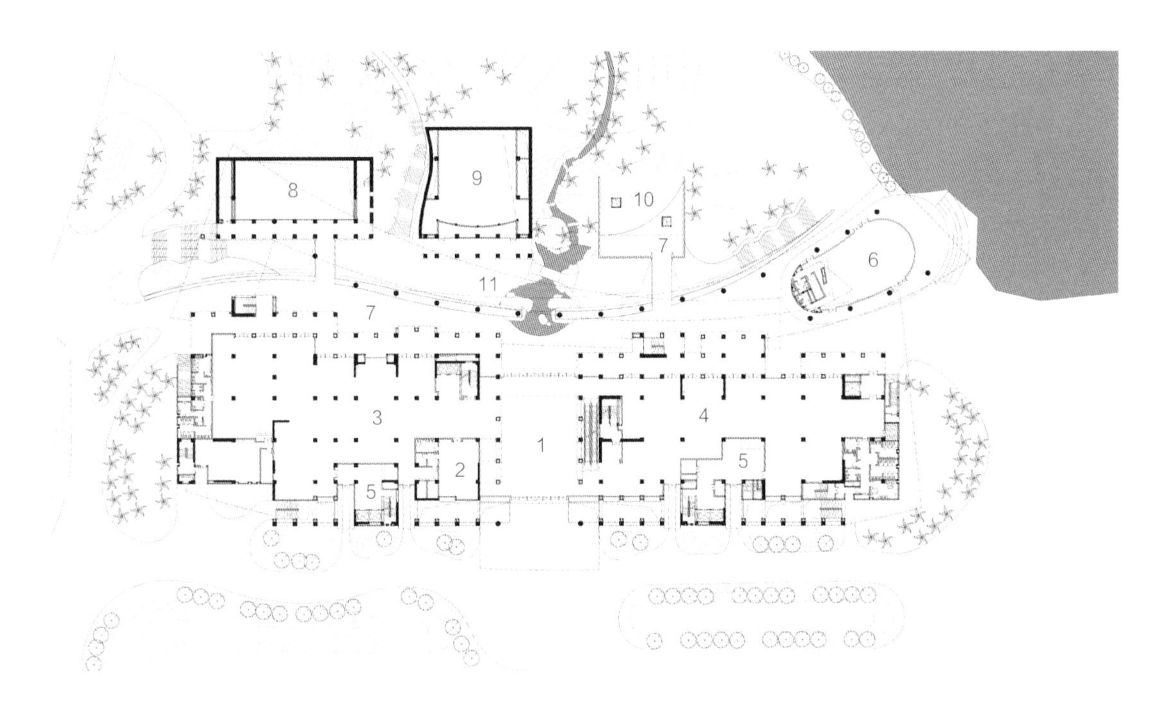

1. 接待大厅 2. 贵宾接待室 3. 智慧城市数字展厅 4. 规划展厅 5. 办公区门厅 6. 多功能厅
7. 室外平台 8. VR展厅上空 9. 报告厅上空 10. 咖啡厅上空 11. 下沉广场上空

首层平面图

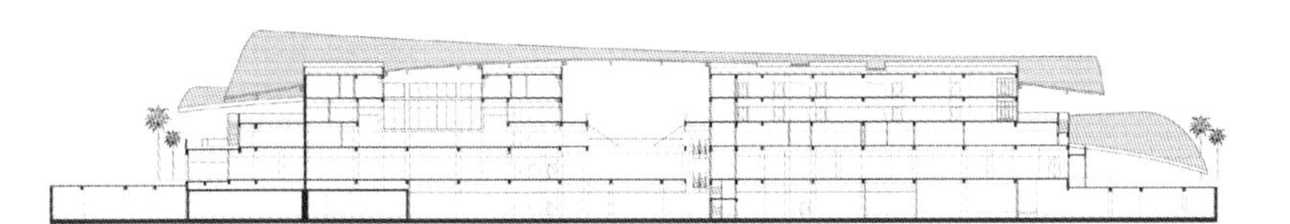

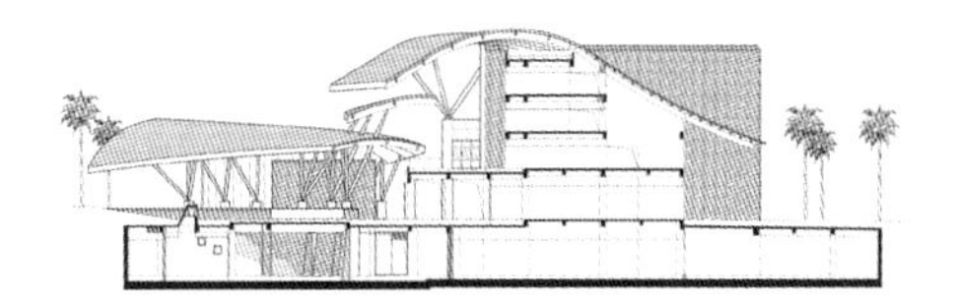

剖面图

建筑上部采用钢木混合结构，支撑构件为钢结构，主梁、次梁、屋面板均采用木结构。屋面则采用色彩变化自然的红雪松木瓦，能够与环境充分融合。

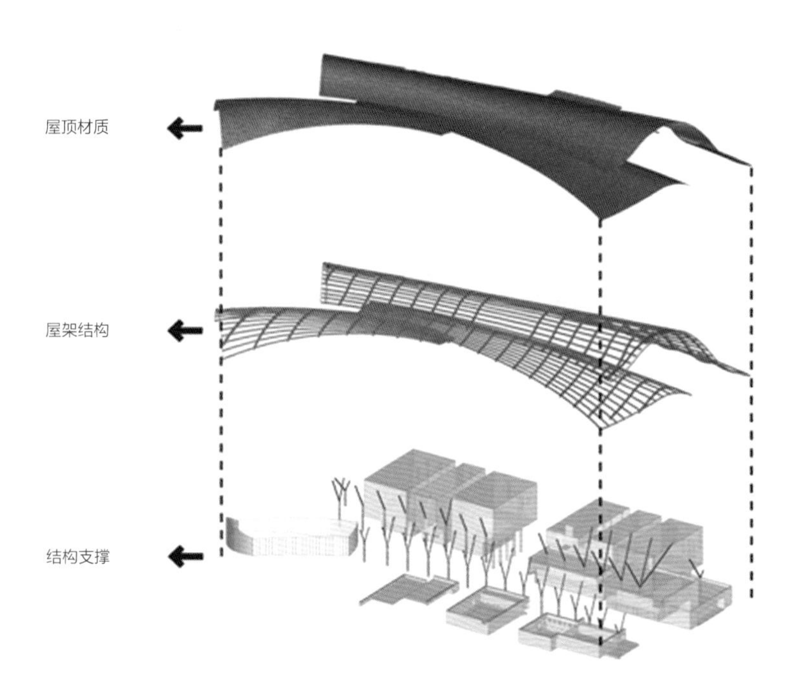

结构分析图

The upper part is a mixed structure of steel and wood. The roof of red cedar tiles, with natural variations in their color, exists in harmony with its surroundings.

玉树康巴艺术中心 Yushu Khamba Arts Center

地点 青海省玉树藏族自治州 / 用地面积 24,563m² / 建筑面积 20,610m² / 高度 28m / 设计时间 2011年 / 建成时间 2013年

方案设计 崔　愷、关　飞、曾　瑞
　　　　 董元铮、高　凡
设计主持 崔　愷

建　　筑 关　飞、曾　瑞
　　　　 董元铮、高　凡
结　　构 王　载、王文宇、陈　明
给 排 水 董　超
设　　备 尹奎超
电　　气 陈沛仁、廖建军
总　　图 白红卫
室　　内 郭晓明、魏　黎

玉树康巴艺术中心是为遭受地震后的结古镇所作的重建工程，汇集了原玉树藏族自治州的剧场、剧团、文化馆和图书馆等多种功能。本着经济、环保的原则，设计将州剧团的辅助功能区与剧院的后台区合并，将排练厅与室外演艺功能合并，尽可能控制规模，并强化各功能区的通用性。平面布局力图通过再现院落空间的组合体现传统藏式建筑的空间精神，并尝试在建筑布局上体现台地特征，建筑在体量上也逐层递减。基于低造价的考虑，建筑通过沿袭传统藏式建筑的色彩形成丰富的视觉效果。

Khamba Arts Center is a reconstruction project in the earthquake-stricken Jiegu Town. The project incorporates various functions including a theater, a troupe, a cultural center and a library. On the basis of cost effectiveness and environmental protection, the design combines the auxiliary areas of the troupe with the backstage, and integrates the rehearsal hall with the outdoor performing area. The layout presents a combination of courtyards to showcase the atmosphere of traditional Tibetan buildings.

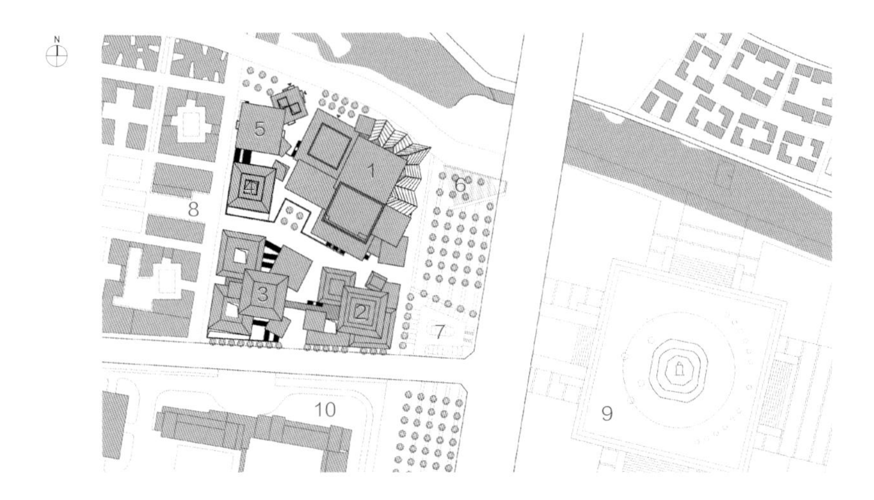

1. 剧场
2. 图书馆
3. 文化馆
4. 剧团
5. 电影院
6. 剧场前广场
7. 停车场
8. 唐蕃古道商业街
9. 格萨尔广场
10. 周边小学

总平面图

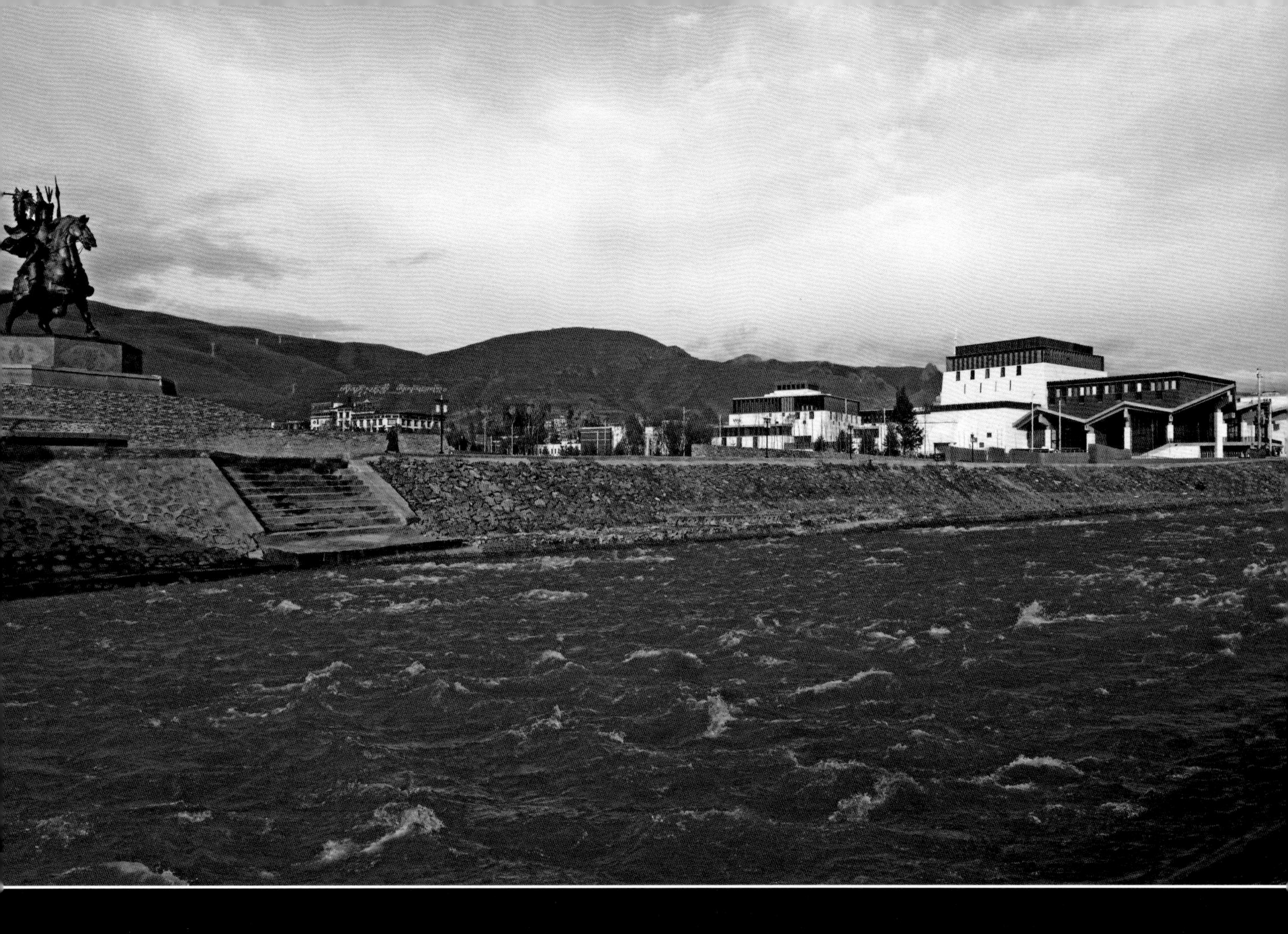

设计从尊重城市文脉的角度出发，总体布局自由松散但错落有致，强调与塔尔寺、唐蕃古道商业街、格萨尔广场等周边城市元素的对位呼应。从建筑的密度上与传统城市肌理相吻合，步行街道的尺度也尽力与唐蕃古道商业街相协调。

With respect for the urban context, the general layout of the arts center is both stretched and well-proportioned, highlighting its compliance to the context of its neighborhood. The density of buildings and the scale of streets are both coherent to Tangbo Passageway Shopping Street.

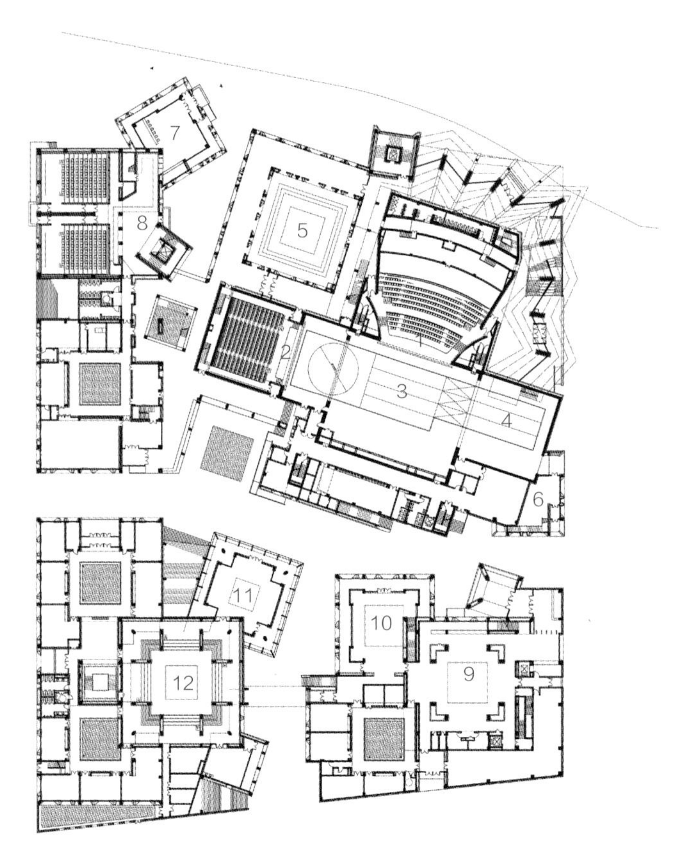

1. 大剧场
2. 多功能剧场
3. 主舞台
4. 侧台
5. 半室外演艺
6. 票务厅
7. 电影院门厅
8. 电影院大厅
9. 期刊阅览
10. 儿童阅览
11. 展厅
12. 共享大厅

首层平面图

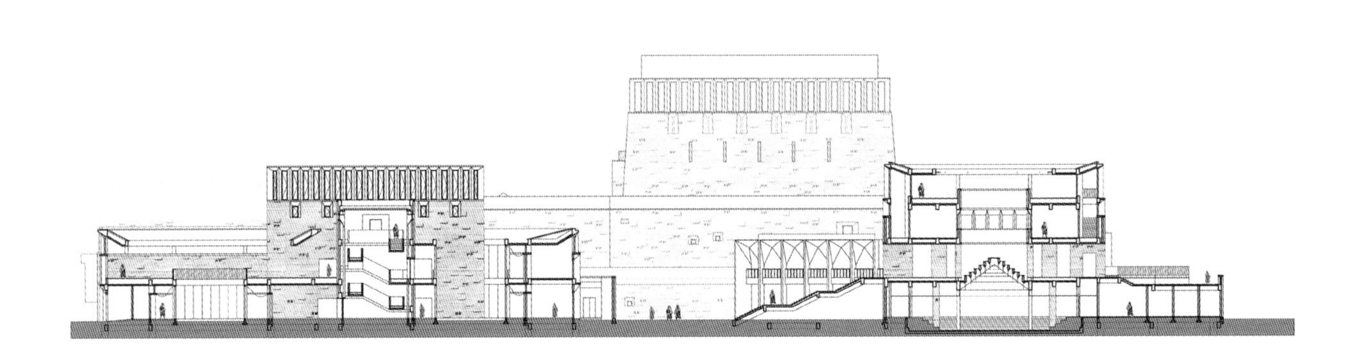

图书馆、文化馆剖面图

元上都遗址博物馆 Site Museum of Xanadu

地点 内蒙古自治区锡林郭勒盟正蓝旗 / 用地面积 6,747m² / 建筑面积 4,997m² / 高度 10m / 设计时间 2009年 / 建成时间 2015年

方案设计 李兴钢、谭泽阳
付邦保、赵小雨
设计主持 李兴钢、付邦保

建　　筑 付邦保
结　　构 王立波、高银鹰、张剑涛
给 排 水 刘　海、何　猛
设　　备 李超英、向　波
电　　气 甄　毅
总　　图 余晓东

博物馆为配合元上都古都城遗址申报世界文化遗产而建，位于平地隆起的草原山峰“乌兰台”。设计充分利用废弃的采矿场布置博物馆主体，以修整被破坏的山体，并结合“乌兰台”蒙语中“红色山岩”之意，采用了红色清水混凝土，犹如从山体中延伸而出。主体建筑设于半山处，充分利用现有废弃的采矿场来布置，既修整被破坏的山体，也将大部分建筑体量掩藏于山体内，遵循了对文化遗产最小干预的原则。半藏半露的长方体部分隐喻遗址城垣，其位置与山体等高线相交并指向元上都遗址的明德门，让建筑与遗址产生轴线上的联系。沿着博物馆的内外参观路径设置了一系列远眺遗址和草原丘陵地景的平台，直至到达山顶敖包。长长的路径和不断停驻的平台是博物馆不可分割的组成部分，将元上都的历史、文化和景观在此串联。

Built for the application of the site of Xanadu for the title of World Cultural Heritage, the museum was located 5 kilometers across a grassland from “Wulantai”, meaning beacon of the red rocks in Mongolian. Mostly embedded in the pit formed from an abandoned quarry, it emerges from the earth in a geometric collection of iron oxide infused concrete, conforming to the principle of minimum intervention to the heritage. Furthermore, the building is oriented to intersect with both the mountain contours and the starting point of the capital city ruins' central axis. A series of platforms for overlooking the ruins and the grasslands are located along both exterior and interior paths, which stretch all the way to the Ovoo on the hilltop that reminds the histery and culture of Mongolian.

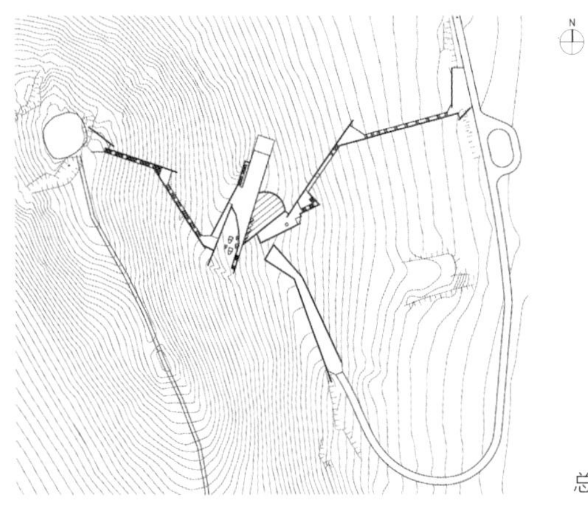

总平面图

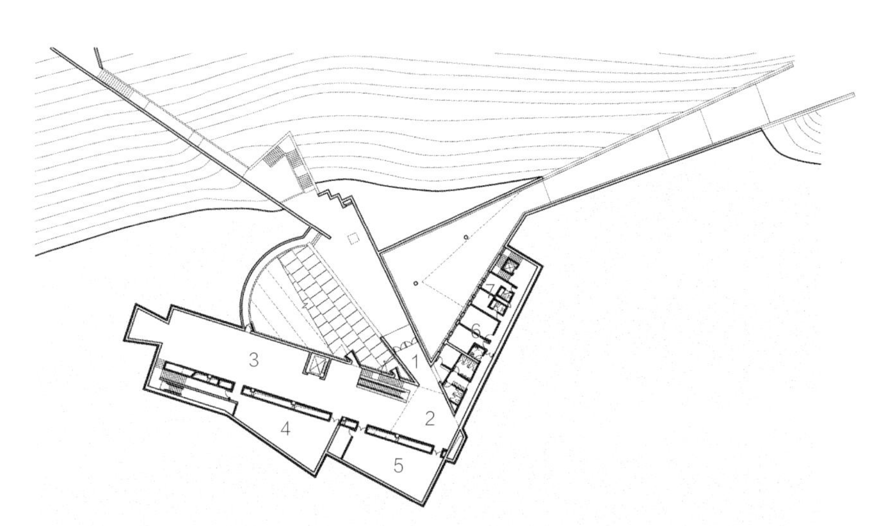

1. 门厅
2. 序言厅
3. 主题展厅
4. 临时展厅
5. 4D 影院
6. 办公室
7. 宿舍

首层平面图

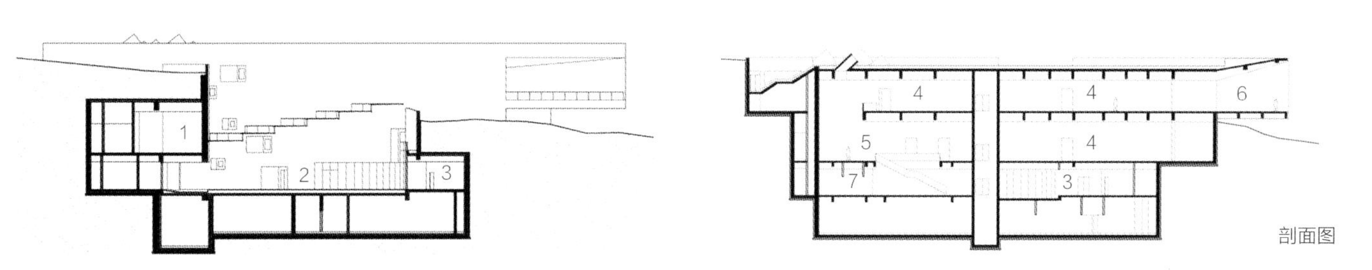

1. 门厅 2. 室外庭院 3. 观众服务 4. 主题展厅 5. 序言厅 6. 观景厅 7. 厨房

敦煌莫高窟数字展示中心 Mogao Grottoes Digital Exhibition Center

地点 甘肃省敦煌市 / 用地面积 40,000m² / 建筑面积 10,440m² / 高度 16m / 设计时间 2012年 / 建成时间 2015年

方案设计 崔 愷、吴 斌
赵晓刚、张汝冰
设计主持 崔 愷

建 筑 吴 斌、赵晓刚
结 构 刘建涛
给 排 水 朱跃云
设 备 王 加
电 气 胡 桃
总 图 高 治
室 内 顾建英、张明晓
景 观 冯 君

莫高窟被誉为"东方艺术宝库"，但庞大的游客数量对遗产的保护和管理造成很大困扰。这座建于绿洲和戈壁之间的莫高窟数字展示中心，即为缓解景区的保护压力而建，集合了游客接待、数字影院、球幕影院、多媒体展示、餐饮等功能。设计伊始，我们最初的感动来自对大自然的敬畏和对古代工匠精美艺术的敬佩。这座建筑，应该是大漠戈壁中的一座小沙丘，造型既如同流沙，如同雅丹地貌中巨舰般的岩体，又类似矗立在沙漠中的汉长城，莫高窟壁画中飞天飘逸的彩带，充满着强烈的流动感。若干条自由曲面的形体相互交错，婉转起伏，巨大的尺度和体量将沙漠地景建筑的特征表达得淋漓尽致。

The digital exhibition center, with functions including reception, film projection, multimedia exhibition and catering services, was built to relieve the pressure of the scenery spot. At the start of the design phase, we were impressed and inspired by both the nature and the ancient craftsmanship. The building resembles the quicksand, the rocks of Yardang landform, the Great Wall in the desert and the flying ribbons in the mural paintings with its dynamic form. Furthermore, the undulant curved roofs are interlaced and interwoven, giving full play to the features of a landscape building in the desert.

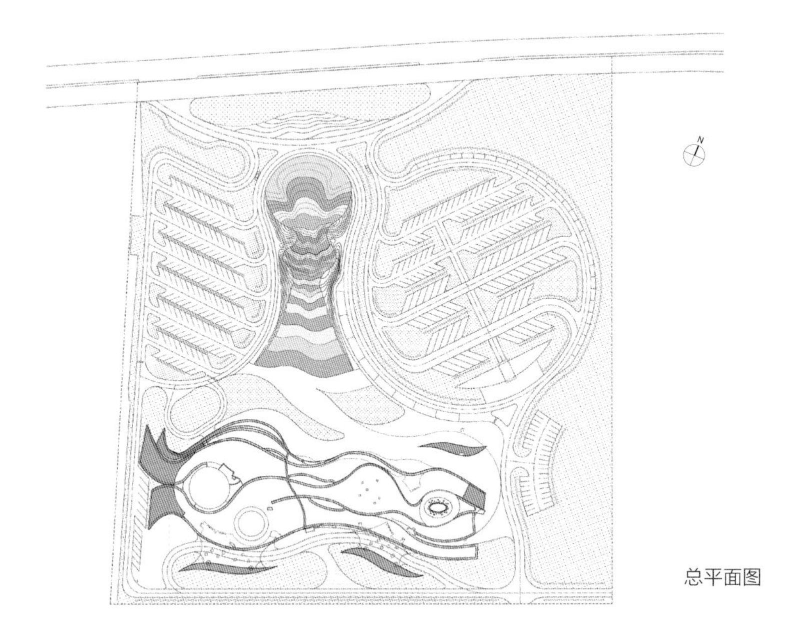

总平面图

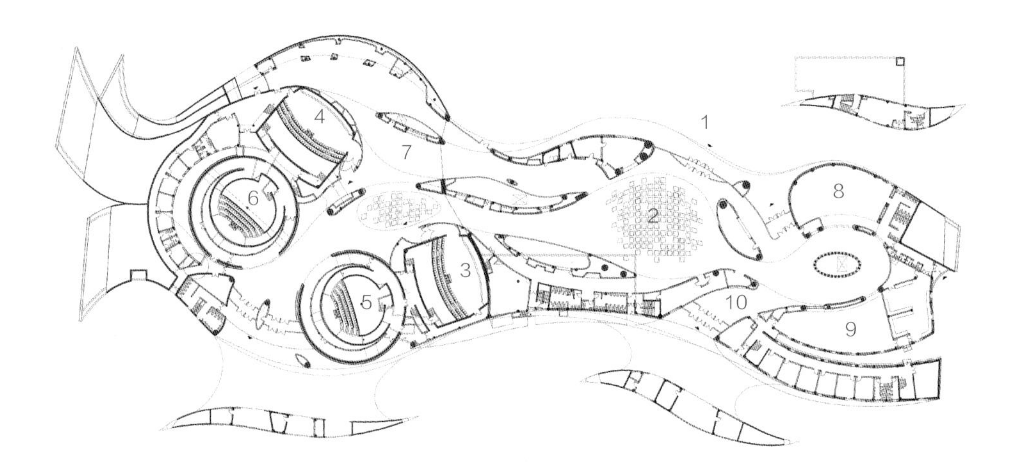

1. 主入口
2. 接待大厅
3. 1号数字影院
4. 2号数字影院
5. 1号球幕大厅
6. 2号球幕大厅
7. 数字展示区
8. 纪念品销售区
9. 餐厅
10. 回程大厅

首层平面图

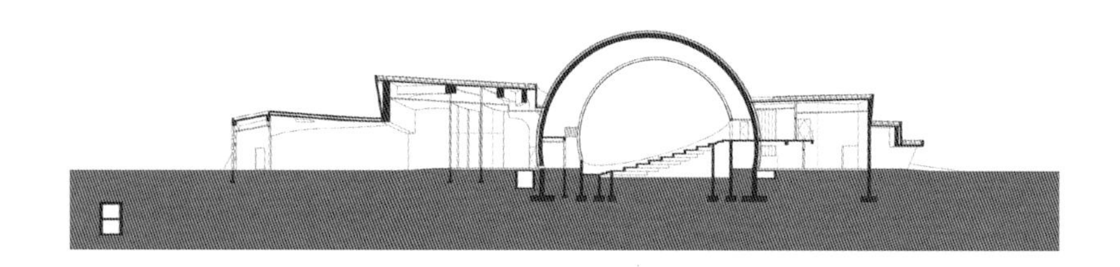

剖面图

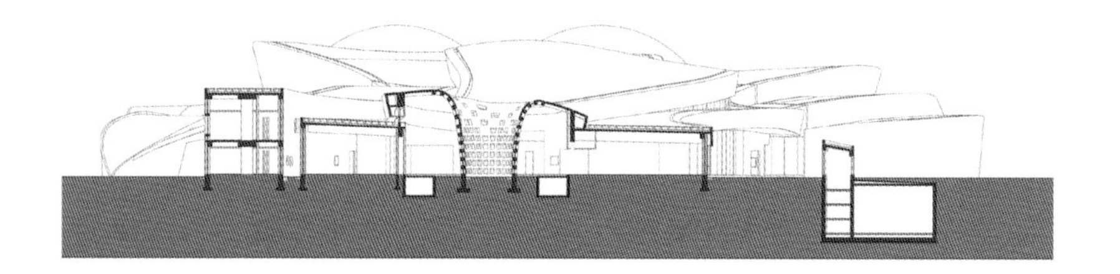

剖面图

充满动感的建筑语汇从室外延续到室内，所有的公共功能均为开放空间，顺应外部形态的变化，室内空间的高度也随之变化。结构支撑体的形态用“墙”的概念，将不同功能、不同高度的空间进行划分，界面清晰明确。

The dynamic features are also applied to the interior of the building, where the height of spaces change with the exterior form. The supporting structure of the building, presented as walls, have served as partitions between spaces of different functions and heights.

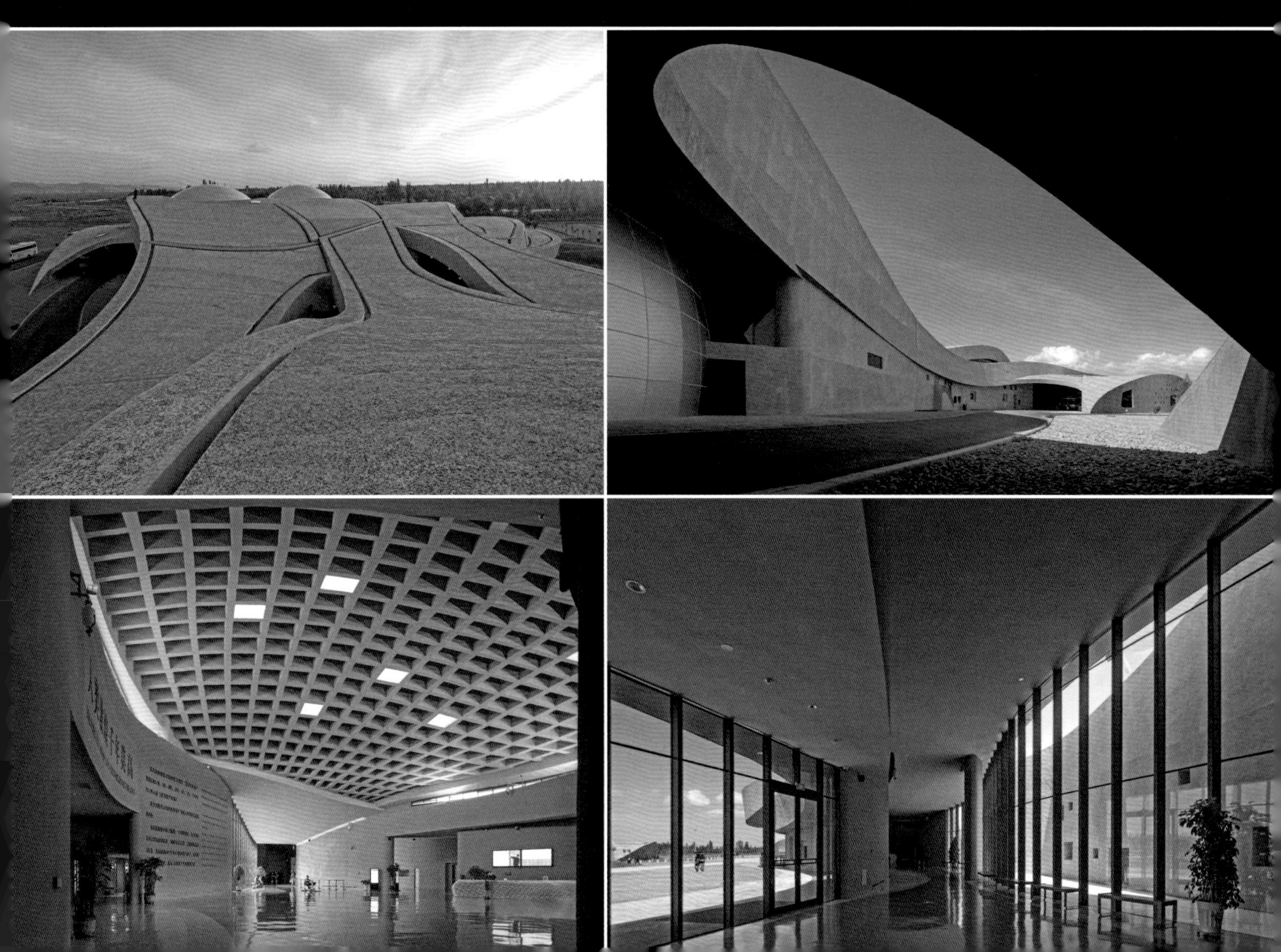

昭君博物馆 Zhaojun Museum

地点 内蒙古自治区呼和浩特市 / 用地面积 404,000m^2 / 建筑面积 15,092m^2 / 高度 15m / 设计时间 2015年 / 建成时间 2017年

方案设计 曹晓昕、梁 力、尚 蓉
宋 涛、范 佳
设计主持 曹晓昕、詹 红

建　　筑 尚 蓉、梁 力
宋 涛、范 佳
结　　构 余 蕾、李 季
董 越、刘会军
给 排 水 安 岩、车爱晶
设　　备 李京沙、屠 欣
电　　气 王 莉、吴 磊
电　　讯 任亚武
总　　图 连 荔、朱秀丽、张 蓉
景　　观 刘 环、刘卓君

昭君博物院的建造，也是一次对王昭君墓地“青冢”所在地整体环境的梳理。博物馆选址位于青冢的神道南侧，与之遥相呼应。博物馆因而以谦逊的姿态出现，控制高度，化实为虚，以突显青冢，并以其外轮廓形成游客进入景区遥望青冢的天然景框。博物馆的主体部分是两个近似的正四棱锥体量，分别作为匈奴与昭君主题馆，由中央连廊衔接起来。外立面为仿夯土混凝土挂板，让建筑仿佛从大地中生长出来。入口雨棚选择重组竹为主材，穿插搭接而成。建筑用新材料、新建构方式来转译古老的传统建构方式，表达对古人智慧的尊重。

The project of Zhaojun Museum features the re-organization of the site where Qingzhong (Zhaojun Tomb) lies. Located south to the tomb passage of Qingzhong, the low-heighted museum takes on a modest look in tribute to Qingzhong, which can be seen from a gap between the two main parts of the building. The two pyramid-shaped volumes serve as exhibition halls for Zhaojun and Xiongnu (an ancient nationality in China) respectively, connected by a central corridor. Finished with rammed earth-like concrete panels, the building seems to have grown form the ground. The canopy at the entrance is made of overlapping and interweaving bamboo bars, paying tribute to the wisdom of ancient people.

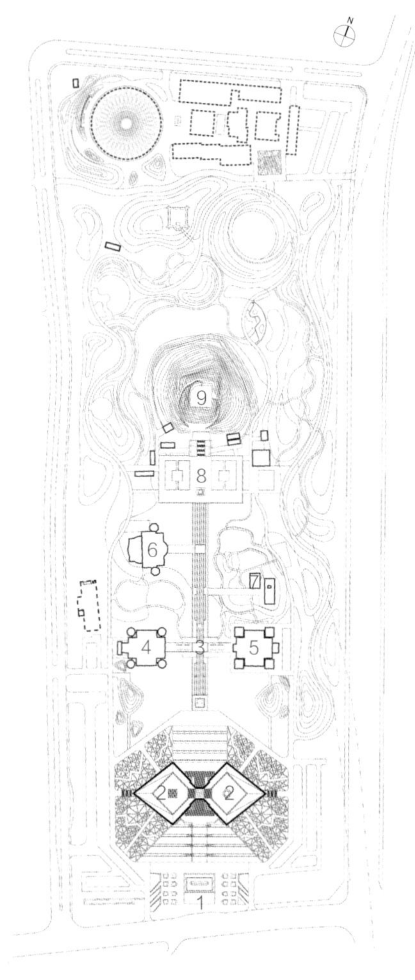

1. 入口广场 2. 博物馆新馆 3. 神道 4. 和亲文化馆 5. 5D 影院 6. 单于大帐 7. 昭君故里 8. 祭祀广场 9. 青冢

总平面图

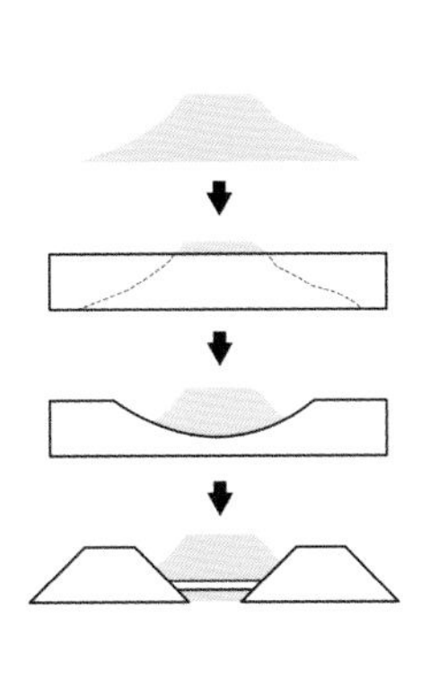

与青冢的关系分析

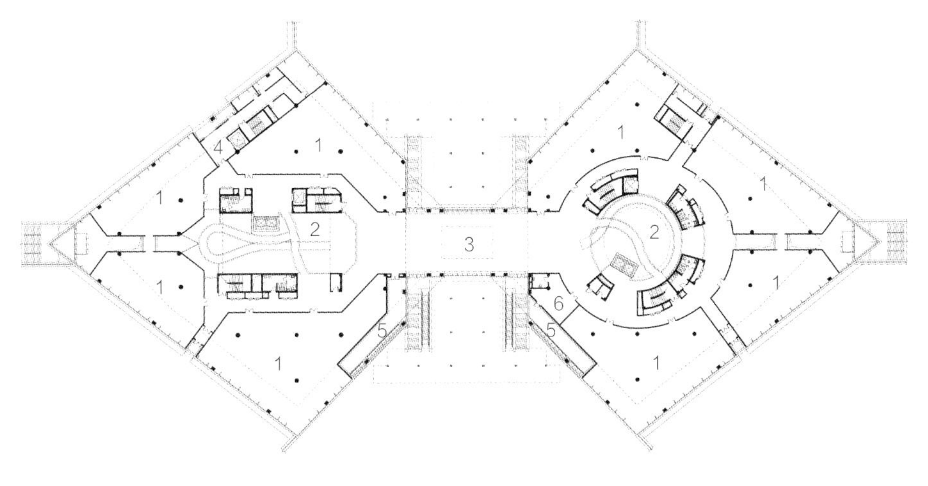

1. 展厅 2. 中庭上空 3. 连廊（序厅） 4. 后勤区门厅 5. 室外平台 6. 贵宾休息室　　首层平面图

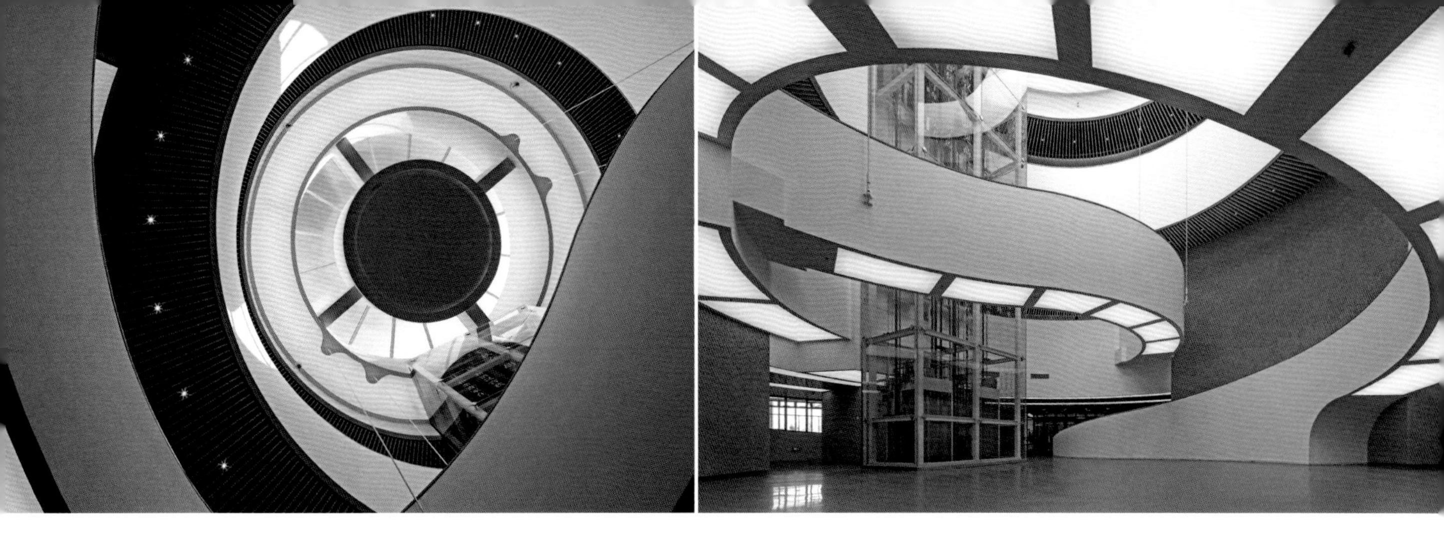

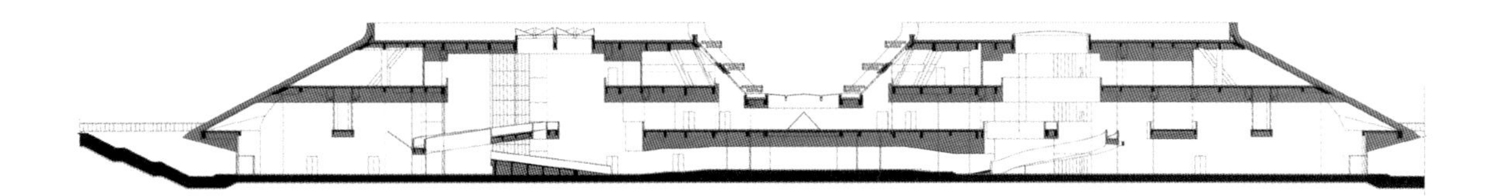

剖面图

大同市博物馆 Datong Museum

地点 山西省大同市 / 用地面积 51,556m^2 / 建筑面积 32,821m^2 / 高度 28m / 设计时间 2010年 / 建成时间 2015年

方案设计 崔 愷、刘 恒、邢 野
设计主持 崔 愷、时 红

建　　筑 刘 恒、邢 野、吴 健
结　　构 朱炳寅、刘 巍
给 排 水 黎 松
设　　备 刘继兴、孙淑萍
电　　气 李维时
总　　图 吴耀懿
室　　内 张 晔、顾大海
景　　观 冯 君

大同博物馆，位于山西大同老城区东、新建的行政文化中心——御东新区的核心位置，与其东侧的音乐厅沿新区南北向轴线对称布置，周边均为各具特色的文化建筑。建筑设计承袭大同深厚的历史文化底蕴，从悠久的龙图腾文化中汲取灵感，并与当地火山群的典型地貌暗合。两个拙朴的弧形体量围绕中庭和庭院盘旋而起，并向外生长出数个分叉，为观展休息区引入光线和风景，而作为主体的展示空间则随着主形体展开，使观者仿佛遁入幽深的石窟之中。遒劲有力的建筑形体在端部直接暴露出功能性断面，其符号化的形态又与汉字笔画产生微妙的联系，使建筑更具力量感和文化内涵。

Datong Museum, located in a central position in Yudong New District, plays an important role in the construction of the area. The design has inherited the essence of the profound historical culture in Datong, while the architectural form draws inspiration from the time-honored dragon totem culture. Two simple arcs rising around the atrium have introduced light into the internal exhibition area. The exhibition space, which serves as the focus of the building, unfolds along the main parts of the building to make visitors feel as if they were going into a deep cave. The section of the building is exposed at the ends of the building.

整体鸟瞰图

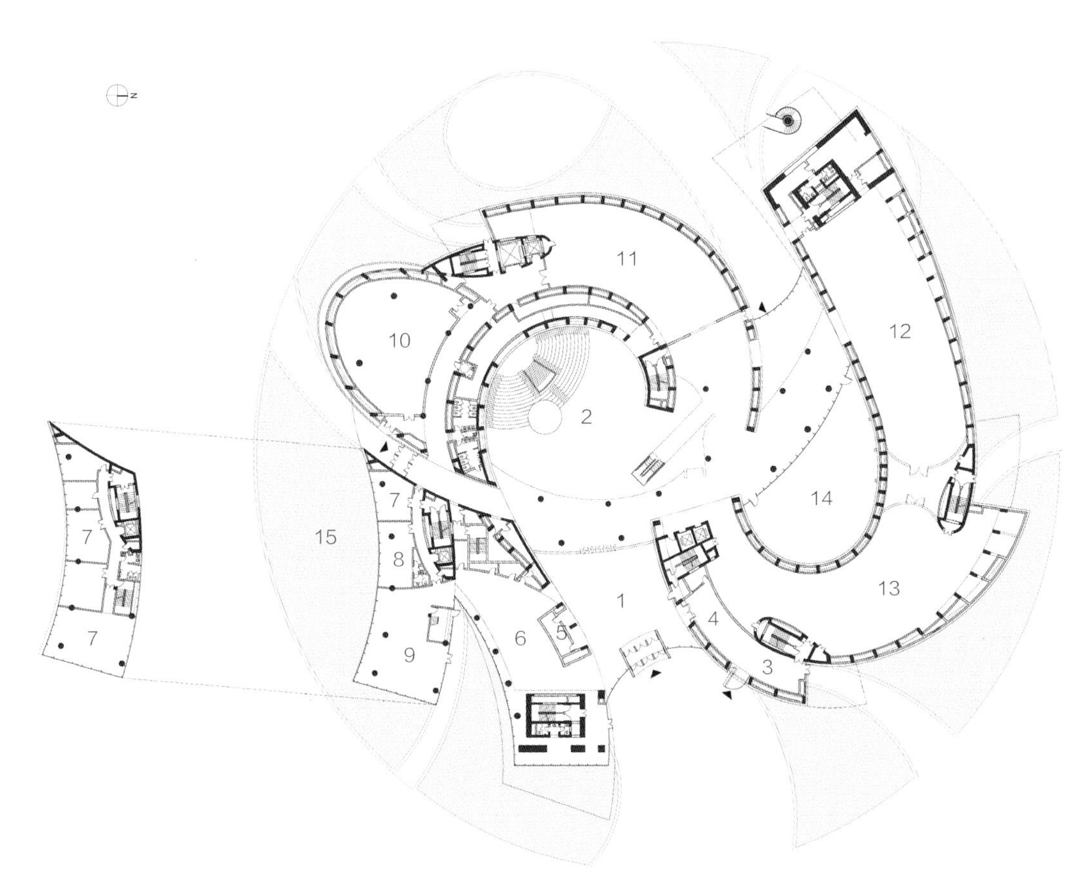

1. 门厅
2. 大厅
3. 安检处
4. 存包处
5. 总服务台
6. 纪念品商店
7. 办公室
8. 接待室
9. 咖啡厅
10. 多功能厅
11. 远古恐龙化石厅
12. 早期历史厅
13. 唐代石刻厅
14. 室外庭院
15. 水池

首层平面图

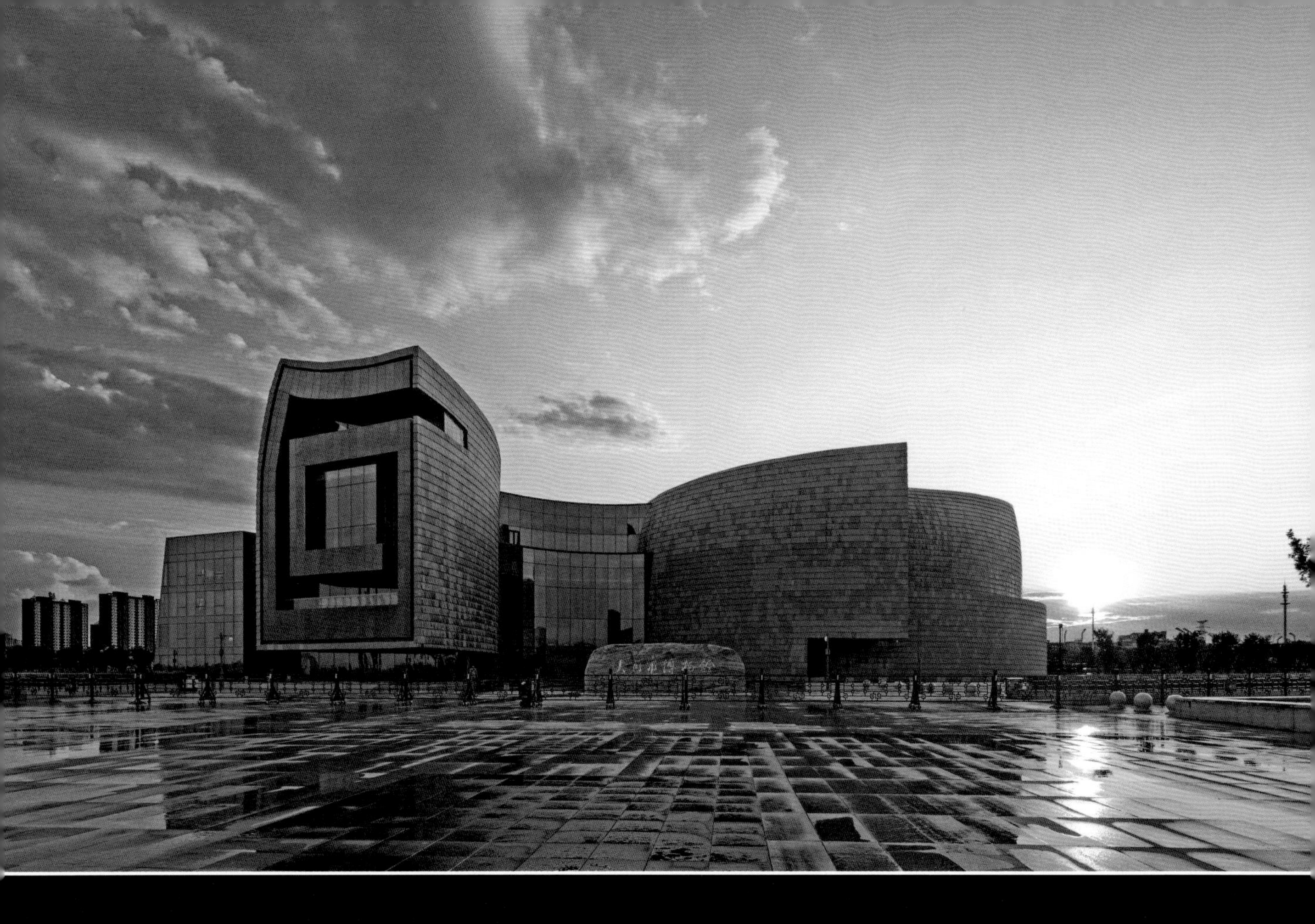

非线性的三维表面上覆盖着上下搭接的花岗石板，同样的石材也蔓延到近似圆壁的水池中，且从下到上呈现逐渐变淡的效果，完整的形态也强化了建筑与大地浑然一体的态势。

The non-linear envelope of the building is covered with granite slabs, which are also applied in a circular pool. The color of the slabs grow lighter from the bottom to the top, merging the building with the land where it belongs.

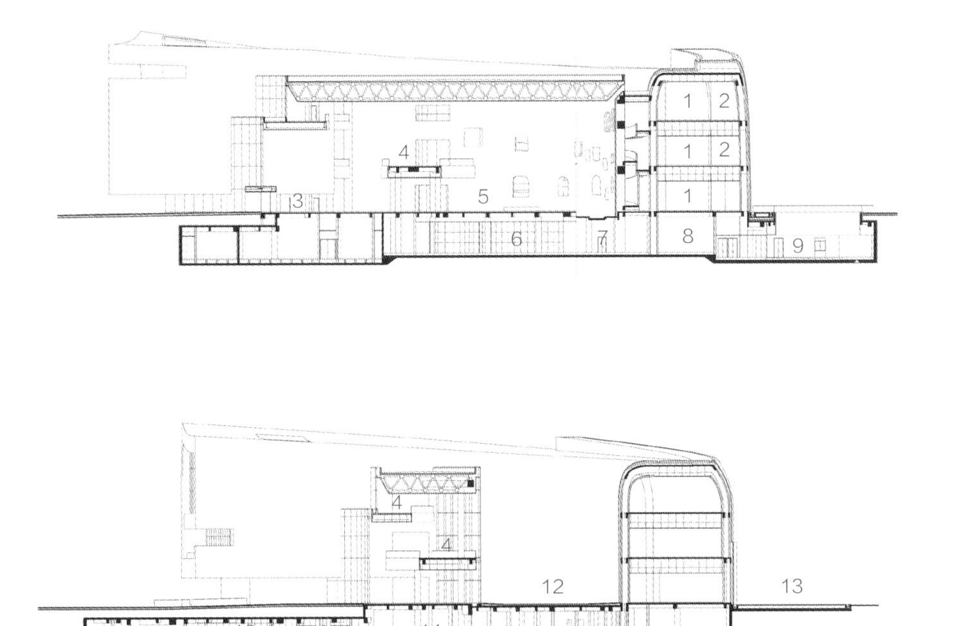

1. 展厅
2. 休息厅
3. 门厅
4. 连桥
5. 大厅
6. 活动室
7. 多媒体室
8. 数字影厅
9. 下沉庭院
10. 餐厅
11. 停车库
12. 室外庭院
13. 室外水池

剖面图

商丘博物馆 Shangqiu Museum

地点 河南省商丘市 / 用地面积 73,613m^2 / 建筑面积 29,613m^2 / 高度 23m / 设计时间 2008年 / 建成时间 2016年

方案设计 李兴钢、谭泽阳、付邦保
郭 佳、张 哲、李 喆
张玉婷、梁 旭、闫 昱
设计主持 李兴钢

建　　筑 张 哲、付邦保
结　　构 张 晔、杨 威
给 排 水 陈 宁
设　　备 刘燕军
电　　气 王 铮
总　　图 连 荔

摄　　影 夏 至

博物馆以商丘归德古城为代表的典型黄泛区古城的形制特征，转化和再现为博物馆的建筑布局、空间序列和形态语言。上下叠层的建筑主体和自下而上、由古至今的展陈布局也喻示了考古学中的“城压城”埋层结构，使其犹如一座被发掘的微缩之城。一系列精心组织的建筑和景观元素，形成了在上下、内外、近远之间往复变化的空间叙事，三层叠加的展厅被水面和庭院环绕，其外又是层层叠落的台地绿植和下设展廊的堤台。参观者游观尽致，通过一系列空间路径完成对这个微缩之城及其所承载延伸的古城历史的完整体验。

The prototype of Guide Ancient City has been transformed and reflected in the layout of Shangqiu Museum, which seems like a miniature of an ancient city, with stacked parts suggesting the “city on city”-type archaeological structure of the buried layers, as well as the layout formed by its transition from the ancient to the present. Stacked exhibition halls are surrounded by waters and courtyards, and through the delicately designed route, visitors will have a comprehensive experience of the miniature city, as well as the history of the ancient city.

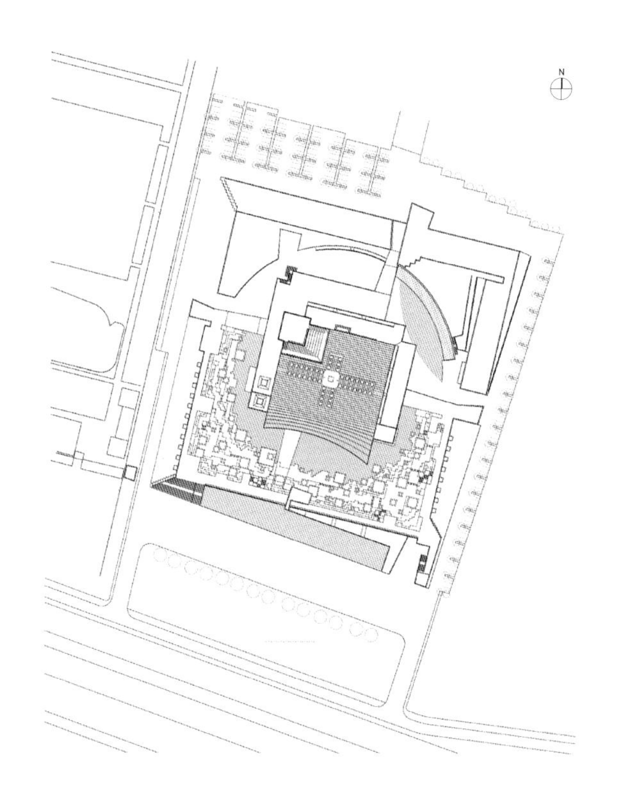

总平面图

参观者由大台阶和坡道登临堤台，凌水沿引桥而入“城”，自下而上沿隐喻城市十字主街的中央十字大厅的坡道；沿途参观各个展厅，最后到达屋顶平台，可由位于建筑各角的眺望台观赏阏伯台、归德古城、隋唐大运河码头遗址等不同方向的古迹。一系列精心组织的建筑和景观元素，形成了在上下内外、近远、之间往复变化的空间叙事。

After climbing up the dam through grand steps and ramps, visitors enters the “city” through a bridge and then visit each exhibition hall along the ramps of the cruciform central hall symbolizing the central cross street of the ancient city. When reaching the roof platform, visitors will see a series of famous sites and enjoy the experience provided by well-designed landscape and architectural elements.

博物馆大量采用了一种廉价的“鲁灰”石材。从博物馆汉画像石藏品受到启发，石材均作“磨切外边 + 中间烧毛”的处理，错缝拼挂，获得精致的细节效果。室内加入了树脂实木面板材，与石材采用统一规格，为室内空间增加了温暖的舒适感。

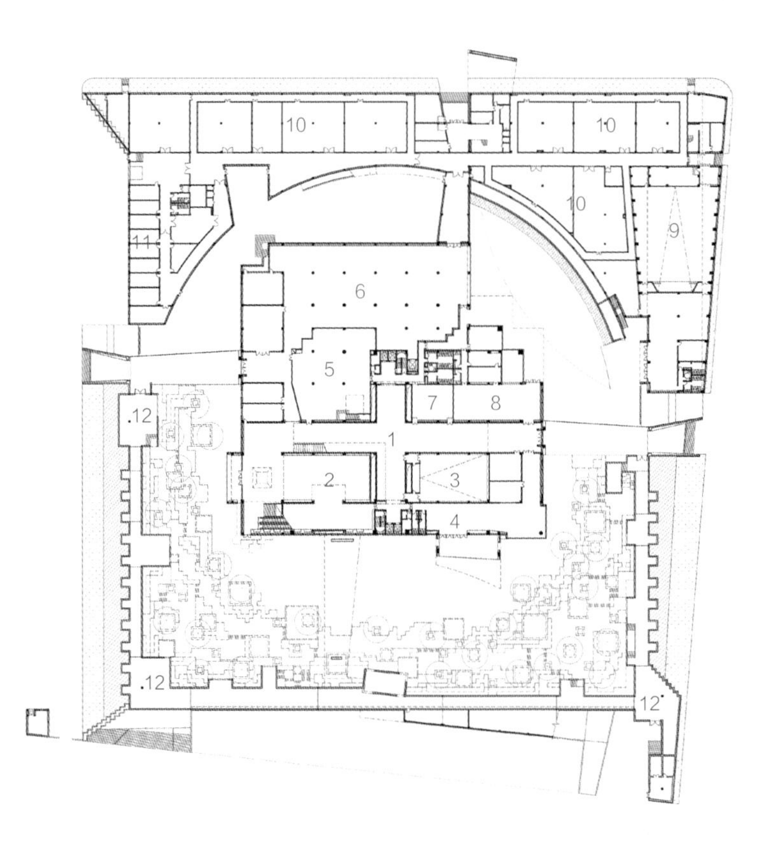

1. 共享大厅
2. 序言厅
3. 放映厅
4. 茶餐厅
5. 展厅
6. 多功能厅
7. 信息厅
8. 贵宾厅
9. 报告厅
10. 展品库房
11. 研究用房
12. 室外展廊

首层平面图

A large quantity of low-cost Shandong grey stones are used in both internal and external spaces. Each stone has been treated with surface grinding and internal singeing, and the staggered joints present elegant details. The resin solid wood panels are applied in the interior space for pleasant visual experiences.

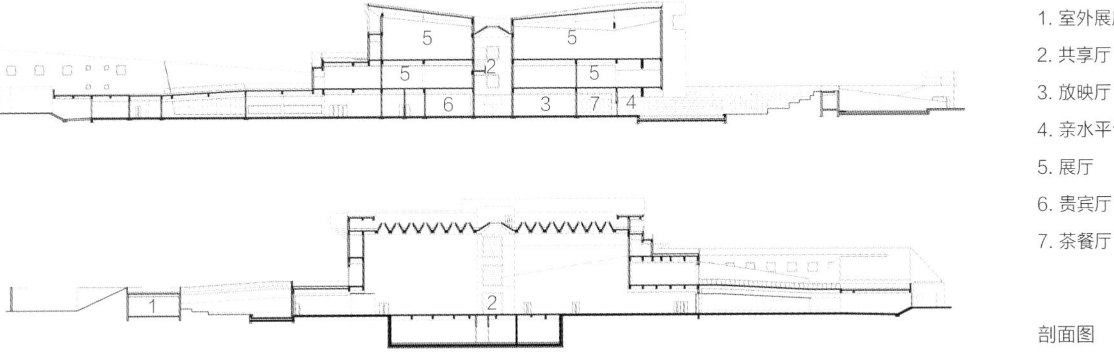

1. 室外展廊
2. 共享厅
3. 放映厅
4. 亲水平台
5. 展厅
6. 贵宾厅
7. 茶餐厅

剖面图

兰州市城市规划展览馆 Lanzhou City Planning Exhibition Hall

地点 甘肃省兰州市 / 用地面积 11,428m² / 建筑面积 16,270m² / 高度 23m / 设计时间 2012年 / 建成时间 2016年

方案设计 崔 愷、康 凯、吴 健
设计主持 崔 愷、康 凯

建　　筑 吴 健
结　　构 张淮湧、王树乐
　　　　 陈 越、尹海鹏
给 排 水 杨东辉
设　　备 郭 然、郑 坤、宋占寿
电　　气 李 磊、沈 晋、姜海鹏
总　　图 高 治、高 伟
景　　观 冯 君
室　　内 上海风语筑文化科技股份有限公司

兰州市城市规划展览馆位于黄河北岸，用地呈不规则状沿黄河展开。建筑取“黄河石”为设计意象——经过切削的体块犹如一块被石头包裹的黄河璞玉，经过河水经年累月的冲刷，呈现出历史和文化的沉淀。立面采用粗犷的现浇清水混凝土，水平向的凹槽自下而上间隔渐宽，南立面还嵌入多处横向的玻璃嵌缝，既表达河水冲刷的感受，也化解了大面积混凝土材料的单调感。这些玻璃嵌缝也成为从建筑内部欣赏黄河美景的窗口。近人尺度的凹缝内还印有黄河卵石肌理，使建筑的地方意味更加浓厚。建筑空间顺应展陈建筑特点，采用回字形的展陈流线围绕城市总规模型展开，外部的清水混凝土墙面也延伸至室内。

Located on the north bank of the Yellow River, the building resembles a stone washed by river for months and years. By means of cutting, it looks like an unprocessed jade covered by stone that indicates the rich culture of the historical city Lanzhou. All exterior walls are cast on site to produce an impressive texture. The horizontal grooves arrayed with varied spacing, especially several horizontal glass strips on the south façade, give its appearance more richness. Those glass strips are also viewing windows of the river. Some grooves with pebbles inside have added to the local characteristics of the building.

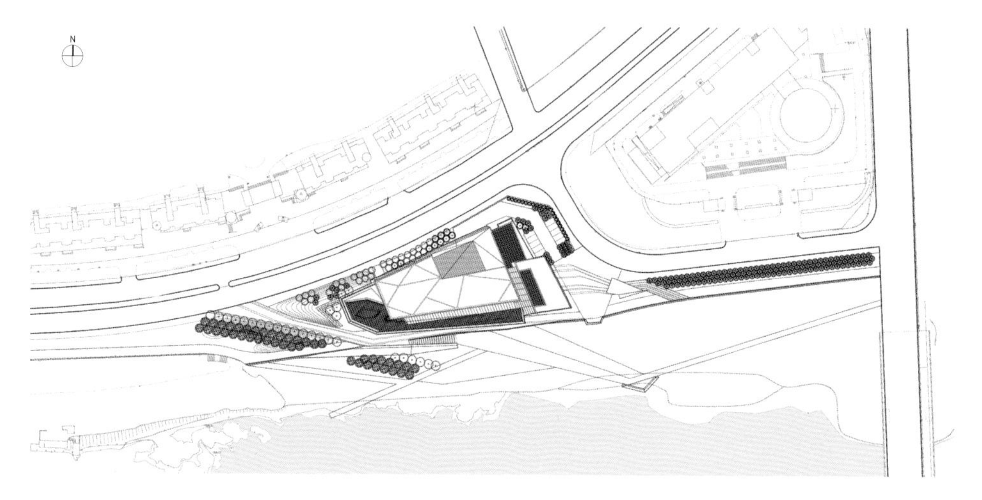

总平面图

1. 展厅
2. 数字展厅
3. 休息区
4. 观看台
5. 多功能厅
6. 办公
7. 办公庭院
8. 室外平台
9. 门厅上空
10. 模型展区上空

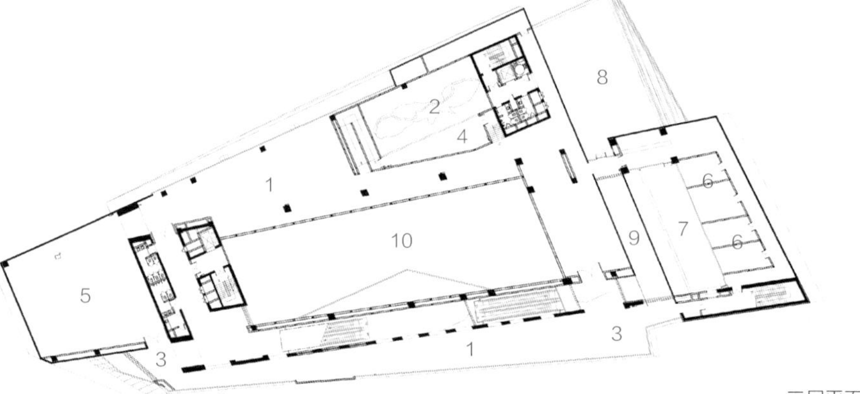

三层平面图

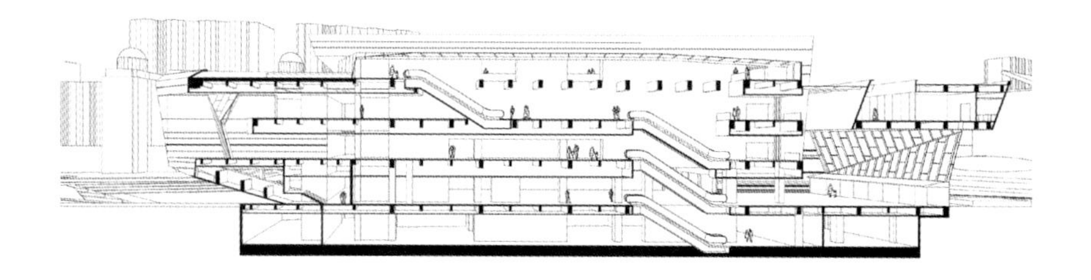

剖透视图

河堤部分保持项目整体的折面形态，与建筑立面的水平向条纹肌理一致。东端的十二面体的宝石形景观挑台，为人们提供了眺望黄河的位置。中部清水混凝土折面向河滩撕开，形成平缓的踏步，为沿河步行的市民提供下至河滩的便利路径。

Folded surfaces are also seen on the river walls, with texture identical to that of the façade. A diamond-shaped landscape platform offers a view of the Yellow River. The folded surfaces have been “split” to accommodate steps down to the bay.

长春市规划展览馆、博物馆 Changchun Planning Exhibition Hall & Museum

地点 吉林省长春市 / 用地面积 73,950m² / 建筑面积 63,172m² / 高度 36m / 设计时间 2010年 / 建成时间 2016年

方案设计 崔 愷、景 泉、杨 磊
设计主持 景 泉、李静威、王更生

建 筑 吴锡嘉、张伟成、杨 磊、杜 捷、王 辰
结 构 孙海林、段永飞、刘会军、高芳华、陆 颖、罗敏杰、王 昊、张世雄、王春圆
给排水 赵 昕、贾 鑫
设 备 汪春华、王春雷、李金双
电 气 丁志强、李 磊、裴元杰
总 图 高 治、吴耀懿
室 内 张 晔、饶 劢、曹 阳、刘璐蕊、王佳旭
景 观 李 力、段岳峰、张 鹏、白雪松、王兆阳、吴 丹、陈鑫文、方 圆、石 宇

建筑坐落于长春城市中轴线人民大街的南端，由规划展览馆、综合博物馆、美术馆、信息服务中心及餐厅等部分组成。设计以长春的城市特色“流绿都市”为出发点，使这座展现城市之美的建筑能够成为流绿都市中绽放的“城市之花”。自由奔放的花朵式建筑形态，造型宛如“如意”，向上展开的趋势体现了城市蓬勃的文化精神。非线性的形体、折板曲面幕墙、菱形钢结构都成为建筑鲜明的特点，同时也是设计和建造中的难点。通过运用参数化技术、全专业BIM设计，并结合先进的绿色技术，建筑最终呈现了完满的建成效果。

Situated at the southern end of Renmin Street of the city's central axis, the Changchun Planning Exhibition Hall & Museum is a landmark of the southern core district of Changchun, showcasing the culture and image of the city. Designed to resemble a "flower in a green city", the building has a form featuring openness and freedom. The building features a non-linear form, folded curved curtain walls and a rhombus-shaped steel structure. Parametric design methods and BIM technologies have been applied in the design process.

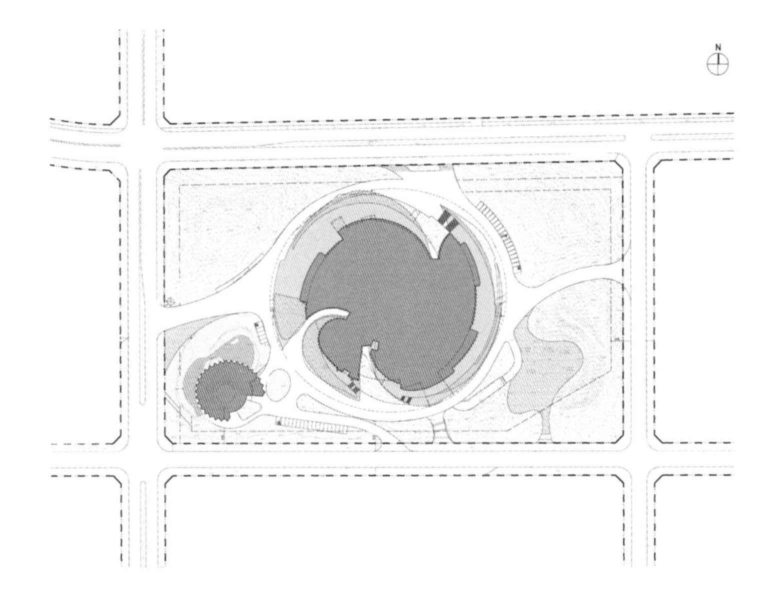

总平面图

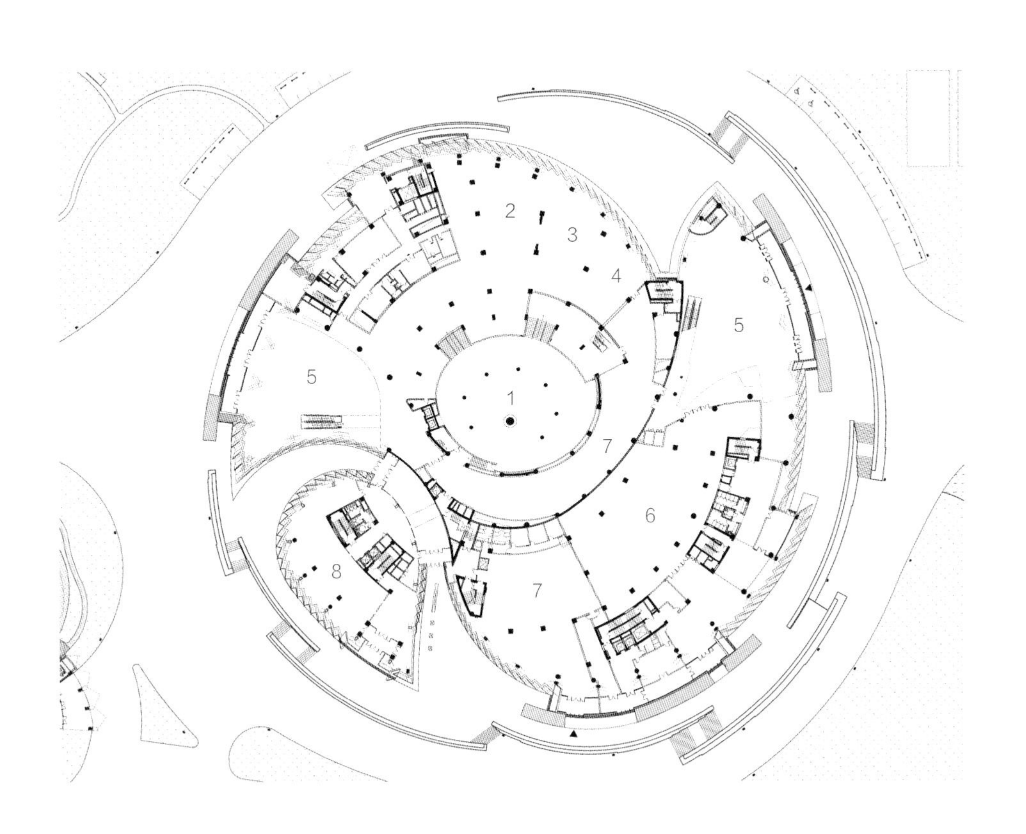

1. 历史展区
2. 历史保护展区
3. 总体规划展区
4. 综合交通展区
5. 礼仪大厅
6. 临时展厅
7. 展厅
8. 信息服务中心大厅

首层平面图

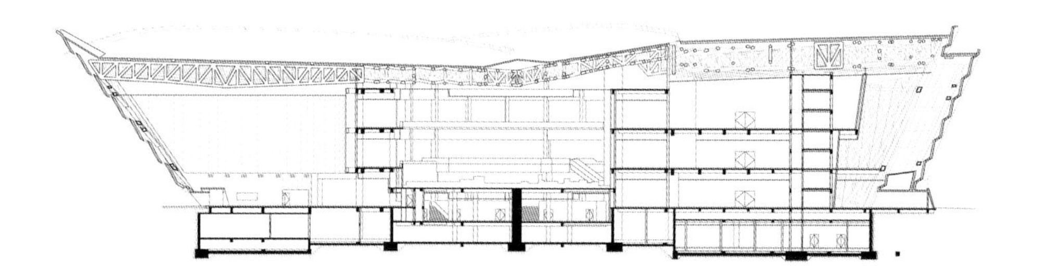

剖面图

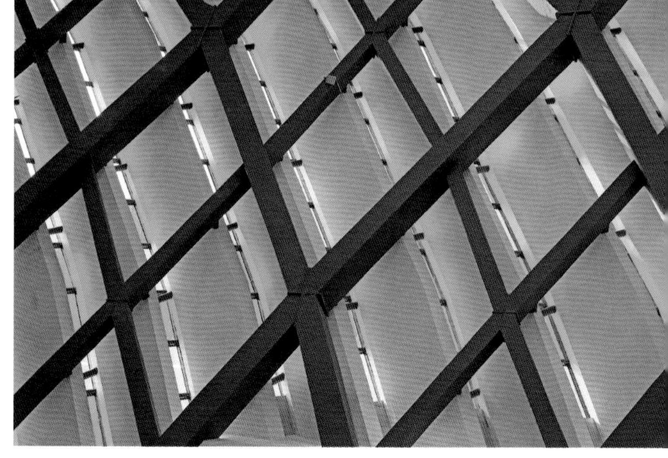

高昌故城游客服务中心 Visitor Center of Gaochang Ruins

地点 新疆维吾尔自治区吐鲁番市 / 建筑面积 4,606m^2 / 高度 13m / 设计时间 2013年 / 建成时间 2016年

方案设计 柴培根、周　凯
　　　　 戴天行、李　赫
设计主持 柴培根、周　凯

建　　筑 戴天行、李　赫
结　　构 王树乐、周　岩
给 排 水 董　超
设　　备 郭晓静
电　　气 李沛岩
总　　图 高　治、李　爽

高昌故城游客服务中心是为配合申报世界遗产而建设的综合游客服务中心。设计的意图是为前来的游客建造一所"视觉过滤器"。游客抵达游客中心伊始，即可感受到以火焰山为背景的建筑全貌，但此时故城是被遮蔽的；逐渐接近入口，主体建筑完全占据视野；进入院落后，游客可登观景塔远眺高昌故城及火焰山，狭长的开窗将自然景观限定于一个宽幅画框之中；离开游客中心，故城形象则成为视野的全部。

The visitor center is a comprehensive service center built for the site application for the title of World Cultural Heritage, bringing people on a tour of varied experiences to highlight the vicissitudes of the ruins. Seen from the entrance, the panorama of the building, with Flaming Mountains as its backdrop unfolds in front of the visitors while the ruins of the city is hidden; only in the courtyard can people have an overlook of the ruins.

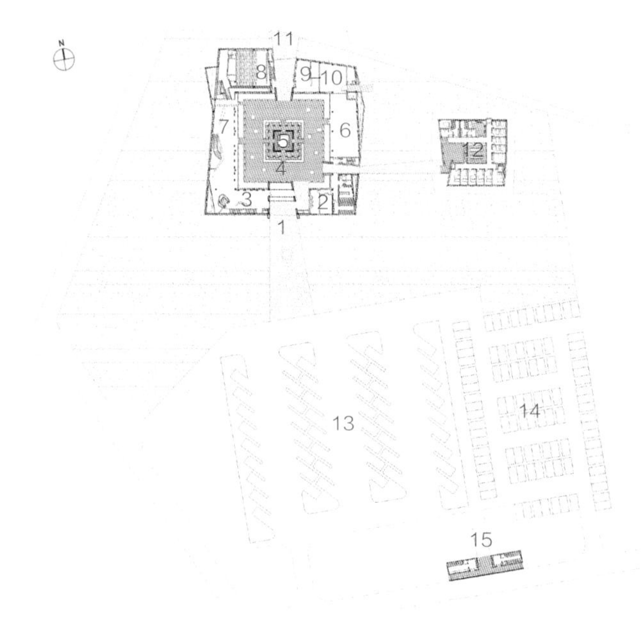

1. 游客中心入口
2. 售票处
3. 咨询处
4. 休息区
5. 沙盘展示区
6. 纪念品商店
7. 书店
8. 影院
9. 邮政电讯服务处
10. 电瓶车修理处
11. 游客中心出口
12. 员工后勤区
13. 大客车停车场
14. 小客车停车场
15. 停车场卫生间

平面图

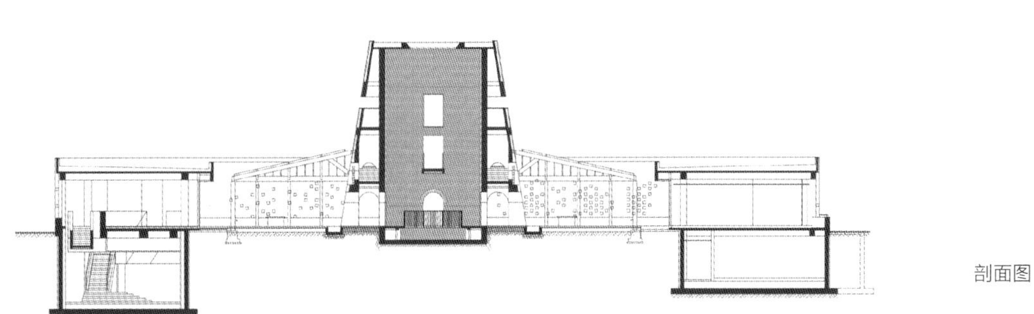

剖面图

略显曲折的建筑边界与故城城墙有所联系，院中央的观景塔也通过倾斜的塔身与墙体上的拱形龛与高昌故城内大佛寺取得意象上关联。龛内的壁画则向游客讲述了高昌的历史文化。

The slightly winding boundary of the building, as well as the tilted viewing tower and its arched niche are all designed to pay tribute to the ruins.

玉门关游客服务中心 Visitor Center of Yumen Pass

地点 甘肃省敦煌市 / 建筑面积 2,584m^2 / 高度 8m / 设计时间 2016年 / 建成时间 2017年

方案设计 牛 涛
设计主持 于海为

建　　筑 谢 悦、牛 涛
结　　构 张淮湧、周 岩
给 排 水 杨向红、董 超
设　　备 郭 超
电　　气 李沛岩
电　　讯 董 超
总　　图 连 荔、王宇恒
景　　观 刘 环、李 旸
室　　内 北京市辛迪森装饰设计公司

游客中心距离玉门关遗址仅200m。此处原有一处供管理人员休息办公的房子，形态与遗址极不协调。改造设计希望呈现出遗址的历史厚重感和大漠的荒凉，尽量缩减建筑体量，将大部分面积设于地下。改造设计重新梳理了停车场和游客中心与遗址本体的关系，借助基地高差将由停车至游客中心的步道分为三段，游客拾级而下，远处的遗址逐渐从视线中消失，游客中心映入眼帘，进入院内的景观塔，又可从高处一览玉门关全景，随后步入景区近距离欣赏遗址。

Aiming to present the profound history of the ruins and the vastness of the desert, the design features a layout with most parts located underground, so that the building poses little disturbance to the site. The path between the parking lot and the visitor center has been divided into 3 sectors by the altitude difference, where visitors' sight shifts from the ruins to the visitor center. In the viewing tower of the courtyard, visitors can have an overlook of the Yumen Pass.

玉门关遗址

尹大擎 摄

外墙材料选用“水洗石”，恰到好处地诠释了建筑的地域特征。景观也以呈现戈壁风貌为主，仅做提示性和遮挡性的墙体、通往入口的小径和几处精心设置的石头和绿化。

The washed granolithic finish on the façade reveal local characteristics of the region. The landscape design remains low-profile to highlight the scene of the Gobi desert.

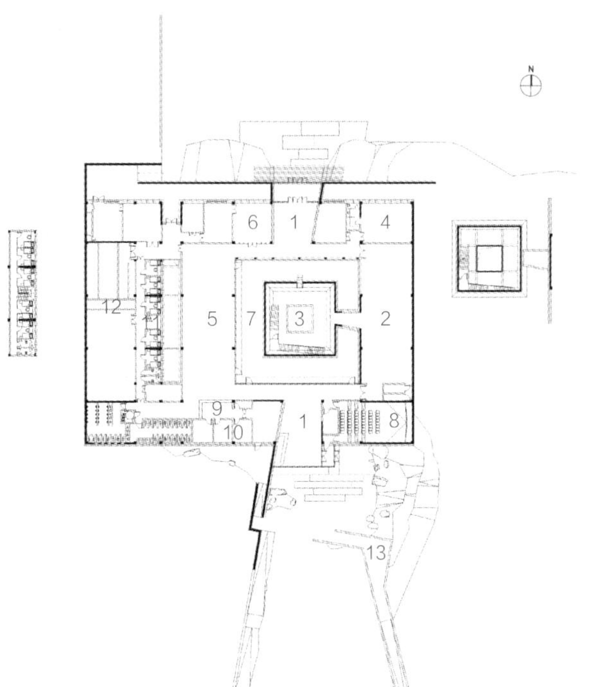

1. 门厅
2. 展厅
3. 景观塔
4. 贵宾接待室
5. 游客服务厅
6. 咖啡厅
7. 庭院
8. 放映厅
9. 问讯处
10. 办公室
11. 员工宿舍
12. 机房
13. 景观步道

首层平面图

可可托海地质博物馆及温泉酒店 Koktokay Geological Museum & Spring Hotel

地点 新疆维吾尔自治区阿勒泰地区富蕴县 / 建筑面积 博物馆 5,700m² 温泉酒店 1,300m² / 高度 10m / 设计时间 2008年 / 建成时间 2012年

方案设计 柴培根、田海鸥
杨文斌、金基天
设计主持 柴培根

建　筑 田海鸥
结　构 孙海林
电　讯 赵 昕

施工图设计 新疆维吾尔自治区建筑设计研究院

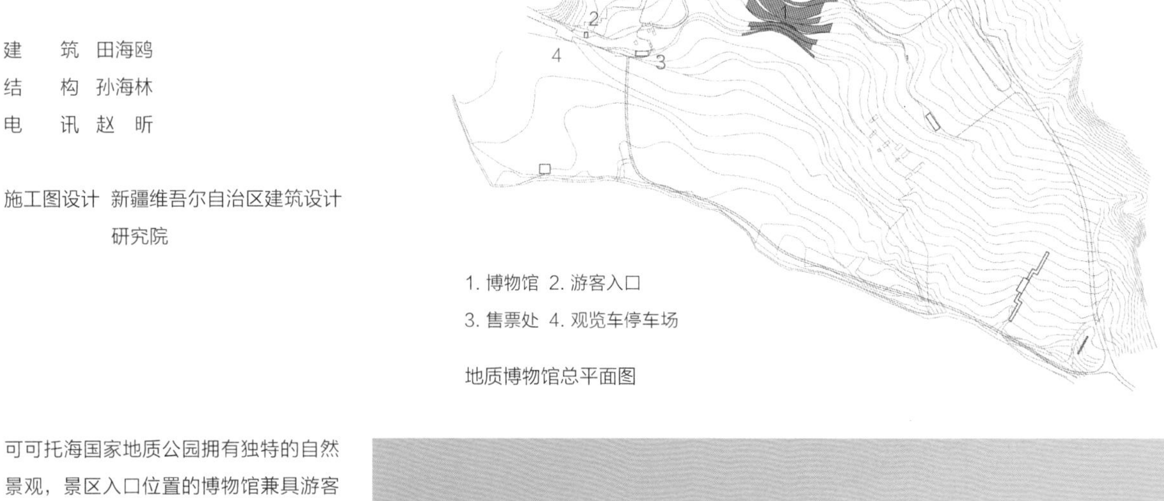

1. 博物馆 2. 游客入口
3. 售票处 4. 观览车停车场

地质博物馆总平面图

可可托海国家地质公园拥有独特的自然景观，景区入口位置的博物馆兼具游客中心的功能。设计将其作为“地景建筑”，在尊重自然的前提下对自然地貌进行抽象再现。覆土的屋顶将地表由周围景观平缓延伸至建筑，从而弱化了建筑形体的边界。人们可以自由行走其间，感受自然、景观、建筑的和谐共生。温泉酒店则位于地质公园的北端，背靠山谷下临河岸，建筑依山就势，与室外温泉池结合为整体，采用底层架空的方式“小心地站在场地”上，维持了原有地表的融雪线。坡屋顶模拟起伏的山势设计，公共空间也充分结合高度的变化形成丰富的层次，客房区则采用简洁平实的方盒子形态与自然景致并置。

Located in Koktokay Geological Park, the museum also doubles as the tourist service center. Designed as a landscape building, the structure is an abstract reproduction of the local landform. The grass-covered roof seems to extend from the land to the building, blurring the boundary of the building. The slightly undulating roof resembles the folds on the land. Going along the contours of the mountain, the spring hotel stands “carefully” on the site with a stilted form, so that the snow-melting line of the terrain is preserved. The sloped roofs resemble the undulating mountains, and the whole building stands in harmony with the nature with a modest look.

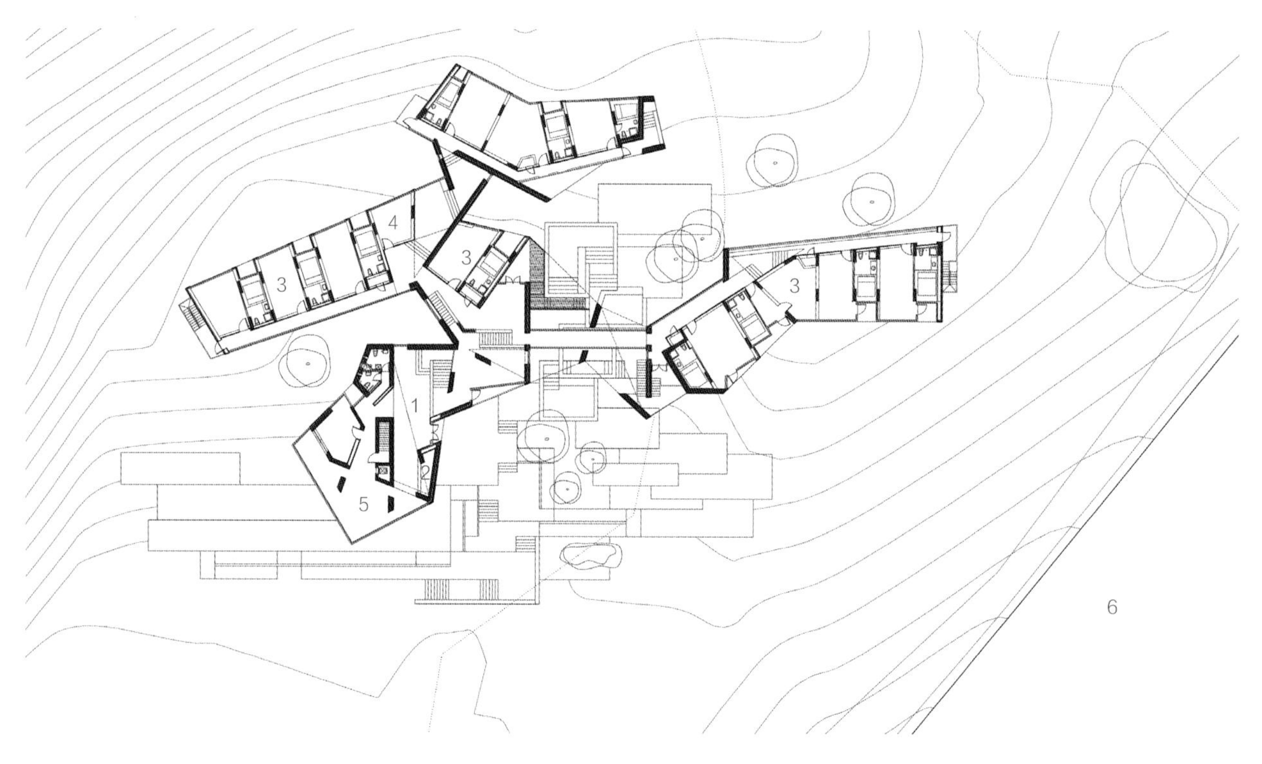

1. 门厅 2. 接待区 3. 客房区 4. 服务间 5. 餐厅 6. 额尔齐斯河

温泉酒店主要平面图

北庭故城国家考古遗址公园配套建筑 Service Buildings in National Archaeological Site Park of Beiting

地点 新疆维吾尔自治区吉木萨尔县 / 建筑面积 12,117m² / 设计时间 2012年 / 建成时间 2013年

方案设计 于海为、刘晏晏、朱起鹏
设计主持 于海为

建　　筑 刘晏晏、朱起鹏
结　　构 周　岩
给 排 水 董　超
设　　备 郭晓静
电　　气 李沛岩

北庭故城是丝绸之路上有千年历史的重要历史名城。考古遗址公园的建设包含多座配套建筑，南门工作站和西大寺保护棚已建成并得到使用。西寺遗址位于北庭故城遗址以东约 1km 处，是高昌回鹘时期修建的主要佛寺之一。原遗址博物馆建于 2008 年，为申报世界遗产，对西寺遗址博物馆又进行了相应的调整改造，按照申遗要求，保护大棚墙面及棚顶颜色采用与北庭故城遗址风貌相协调的防水涂料，入口大厅与中间展厅的室内设计也尽量突出遗址的质感。

Located in Beiting Ancient Town, an important historical city on the Silk Road, the archaeological site park of Beiting consists of a series of service buildings, among which the South Gate Workstation and the West Temple Shed have been completed. To meet the demands for Beiting's application for the title of World Heritage, the museum of West Temple site have been renovated, where the walls of the shed have been preserved, and the colors for the waterproof paintings on the shed exists in harmony with Beiting site.

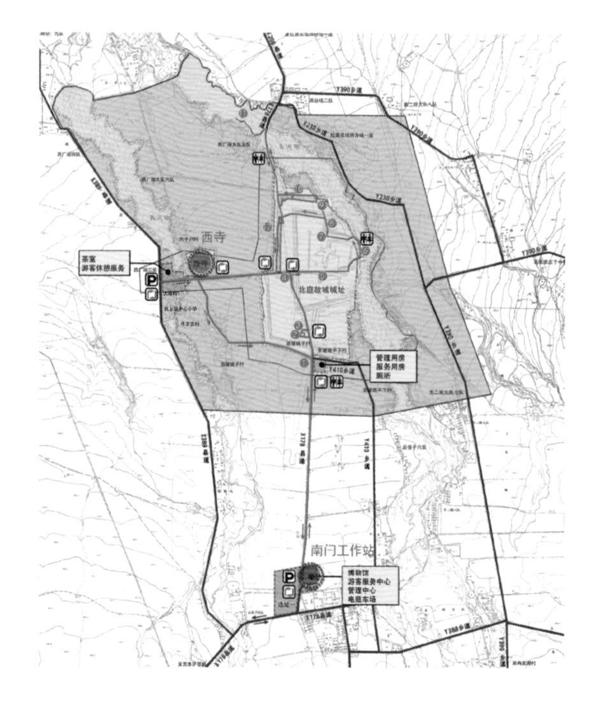

两处建筑在遗址保护区中的位置

地域特色

北庭高昌回鹘佛寺遗址博物馆

南门工作站位于北庭故城遗址南城墙外，是游客进入遗址前乘车和休憩的场所。设计将功能房间掩藏在南北向干道东侧的农田之中，充分利用现状成年乔木和场地的自然高差。以之字形的路径形成游人进入遗址的停顿点，通过控制人在不同高程的空间感受，将遗址徐徐展示在游人眼前。外墙采用当地石材并以当地工艺砌筑，其色调质感与场地浑然一体。

The South Gate Workstation is located outside the southern city walls of the Beiting Site. Taking advantage of the grown arbors and the natural altitude difference of the site, the design has hidden the rooms of the workstation behind the landscape of the fields. Staying spots are defined by the Z-shaped route, which reveals the ruins to the visitors as they step onto various altitudes. The exterior walls of the workstation are built of local stones with local techniques, merging into the site with its simple colors and textures.

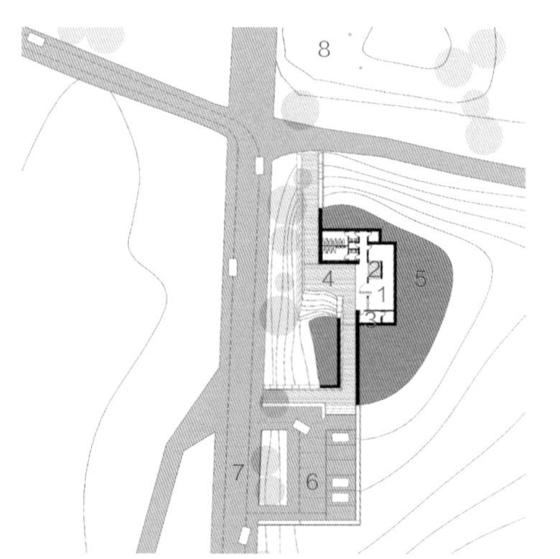

1. 服务区
2. 天井
3. 设备间
4. 庭院
5. 填土
6. 停车场
7. 道路
8. 遗址

平面图

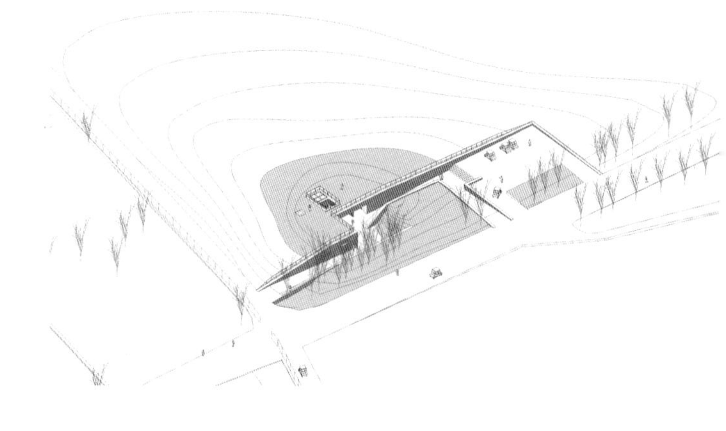

轴侧图

遵义海龙囤展示中心 Zunyi Hailongtun Exhibition Center

地点 贵州省遵义市 / 建筑面积 1,880m^2 / 设计时间 2013年 / 建成时间 2014年

方案设计 于海为、朱起鹏、刘晏晏
穆小贺、魏亚文
设计主持 于海为、刘晏晏
建　　筑 朱起鹏、刘晏晏
穆小贺、魏亚文
结　　构 王　奇
景　　观 刘　环
室　　内 博溥（北京）建筑工程公司

海龙屯遗址始建于宋，由明代土司杨应龙最终完成，是我国保存最为完整、壮观的古代军事城堡遗存。海龙囤展示中心由谷底内原有的三层瓷砖饰面建筑改造而成，集文物展示、游客服务及办公等功能为一体。为了应对申报世界遗产的紧迫时间要求，改造以简约而轻巧的方式，削弱原有巨大的建筑体量，而不减少所需的建筑面积。通过竹子这种天然材料的层层包裹，在视觉上让生硬的建筑消隐，匍匐于山谷之中。在建设资金和地方施工水平有限的条件下，实现了充满意境和吸引力的效果。当地盛产的原竹材料，以简单的金属扣件栓接，便于安装，也抹去了人工造作的痕迹。

Built in Hailongtun, a place with well-preserved ancient remains of military fortifications, the exhibition center is a renovation project for a three-story building finished with tiles, incorporating functions including exhibition, service and offices. The renovation design took a simple and light approach, which breaks down the huge volume without reducing the floor area. Wrapped in bamboo, the building seems to have faded into the valley, presenting an appealing image. Pieces of local bamboo, bolted by simple metal fasteners, are both easy to assemble and free of signs of artificial interference.

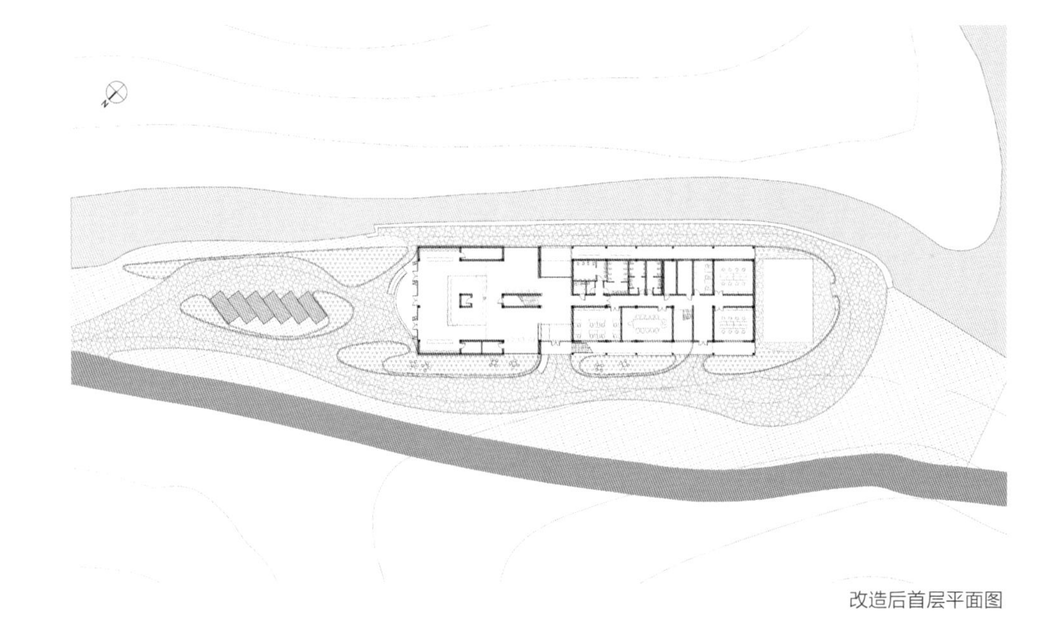

改造后首层平面图

地域特色

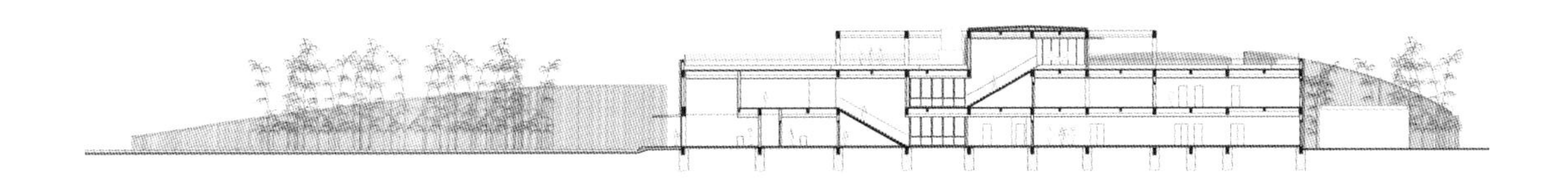

剖面图

永顺县老司城博物馆及游客中心 Museum & Visitor Center of Yongshun Ancient Tusi Castle

地点 湖南省永顺县 / 用地面积 87,314m² / 建筑面积 5,950m² / 高度 15m / 设计时间 2013年 / 建成时间 2015年

方案设计 崔 愷、张 男、朱 巍
设计主持 崔 愷、张 男

建　　筑 朱 巍、李 喆
结　　构 王 玮、王 震、王 超、何相宇、曹永超
给 排 水 董 超
设　　备 杨向红、钟晓辉、马任远
电　　气 李唯时
总　　图 高 治、李 爽
室　　内 顾建英
景　　观 刘 环、张景华

摄　　影 孙海霆

作为老司城遗址的接待和展示设施，这一项目由博物馆和游客中心两部分组成，均选址于高山耸峙的山谷之间。博物馆分为上下两层，重叠倚靠于山坡上，呈带状布局，略带转折而与地势契合，也便于功能流线的布置。外墙由混凝土与河谷卵石垒砌墙体组合而成，入口、观景厅等位置由木构或仿木材料装饰。厚重的墙体仅在必要位置开设洞口，休息厅的庭院和两个下沉庭院则为建筑提供了自然通风采光，覆土植草屋面和木构架的爬藤与周边树木一同构成更大范围的"景观立面"。室外地面也采用卵石铺地，就地取材，与环境更为协调。游客中心是博物馆的后续项目，延续了博物馆对于地域性材料与传统手工艺的利用，除卵石墙体和竹格栅外，尝试了屋面小青瓦和立面竹板的做法。钢结构既能满足施工期限要求，也以清晰的结构形式致敬当地民居的木框架结构传统。

The project consists of a museum and a visitor center, which are both located in the valley. The 2-story museum backs on the mountain, conforming to the terrain with a belt-like layout with several turns. Its exterior walls are dominated by concrete and cobble stones. A larger range of "landscape façade" is formed by the green roof, the veins on the timber frames and the trees on site. The visitor center also features the utilization of local materials and craftsmanship. Besides cobble stone walls and bamboo gratings, small black tiles and bamboo boards are also applied. The steel structure, which can be easily built under a clear and simple logic, pays tribute to the wooden structure of local folk house.

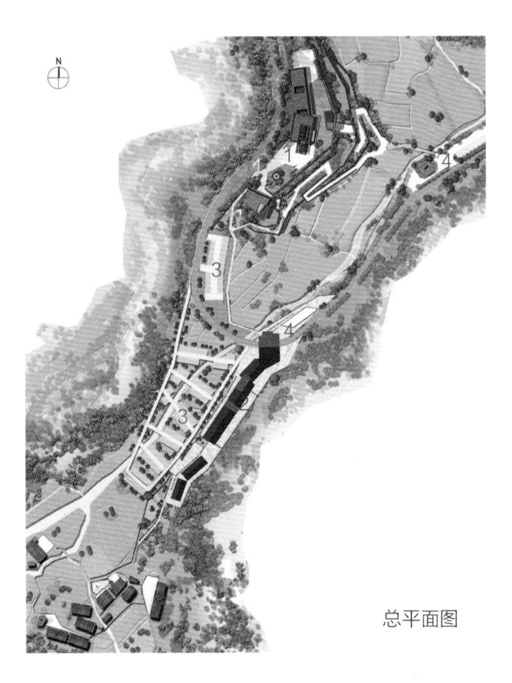

总平面图

1. 博物馆 2. 游客中心 3. 停车场 4. 电瓶车场

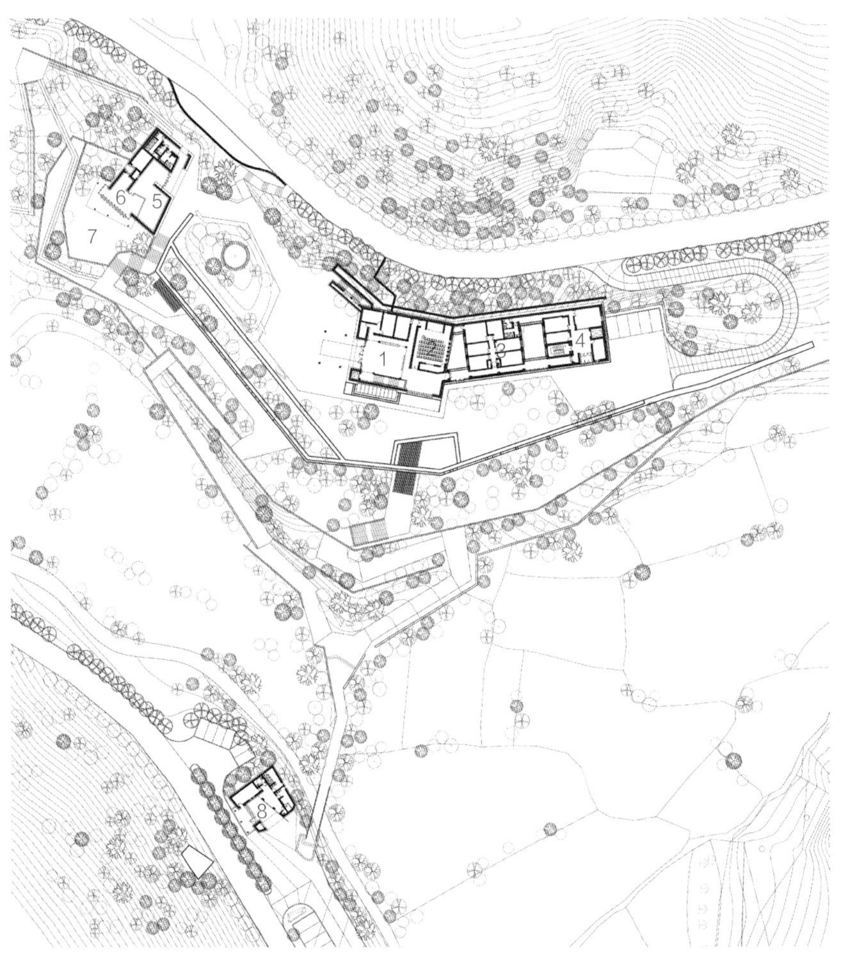

博物馆二层平面图

1. 门厅 2. 报告厅 3. 办公区 4. 接待、会议区 5. 纪念品商店 6. 咖啡厅 7. 室外景观平台 8. 售票区

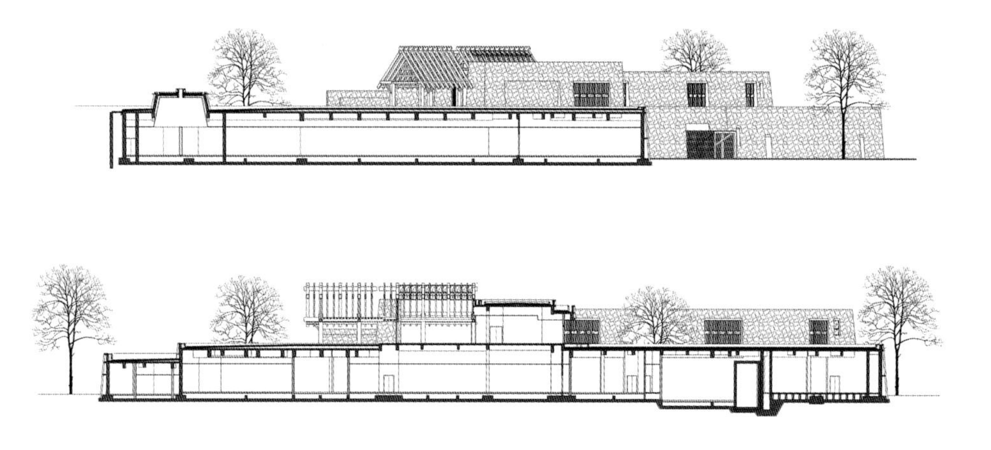

博物馆剖面图

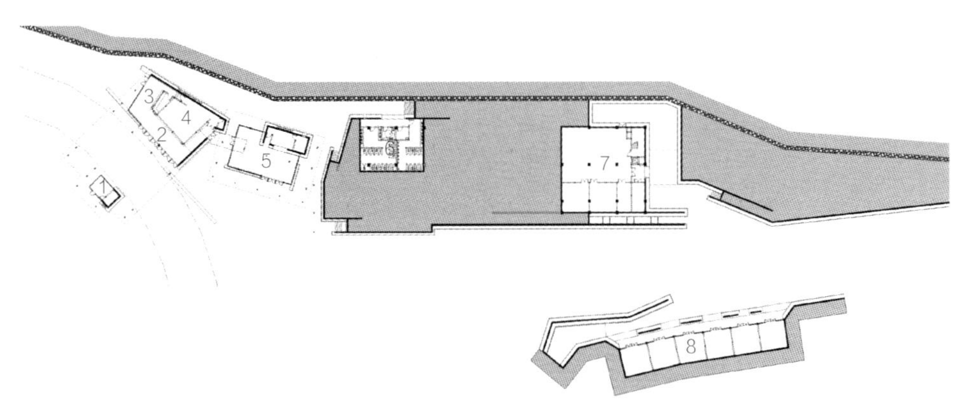

1. 检票室　2. 候车厅　3. 茶室　4. 放映厅　5. 售票大厅　6. 卫生间　7. 厨房　8. 商店

游客中心首层平面图

老西门窨子屋博物馆 Old West Gate Cellar Museum

地点 湖南省常德市 / 用地面积 1,380m² / 建筑面积 2,165m² / 高度 9m / 设计时间 2013年 / 建成时间 2015年

方案设计 何 勍、曲 雷
设计主持 曲 雷

建　　筑 何 勍、王 强
结　　构 李忠盛、张 慧
设　　备 范向国、李 严
电　　气 姜晓先、刘 涛
总　　图 王雅萍

“窨子屋”是常德一带具有传统特色的民居形式。历经战火劫难及时代变迁，时至今日传统的窨子屋基本已消失殆尽。“窨子屋博物馆”因此成为常德老西门20万平方米棚户区改造项目中传承地方传统文化的精神标签，既是博物馆，建筑本身也成为其中最重要的展品。整个建筑自东向西分为三个部分，依次为全榫卯木结构、中厅及框架砌体结构。一座不锈钢桥，将旧砖墙与现代镜钢墙面连为一体。新与旧、传统与现代相互穿插于建筑的各个角落，方寸之地，步移景异。新石与旧瓦、幕墙与花窗、涂料与铜板、光影与建筑，以对比形成丰富的对话。屋顶由精心收集而来的老瓦与新瓦复合而成，64个各色柱墩支撑着老工匠精心树立的榫卯木架，现代化设施如照明管线、保温、防水、空调等则被细心暗藏于构造细节之中。

"Yinzi" house, or cellar house, is a form of folk house in Changde, Hunan Province. The Old West Gate Cellar Museum takes the lead in inheriting the local culture of the Old West Gate region of Changde, serving as both a museum and a most important exhibit on its own right. The building consists of three parts with various structural forms. A bridge made of stainless steel connects an old brick wall and a newly built wall of mirror-finished stainless steel. In every corner of the building, contrasts are formed between the old and the new. The roof is covered with new and old tiles. The meticulously erected wooden structure of mortise and tenon are supported by 64 column piers, while pipelines are hidden through detailed design.

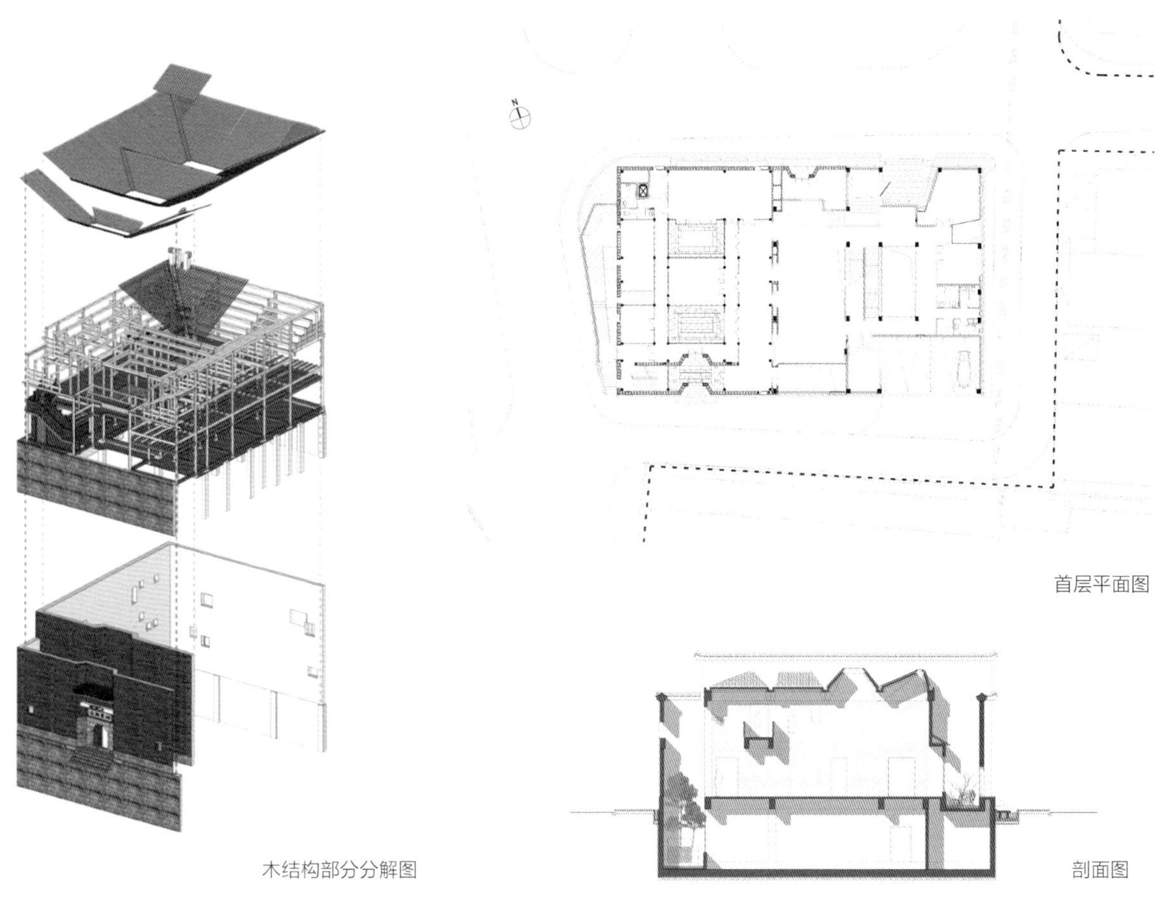

木结构部分分解图

首层平面图

剖面图

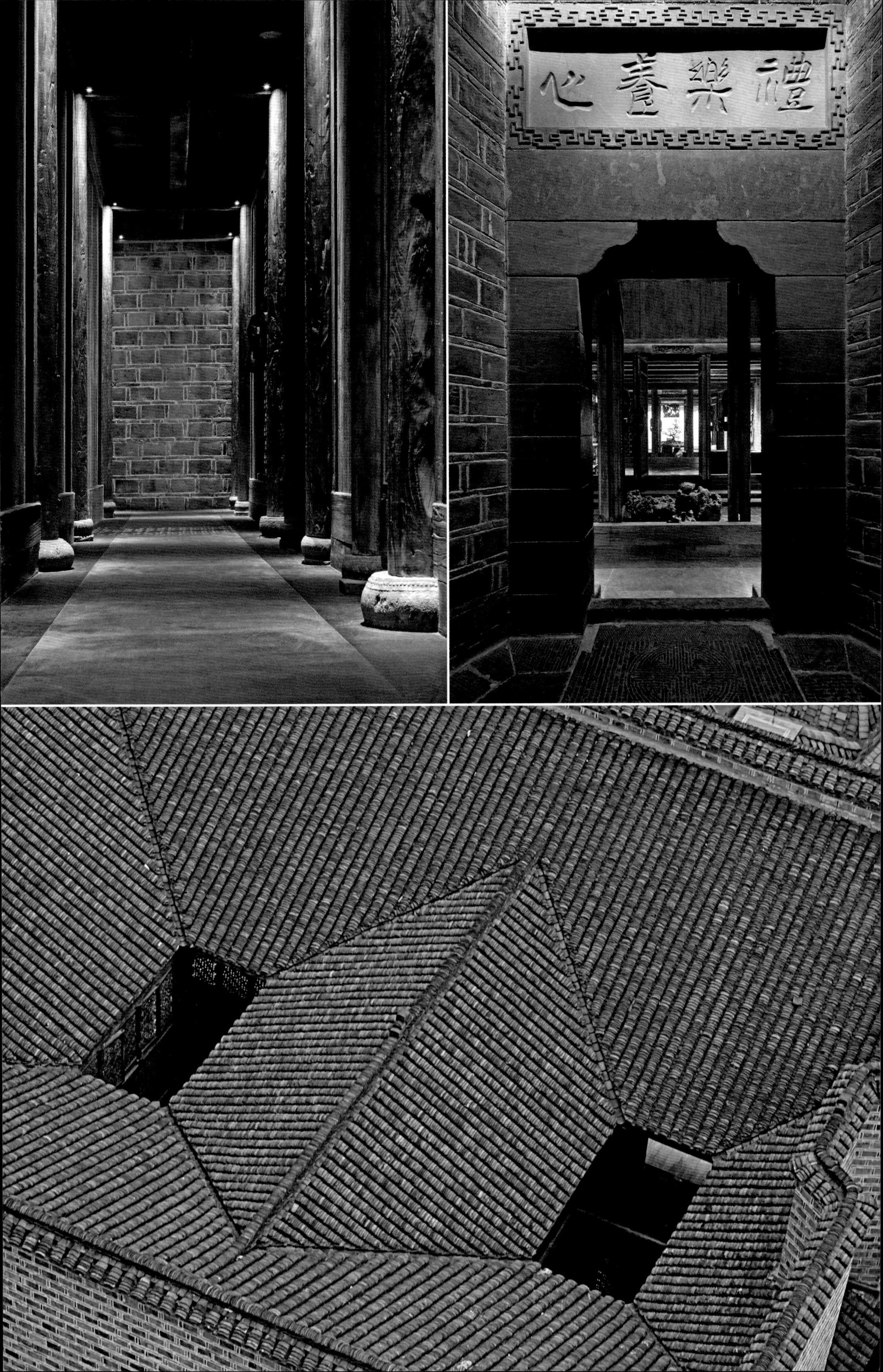
禮樂養心

老西门丝弦剧场 Old West Gate Silk String Theater

地点 湖南省常德市 / 用地面积 3,000m² / 建筑面积 5,000m² / 高度 24m / 设计时间 2015年 / 建成时间 2017年

方案设计 曲 雷、何 勍
设计主持 曲 雷

建 筑 何 勍、王 强
童佳明、冀文涛
结 构 李宗盛、郭 伟、张 慧
给排水 李 严、贺立军
设 备 范向国、张英男
电 气 叶 劲、刘 涛
总 图 王雅萍

丝弦剧团是常德丝弦和花鼓戏两项非物质文化遗产的所有者，丝弦剧场则是老西门棚户区改造中最大的公共建筑，改造促使其成为点亮城市片区的新地标。
休息厅采用悬挂结构，最大出挑达 10 余米，通高的无窗框结构式隐索幕墙将拉索隐藏在 5cm 玻璃的胶内部。小剧场内部每块栅顶和侧墙都可依演出需要变换布置。较低的舞台使观众获得了亲近的观感。
外立面的紫铜色铝板和碳化竹钢，暗示了现代建造与传统材料的关联，带着时光纹理的灰色页岩板，对应着明清古城墙遗址。粗糙而参差的陶粒混凝土基座则与近旁著名的常德会战留下的抗战碉堡相呼应。直径 30mm 的遮阳铝管仿佛琴弦，满布曲面，自成韵律。

The Silk String Theater boasts two intangible cultural heritages. The lounge has a maximum overhanging length of 10 meters. Each part of the ceiling can be modified on demand, and the low stage has drawn the audience close to the performers. The purple-bronze colored aluminum panels and bamboo-based steel bars have implied the connection between modern building technology and traditional materials. Gray shale slabs resemble the texture of the ancient city walls, while the coarse ceramsite concrete base pays tribute to the fortifications built during the Changde Battle against Japanese aggression.

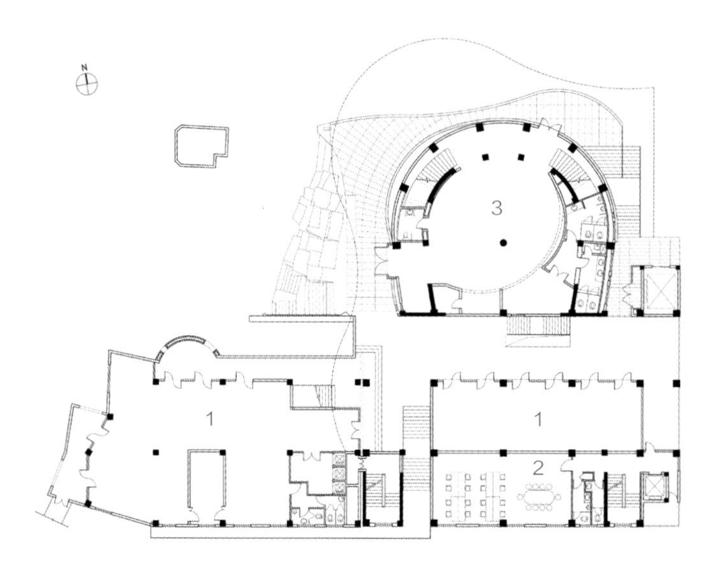

1. 止间书屋
2. 办公
3. 售票大厅

二层（剧院入口层）平面图

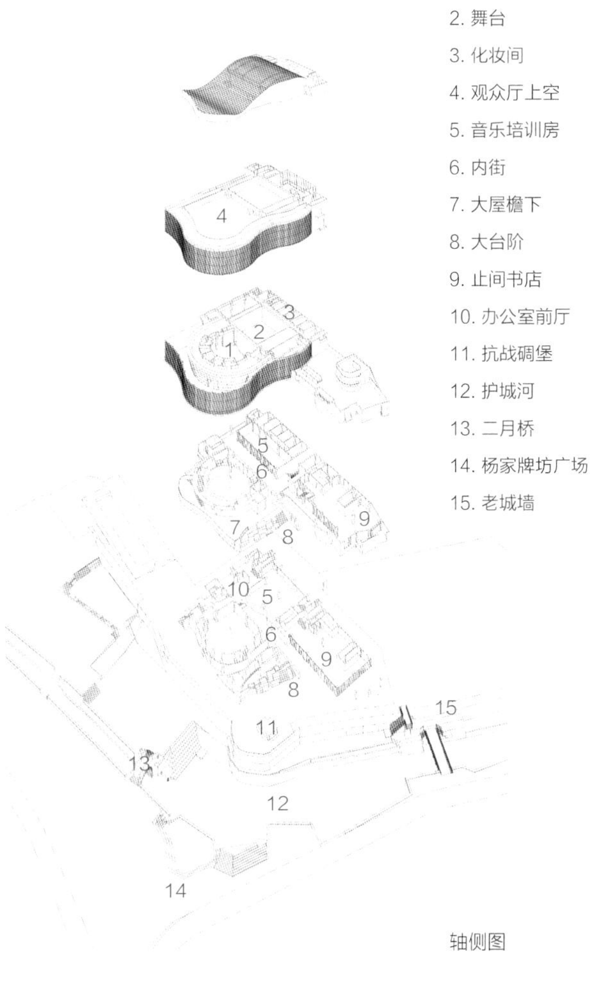
1. 观众厅
2. 舞台
3. 化妆间
4. 观众厅上空
5. 音乐培训房
6. 内街
7. 大屋檐下
8. 大台阶
9. 止间书店
10. 办公室前厅
11. 抗战碉堡
12. 护城河
13. 二月桥
14. 杨家牌坊广场
15. 老城墙

轴侧图

老西门剧场

昆山大戏院 Kunshan Grand Theater

地点 江苏省昆山市 / 用地面积 20,726m² / 建筑面积 50,553m² / 高度 29m / 设计时间 2011年 / 建成时间 2017年

方案设计 崔　愷、刘　恒、叶水清
　　　　 陈梦津
设计主持 崔　愷、刘　恒

建　　筑 叶水清、梁世杰
结　　构 王义华、程立新
给 排 水 高　峰、尹　华
设　　备 金　跃、顾艳娜
电　　气 陈　琪、崔振辉
总　　图 白红卫
室　　内 张　晔、韩文文

施工图设计 中旭建筑设计有限责任公司

项目在原昆山大戏院位置进行重建，并与西侧图书馆、南侧游泳健身中心及西南侧体育公园共同组成一个集文化、商业、休闲、展览、展示等功能于一体的城市文化综合体。建筑外部延续周边街区的边界，内部则为宛转流动的曲线体量，结合多样化的楼梯设计，使层层错落的室外平台形成连续空间。入口处的斜向轴线延续了原有的场地肌理，形成面对城市街角的入口广场空间，建筑之间以完整的屋盖相连。斜向轴线上设有下沉庭院、广场，结合河边景观、滨水平台，可将人流自然引向南侧的市民广场。剧场休息厅外侧不同色彩和尺寸的红色铝管，在灯光作用下宛若柔婉的水袖。电影院墙体通过参数化设计从室外的红色渐变到室内的蓝色调，外侧用不锈钢丝网装饰，形成如织物般虚幻的效果。

As a reconstruction project of the former Kunshan Theater, the project has become a culture complex that consists of a theater, a library, a swimming pool and a sports park. Curves, as a prominent feature of the building, have not only set the theme for the interior space, but also defined the feature of the building's exterior platforms. The artistry of the building is fully represented through the combination of decoration and illumination. The exterior wall of the theater lounge is decorated with red aluminum tubes, resembling the iconic fluttering sleeves in Kunqu opera, and the walls of the movie theater have a finish of Malay paint coated with stainless steel net.

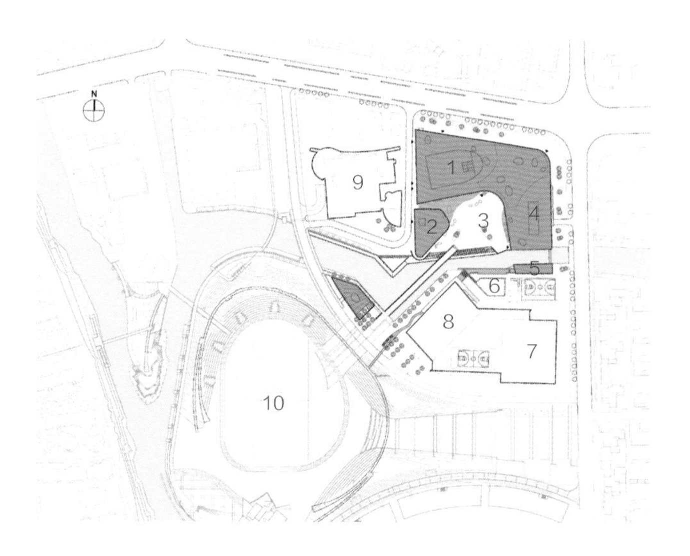

1. 多功能剧院
2. 电子阅览室
3. 中心广场
4. 电影院
5. 茶室
6. 室外游泳池
7. 篮球馆
8. 游泳馆
9. 图书馆
10. 体育公园

总平面图

大屋盖底面的三角形镜面不锈钢单元和不锈钢在白天和夜间反射出不同的城市景象，让城市空间与内部空间有机结合。

Reflections of the urban space, varying from day and night, are introduced into the building by the stainless steel mirror units under the huge roof.

1. 剧院门厅
2. 舞台
3. 侧舞台
4. 化妆
5. 电子阅览室
6. 餐厅
7. 厨房

二层平面图

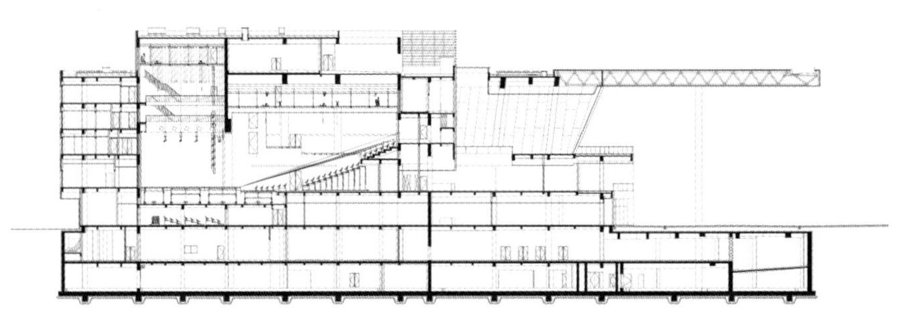

剧院剖面图

富强 民主 文明 和谐
自由 平等 公正 法治
爱国 敬业 诚信 友善

临沂大剧院 Linyi Grand Theater

地点 山东省临沂市 / 用地面积 82,166m² / 建筑面积 66,035m² / 高度 37m / 设计时间 2011年 / 建成时间 2015年

方案设计 崔 愷、康 凯、吴 健
设计主持 崔 愷、康 凯

建　　筑 朱 巍、吴 健、彭 彦
结　　构 赵剑利、邵 筠、许 庆
　　　　 尹胜兰、田京涛
给 排 水 杨东辉、董新淼、唐致文
设　　备 刘燕军、路 娜、韩亚征
电　　气 李俊民、陈双燕
　　　　 丁志强、沈 晋
总　　图 郑爱龙
景　　观 冯 君

临沂大剧院流畅的形态取"蒙山沂水"之意，两个似合似分的曲线形屋面将大剧院、音乐厅、电影院等功能体聚合其中，并在单体之间形成一条蜿蜒的商业街，将人流引入建筑内部，使剧场同时也成为市民休闲娱乐的开放性场所。屋面盘旋而上，既有中国传统建筑的意味，又有无限上升之感。建筑外饰面材料透明部分采用通透的玻璃幕墙，实体部分采用厚重干挂花岗石板。以竖向构件组合而成的立面具有流动的韵律感，也暗合临沂丰富的汉代竹简遗存的形式。

Inspired by the mountains and waters of Linyi, the design of the theater has integrated a theater, a concert hall and a movie theater. Furthermore, a winding street among the spaces provides access into the building, making the theater an open space for daily activities. The roof ascends in a spiral way, presenting traditional features and a sense of infinite ascending. Dominated by transparent glass curtain walls and dry-hanging granite plates, the façade reveals a sense of mobility while resembling bamboo slips in Han Dynasty.

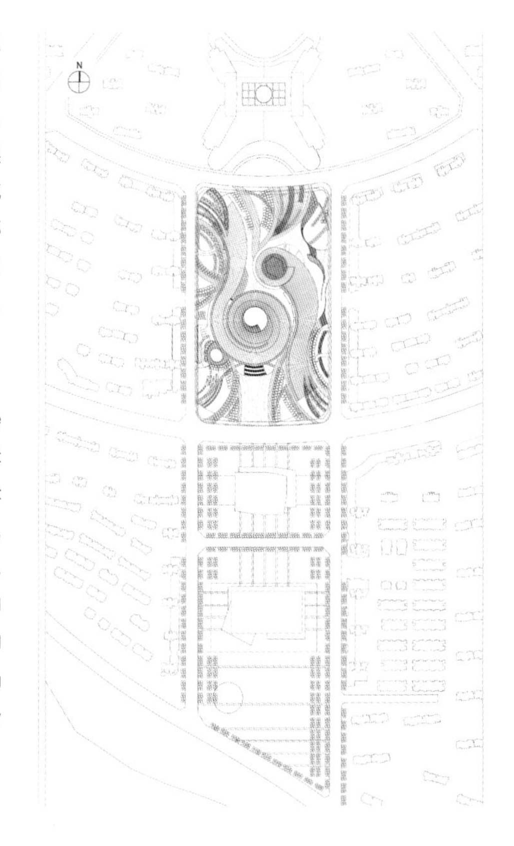

总平面图

黄河口大剧院 Yellow River Estuary Grand Theater

地点 山东省东营市 / 用地面积 294,900m² / 建筑面积 45,094m² / 高度 36m / 设计时间 2011年 / 建成时间 2015年 / 座位数 大剧场1300座

方案设计 张 祺、宋 菲、杨鸿霞
设计主持 张 祺、刘明军

建　　筑 姚文博、胡 斯
结　　构 鲁 昂
给 排 水 陶 涛、周 博
设　　备 金 跃、王 佳
电　　气 杨宇飞、熊小俊
总　　图 余晓东

黄河口大剧院坐落于山东省东营市的黄河入海口，由风景优美的清风湖环抱，周边还将建设多所城市文化设施。剧院本身容纳了一个大剧场、一个多功能小剧场、一组电影厅和餐饮设施。建筑造型取意“水城雪莲”，以黄河源头的巴颜喀拉山雪莲为寓，强调黄河首尾之间的联系，表达出诗意的自然和文化意境。建筑外立面因而采用玻璃与钢框架的组合，悬挑而成一片片花瓣，以当代的先进技术体现建筑的设计构思。

Situated at the Yellow River's estuary, the grand theater borders the beautiful Qingfeng Lake, as well as several cultural buildings to be built. The cultural complex consists of a large theater, a multi-functional small theater, a group of movie halls and restaurants. The form of the building symbolizes “a snow lotus in a riverside city” to present a romantic style in both natural and cultural context. The façade composed of glass and steel frames resembles petals while revealing the power of advanced building technology.

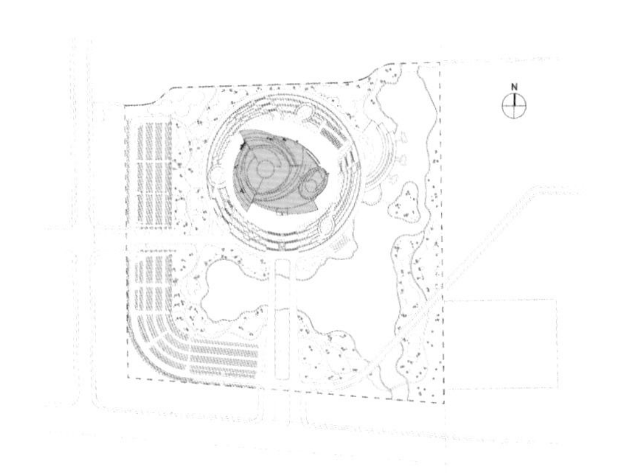

总平面图

幕墙节点效果图

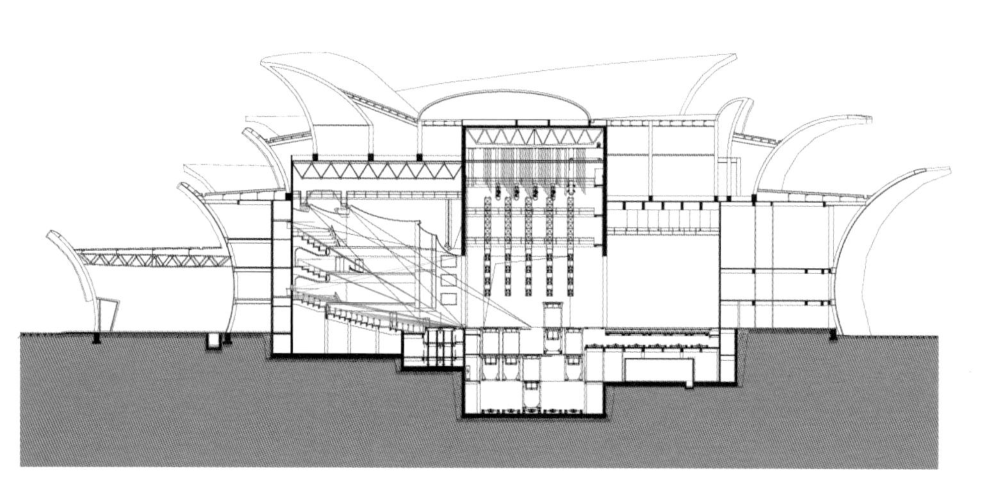

剖面图

地域特色

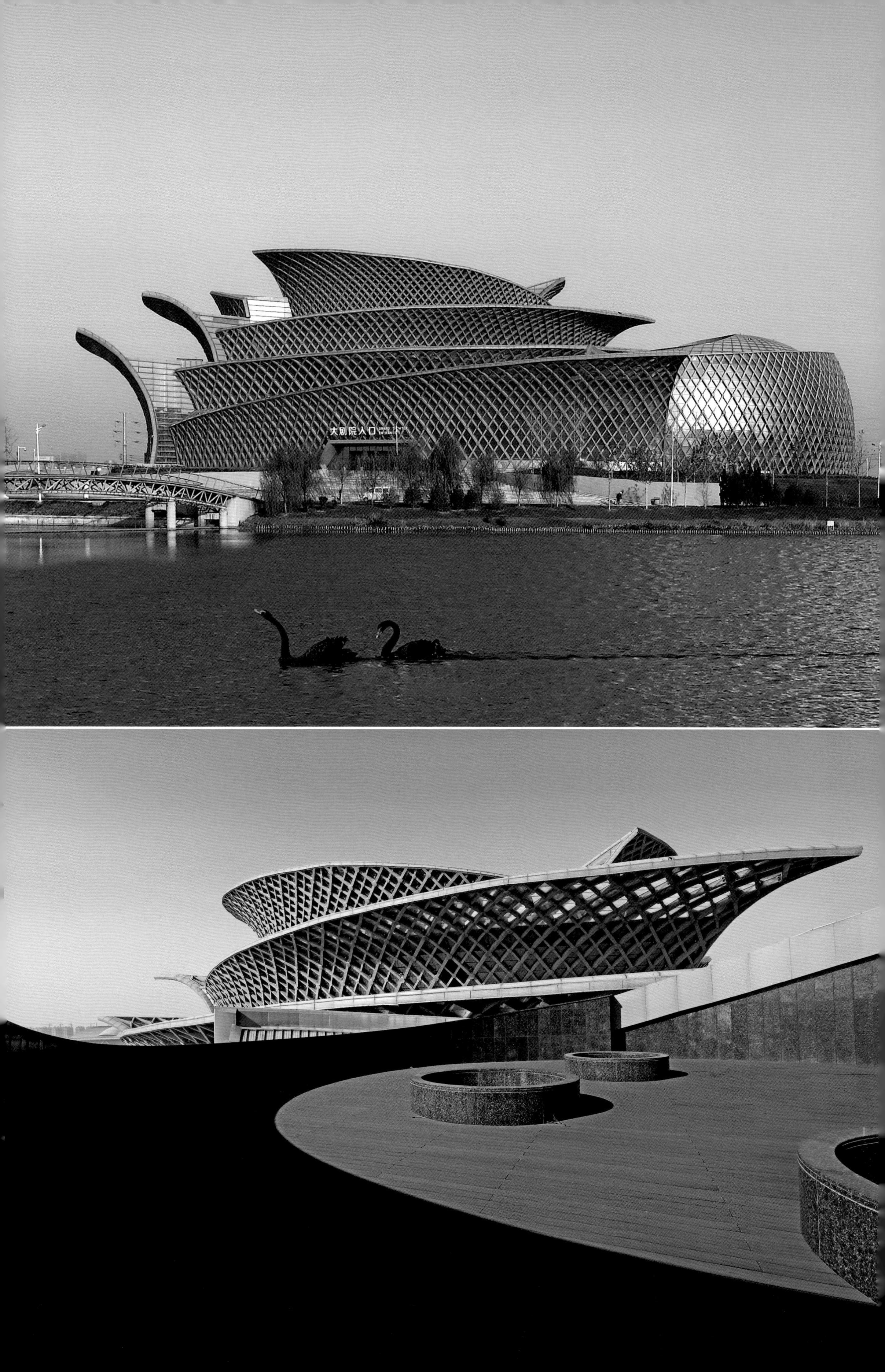
大剧院入口

敦煌市公共文化综合服务中心 Dunhuang Public Culture Comprehensive Service Center

地点 甘肃省敦煌市 / 用地面积 20,000m² / 建筑面积 19,936m² / 高度 22m / 设计时间 2008年 / 建成时间 2014年

方案设计 崔 愷、吴 斌
郑 虎、崔 剑
设计主持 崔 愷、吴 斌

建 筑 辛 钰、崔 剑
结 构 魏丽红
给 排 水 裴黎君
设 备 赵 琪
电 气 张 謇
总 图 王雅萍
景 观 李 莉

在“聚落”理念和统一模数的控制下，博物馆、图书馆、文化馆、档案馆等多种功能被整合到同一个建筑之中。如同当地民居般的院落布局，既有利于自然采光和通风，又增加了空间的层次。建筑体块的高低错落，形成不同标高的活动平台，为使用者提供了开放的交流、沟通场所。根据内部功能需要，外立面尽量减小开窗，整齐排列的小方窗如同当地葡萄晾房镂空的方洞，有着深邃的阴影，同时强化了立面的秩序感。立面材料为土黄色的砂岩，如同鸣沙山、戈壁之色，表面如砂砾般粗犷，在阳光下更显肌理。整组建筑如台地般，呈现出高低错落的“聚落”形态，形成尺度亲切、层次丰富的公共空间。

With the concept of clusters and the framework of modules, various facilities, including a museum, a library, a cultural center and an archive are integrated into one building. A layout dominated by courtyards can facilitate natural lighting and ventilation while adding to the diversity of the space. The different heighted volumes has endowed the building with a proper scale and platforms on different altitudes for communication. The neatly arranged small windows, like the holes on the walls of local grape-drying houses, cast long shadows. The sandstone with earthy color on the façade gives the building a rough texture.

总平面图

阶梯状的体形串联起一条向上引导的空间线索，通往不同高度的屋顶平台。在中部的美术馆及公共展厅部分围着主体拾级而上，可以到达最高处平台的空中室外展场，北眺城市小镇风光，南望鸣沙山之曼妙曲线。

The stepped volumes form an upward route to rooftop platforms at various altitudes. The outdoor gallery dominates the highest point, where visitors can have a panorama of both the city and the undulated Mingsha Hill.

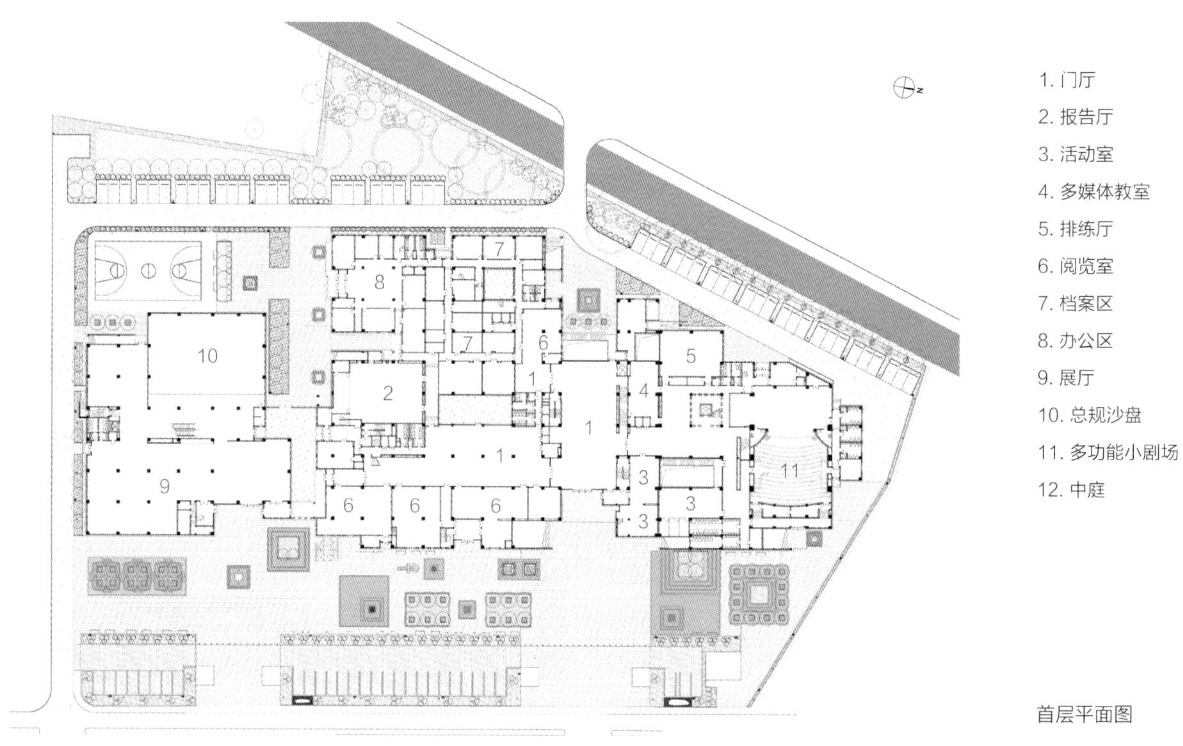

1. 门厅
2. 报告厅
3. 活动室
4. 多媒体教室
5. 排练厅
6. 阅览室
7. 档案区
8. 办公区
9. 展厅
10. 总规沙盘
11. 多功能小剧场
12. 中庭

首层平面图

剖面图

兴安盟图书馆、科技馆 Hinggan League Library & Science Hall

地点 内蒙古自治区乌兰浩特市 / 用地面积 21,054m² / 建筑面积 18,530m² / 高度 24m / 设计时间 2014年 / 建成时间 2017年

方案设计 丁利群、周绮芸、范高萁
设计主持 丁利群

建　　筑 周绮芸、范高萁、张　硕
结　　构 施　泓、王　震、刘雪冰
给 排 水 朱跃云、张庆康
设　　备 金　健、陈高峰、全　巍
电　　气 曹　磊、高　洁
总　　图 白红卫

作为一座整合了图书馆、科技馆、城市规划展览馆、科技书店等多项功能于一体的综合文化展览建筑，建筑的选址具有得天独厚的自然景观，南侧为河道和城市景观带，基地内也有多株茂密的大树。设计因而从尊重现状环境出发，围绕保留的树木设置室外庭院，继而以此组织室内空间和不同的建筑体量，形成“五院五庭五重山”的空间格局。建筑同时响应不同的城市界面。面向城市主干道的北侧尽量保持完整的体量，与道路尺度呼应；南侧则向河面开放，展现错落有致的建筑形态，并在二层设置室外休憩平台。二层以上的结构体系由柱网转化为密肋柱，在外观上呈现为规则的立面，与室内空间保持一致，呈现出均匀而优雅的效果。五个建筑体量内部的中庭空间和室内功能结合，展示不同的空间个性，与各自的空间氛围相得益彰。

As a comprehensive cultural center incorporating a library, a science hall, a city planning exhibition hall and a science bookstore, the complex boasts unique natural landscape thanks to its adjacency to a river and a landscape belt. With respect for the natural resources, the design features a series of courtyards, the locations of which were based on the preservation of trees on site, forming a layout of “5 courtyards and 5 buildings”. The northern part of the building, which faces a main urban road, has a complete form that conforms to the scale of the road; the southern part takes on an open gesture with an outdoor platform at the 2nd floor. The column grid are transformed into a system of ribbed columns at the 2nd floor and above.

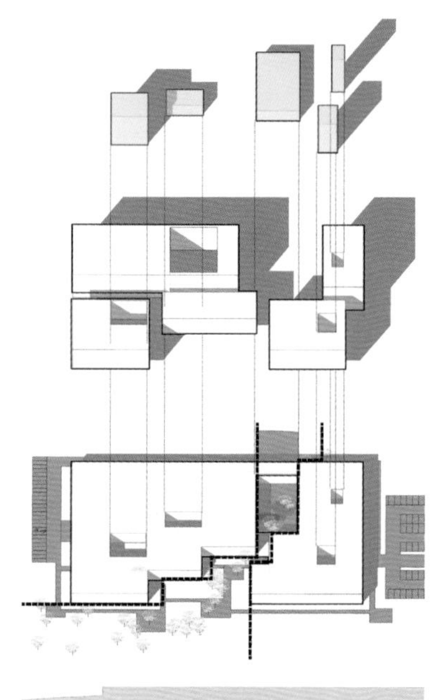

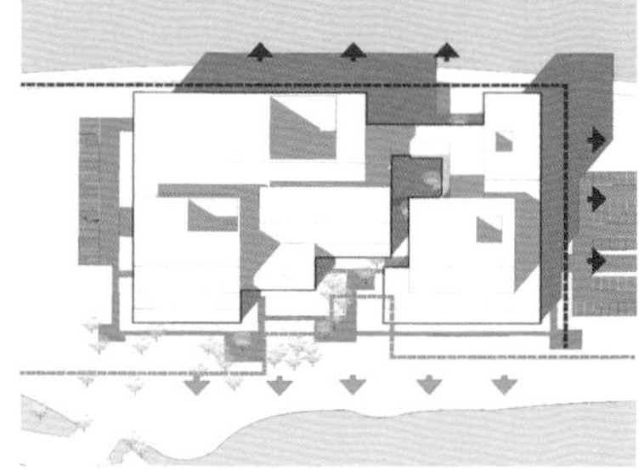

设计意向图

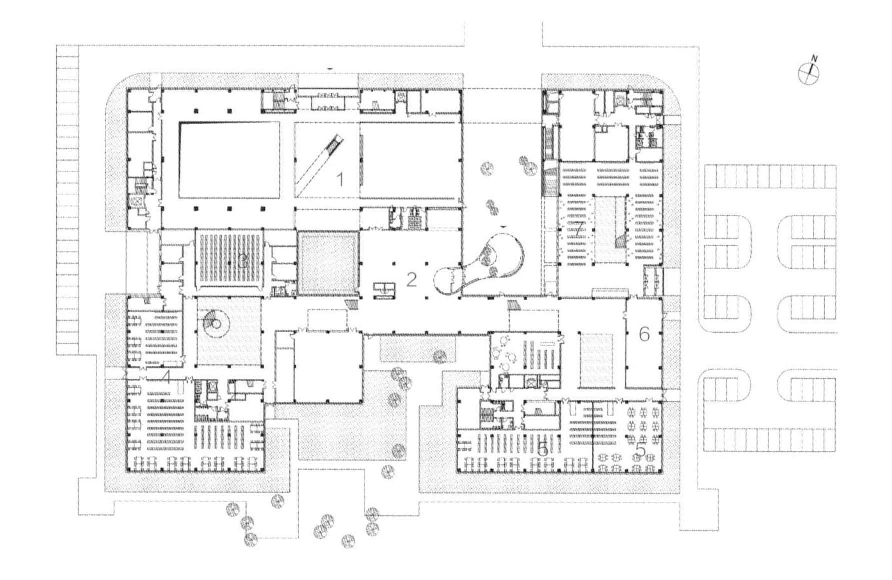

1. 城市规划展览馆
2. 图书馆大厅
3. 书库
4. 视听阅览室
5. 期刊阅览室
6. 咖啡厅
7. 书店

首层平面图

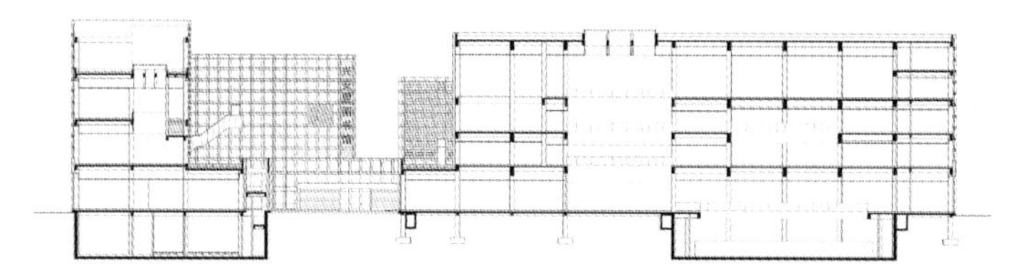

剖面图

兴安盟图书馆

吴桥杂技艺术中心 Wuqiao Acrobatic Art Center

地点 河北省吴桥县 / 建筑面积 15,000m² / 高度 15m / 设计时间 2009年 / 建成时间 2014年

方案设计 徐 磊、刘小枚
孟海港、弓 蒙
设计主持 徐 磊

建 筑 刘小枚、孟海港
结 构 朱炳寅、宋 力、周 岩
刘 巍、王 奇、张 路
给排水 陶 涛
设 备 劳逸民
电 气 曹 磊、李建波
总 图 连 荔
室 内 北京筑邦建筑装饰工程有限公司

摄 影 王洪跃

地处杂技之乡的吴桥杂技艺术中心，拥有优质的表演资源，但有限的城市环境资源和造价成为设计需要克服的困难。设计者把对杂技的领悟体现于建筑之中。杂技演员的姿态是普通人的姿态的延伸和异化，建筑的形式逻辑也遵循这样一个原则：朝向城市一面，采用完整、规则但有构成感的界面；内部院落空间的界面则在延伸的过程中向不规则的方向异化，运用简单的操作，将外层构成式界面经过多次折叠，然后在不同的高度推出或推进，空间的不确定性，自然形成了一定的魔幻的感觉。建筑的构造做法在造价控制的前提下体现较好的品质。深红色陶土砖为北方空旷的城市环境增加暖意且无需维护；减少开窗以利节能，也减少了型材和玻璃的变化；室内则以石膏板吊顶、涂料墙面和地砖为主，仅在重要公共空间使用石材。在观众厅，将演员出现和跑台路径设计得更具趣味，从空间上增加了表演的吸引力。

Located in Wuqiao, the hometown of acrobatics, the design was challenged by limited resources and budget. Since the posture of acrobats seems like extensions and variations of those of ordinary people, the design has adopted a similar approach that features a regular and complete boundary along the street and an interior courtyard with irregular extensions to various directions. Through the “folding” of the envelope of the building and protruding volumes at various heights, a sense of uncertainty is created to reflect the magic of acrobatics. The detailed design of the building has guaranteed high quality with a limited budget. The dark red terracotta brick is a maintenance-free material that adds warm colors to the city in Northern China, and various approaches for energy conservation are applied to the project.

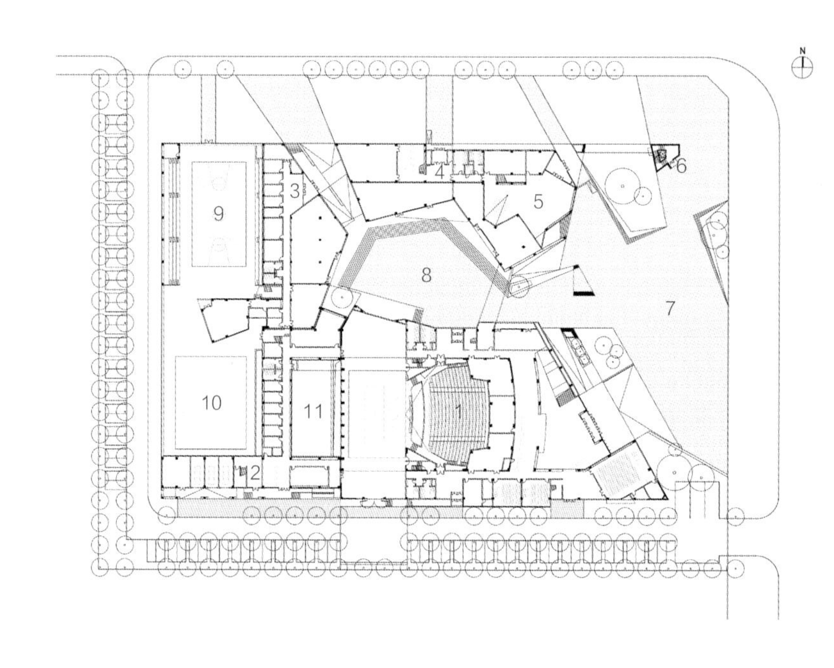

1. 剧场
2. 文化局
3. 全民健身中心
4. 文化图书馆
5. 博物馆
6. 标志塔
7. 广场
8. 梦舞台
9. 篮球场
10. 门球场
11. 庭院

首层平面图

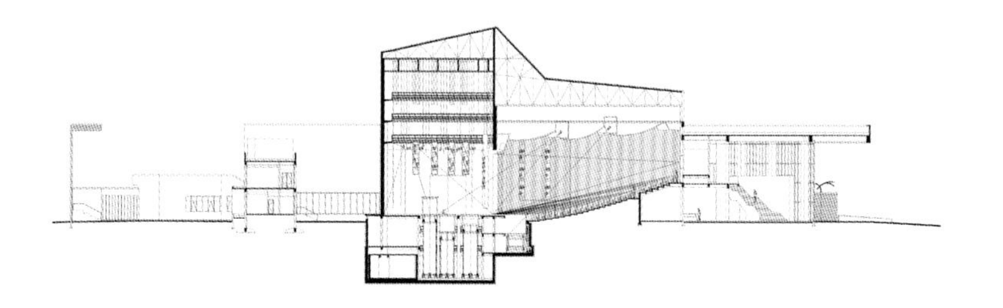

剖面图

桥杂技艺术中心
QIAO OPERA HOUSE

吴桥全民健身中心

葫芦岛文化中心 Huludao Cultural Center

地点 辽宁省葫芦岛市 / 用地面积 56,700m^2 / 建筑面积 41,888m^2 / 高度 24m / 设计时间 2009年 / 建成时间 2016年

方案设计 张 波、王 钊、金海平
张婷婷、靳树春
设计主持 张 波

建　　筑 王 钊、金海平
张婷婷、靳树春
结　　构 王 鑫
给 排 水 郭汝艳、石小飞
设　　备 劳逸民
电　　气 李俊民、苑海兵
总　　图 白红卫

建筑群沿一条城市环路展开，自西向东分别是图书馆、博物馆和文化馆，由长廊连为整体，博物馆与文化馆之间设置公共演艺广场。完整又略有变化的弧形建筑体量，成为沿路及滨河景观中的标志性场景。建筑分三个层次对葫芦岛著名的历史遗存“水上长城”进行了诠释：连续的弧形柱廊延续水上长城的空间序列；空心柱、灯台、花坛等构件组合，对水上长城构成语言进行了抽象提炼；菱形的建筑纹样也源于水上长城独特的平面形式，并以格构、丝印的形式体现在建筑中。

The building cluster consists of a library, a museum and a cultural center. The renowned historical remains of Huludao are reflected in the buildings in three ways: The hierarchy of “the Great Wall on waters” can be seen in the colonnade; components including lamp stands and flowerbeds, serve as abstraction of “the Great Wall on waters”; the rhombus patterns on the buildings are inspired by the unique feature of the Great Wall on waters.

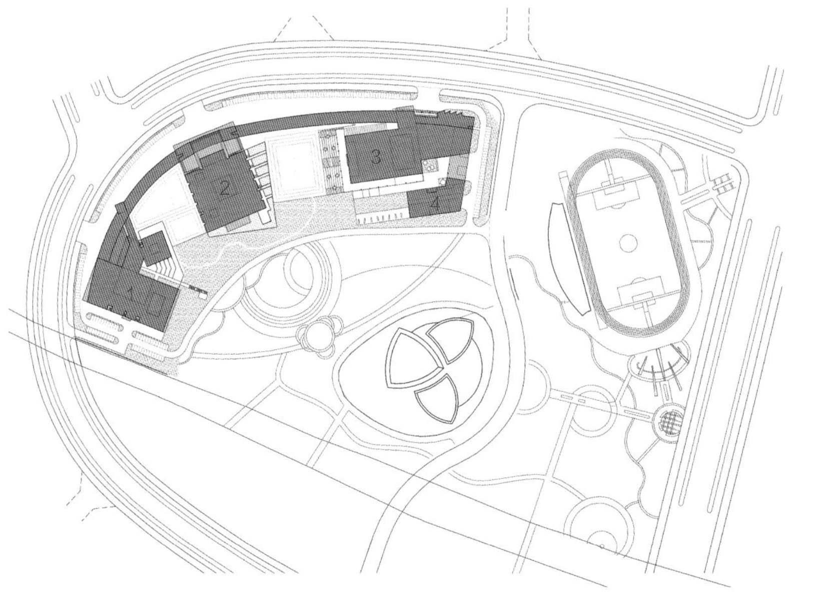

1. 图书馆
2. 博物馆
3. 文化馆
4. 附属办公楼

总平面图

鸟巢文化中心 Bird's Nest Cultural Center

地点 北京市朝阳区 / 建筑面积 15,396m² / 高度 10m / 设计时间 2012年 / 建成时间 2015年

方案设计 李兴钢、张玉婷、谭泽阳
唐 勇、张司腾
设计主持 李兴钢、谭泽阳

建 筑 张玉婷
结 构 王大庆
给 排 水 郭汝艳
设 备 胡建丽、高 强
电 气 曹 磊

鸟巢文化中心是一个在保护奥运遗产的前提下，按照赛后经营计划，对国家体育场局部空间进行改造形成的文化艺术交流场所。由体育场东北入口处，经由向下延伸的引道，即可到达北侧下沉庭院和大厅。庭院中是竖向层叠的“片岩”假山和水平拼合的水面。池岸、浮桥、平台、亭榭均由统一模数的清水混凝土单元板块组成，与草木配合，营造出兼具古意和今风的山水园林。同样模数的水平片状单元则蔓延到大厅室内，形成叠落的混凝土台地，兼具展示和观演功能。大厅通高近 10m，上圆下方的混凝土柱规则排布，而从地面延伸下来的体育场“鸟巢”的钢结构柱则因随意的倾斜，显得如同室内巨大的雕塑装置。

The project is a cultural and arts center after partial renovation of the National Stadium (the Bird's Nest) in line with the post-match operation scheme, and the renovation has been carried out under the premise of preserving Olympic heritage. A downward path provides access to the sunken courtyard and the lounge. In the courtyard with a central pond, the pond banks, bridges, platforms and pavilions are all made of modular components of fair-faced concrete, which has formed a garden of both ancient and modern features. Horizontal concrete plates with identical modules are also applied to the interior.

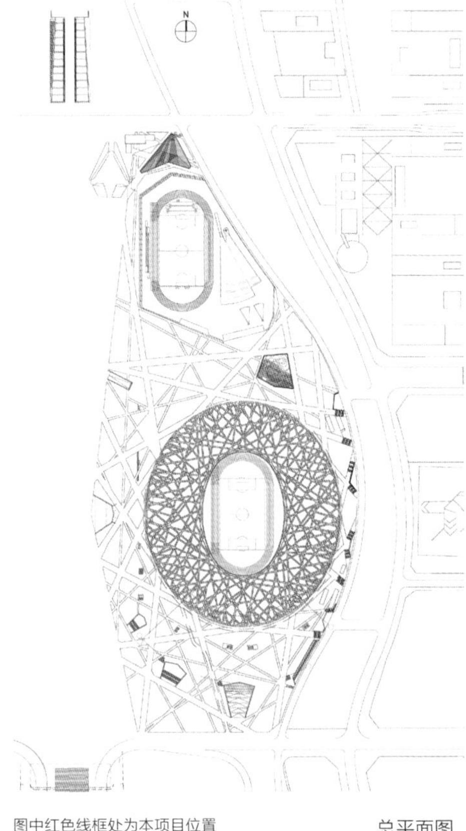

图中红色线框处为本项目位置

总平面图

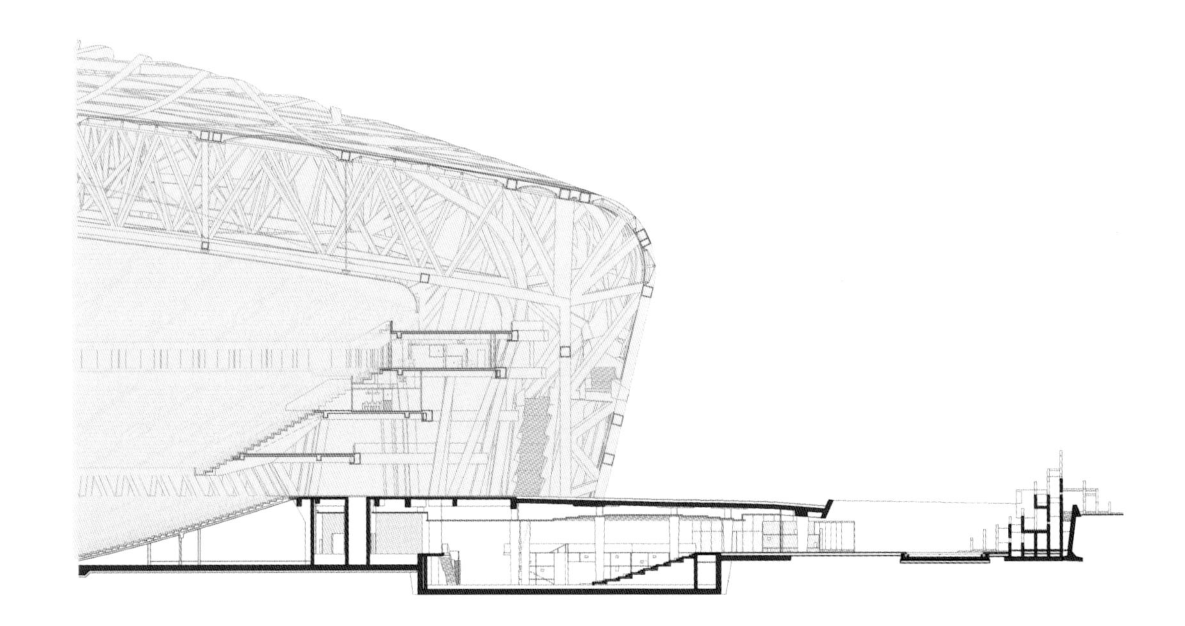

整体剖面图（呈现与国家体育场的关系）

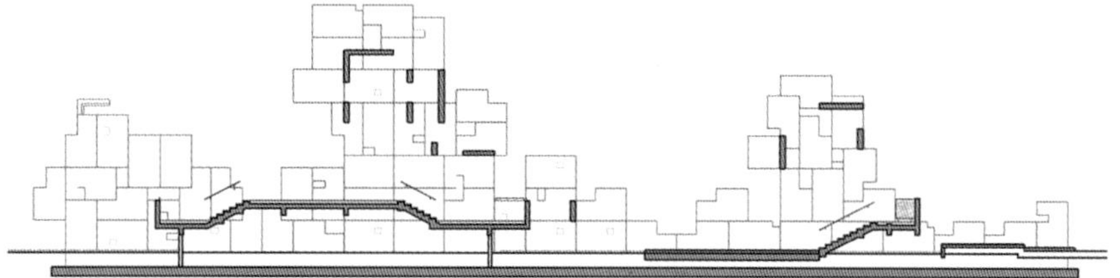

景观庭院剖面图

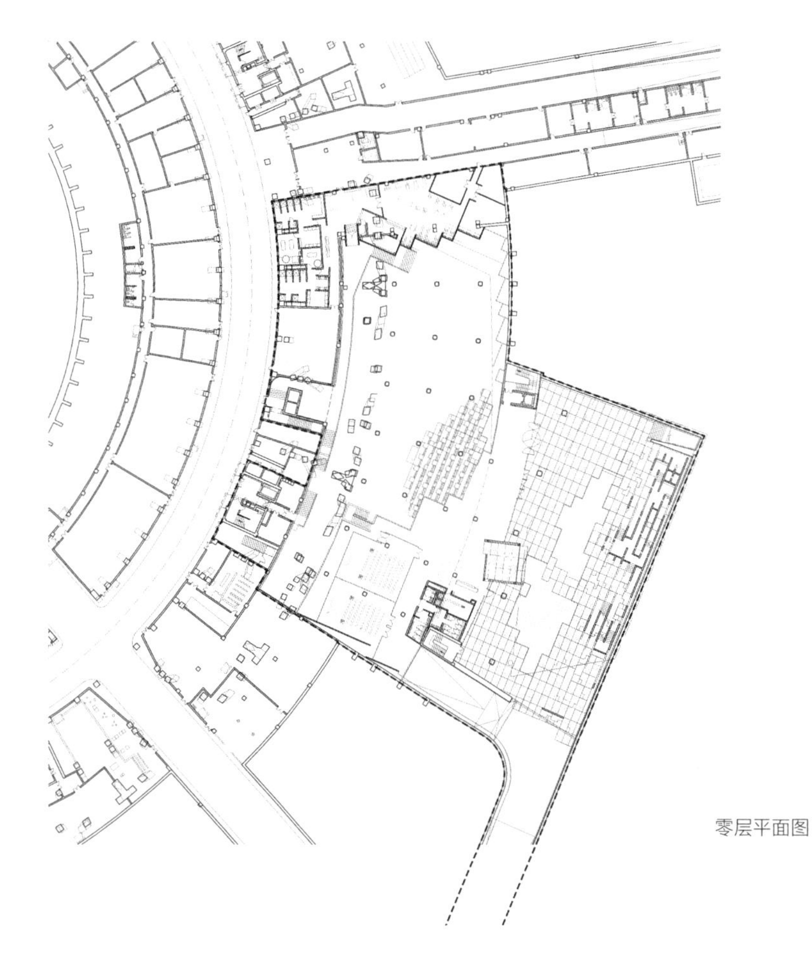

零层平面图

青海省图书馆二期、美术馆、文化馆 Library Phase II, Arts Gallery and Cultural Hall of Qinghai

地点 青海省西宁市 / 用地面积 24,786m^2 / 建筑面积 44,570m^2 / 高度 24m / 设计时间 2013年 / 建成时间 2016年

方案设计 崔 愷、李 峰、刘 洋
设计主持 崔 愷、时 红

建　　筑 李 峰、刘 洋、石 磊
结　　构 朱炳寅、杨 婷、徐宏艳
给 排 水 洪 伟、陈 宁
设　　备 牟 璇
电　　气 何 静、陈沛仁
总　　图 齐海娟

作为西宁市的重要市民文化生活场所，该项目不仅包括新建的图书馆二期、美术馆和文化馆，也对图书馆原有建筑进行了改造。考虑到新宁广场周边混杂无序的建筑现状，设计在高度和形态上与近旁的青海省博物馆取得关联，以方整、敦实的体量，形成文化建筑对广场的围合感，也符合西部地区粗犷豪放的地域特征。完整的建筑形体中，既有微微探出而增加的与广场的互动，也有结合入口、门厅设置的开洞，加之灵活布置的方洞和水平带窗，形成如山岩般切削、劈裂的整体形象。由博物馆浅红色石材引申而来的深浅变化和表面处理层次，通过随机拼贴，组合出细节丰富而整体统一的质感，是对广袤原野上民居石材堆叠形式的呼应。

As a key cultural venue for the citizens of Xining, the project consists of a series of facilities, including the newly-built Library Phase II, Arts Gallery and Cultural Hall, as well as the renovated library. With a regular and solid form, the buildings have enclosed the square while revealing the local features of Western China. The openings on the façade are made in a flexible way. The randomly collaged light-red stones on the façade resemble the stones of local folk houses, showcasing the details and texture of the building, paying tribute to the piling-up of stones of folk houses on the grassland.

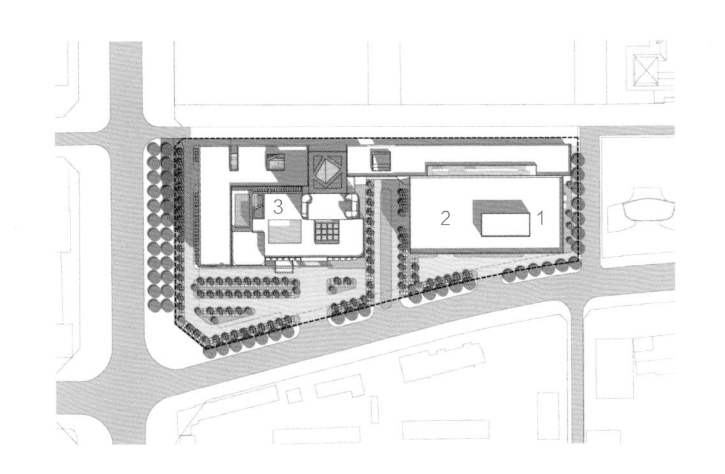

1. 文化馆
2. 美术馆

3 . 图书馆

总平面图

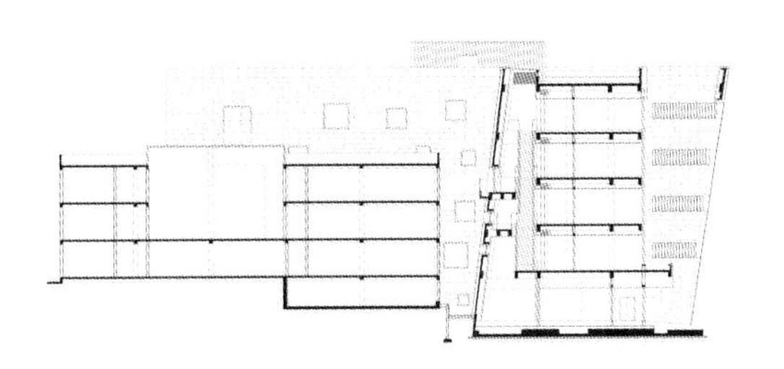

剖面图

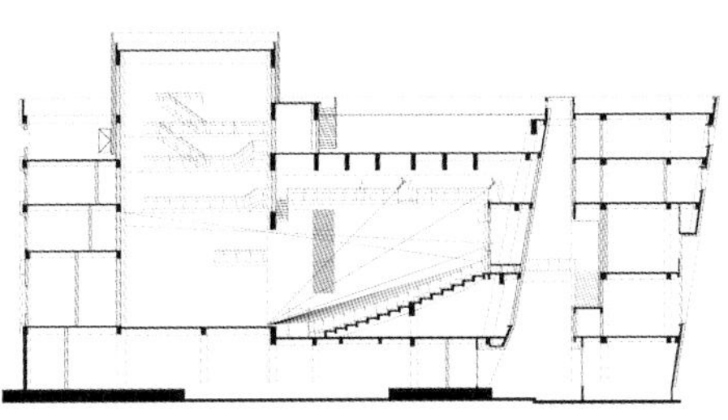

剖面图

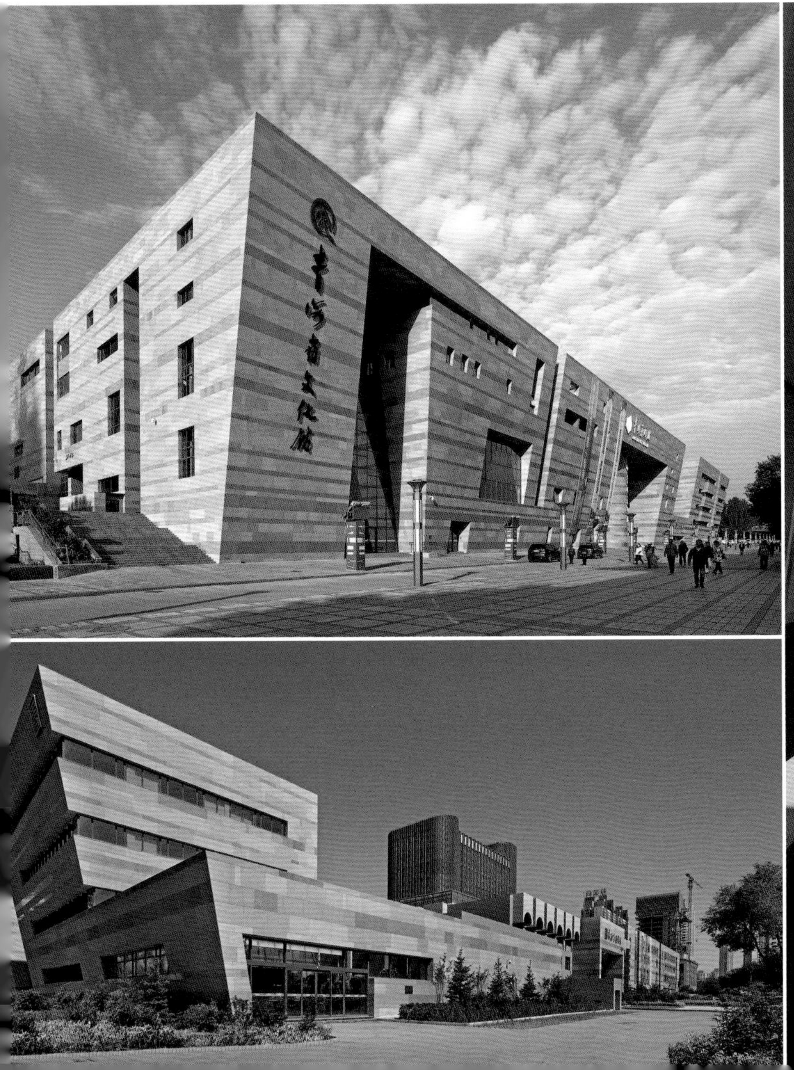

中铁青岛世界博览城 Conference & Exhibition Center of CREC Qingdao World EXPO City

地点 山东省青岛市 / 用地面积 300,000m² / 建筑面积 266,900m² / 高度 35m / 设计时间 2015年 / 建成时间 2018年

方案设计 徐　磊、高庆磊、李　磊
范高萁、孙煦暄、王南珏
金　星、赵　迪
设计主持 徐　磊、高庆磊

建　　筑 李　磊、赵　迪、范高萁
王南珏、金　星、孙煦暄
结　　构 梁　伟、孙海林
董　越、刘会军
给 排 水 匡　杰、车爱晶
设　　备 宋孝春、李　娟
电　　气 贾京花、王　莉
电　　讯 刘畅旸
总　　图 高　治、高　伟

青岛世界博览城拥有近临海景的优势。建筑形式通过中央十字展廊将 12 个独立展厅相连，同时展廊空间与周边道路、环境共享，形成开放式布局，创造出新的城市景观。气势恢宏的中央展廊，在减少能耗的基础上，具有鲜明的形象标志性。中央展廊能够为东亚海洋合作论坛及海事防务展提供适宜的大型展示空间，具备良好的天气适应性，同时与青岛独特的人文历史相契合。各展厅简洁、易用，与展廊保持着高度的融合性，同时也在阐述以海洋文化为背景的社会共识。展馆近人尺度采用了幕墙砖系统，与青岛独有的建筑风格呼应，延续青岛宜居城市所表达的城市气质。展览与停车功能本身决定了形象的简洁实用，建筑内在的空间结构形成了外部形态的美学韵律和视觉冲击力，使建筑形象真实而动人。

As the core sector of Qingdao World Expo City, the project borders the sea and takes on an overall look that is simple and practical, matching with its main functions of exhibition. 12 independent large exhibition halls are connected by a huge cruciform gallery in the center of the building, presenting a layout that is both iconic and climatically adaptive. The building is open to its neighborhood, standing out as unique urban landscape. A system of curtain wall of bricks is applied in consistence with the city's unique architectural style. Technological measures, including modular structures, natural ventilation and natural smoke exhaust, are integrated with other measures for energy saving and cost effectiveness.

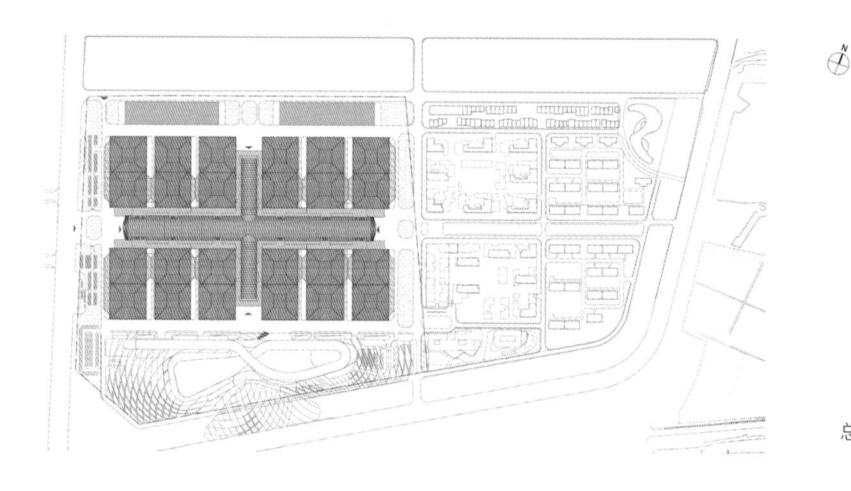

总平面图

N3
2018首届青岛军民融合科技创新成果展
S3
融合创新
S6
N

展廊结构采用了一种新型的预应力索拱，其截面高度仅为 500mm，可实现长达 48m 的跨度。索拱结构是索和拱组成的一种杂交结构。利用索的拉力或撑杆提供的支承作用调整结构内力分布并限制其变形的发展，进而有效提高结构的刚度和稳定性。

To maintain the lightness of structures a novel prestressed cable arch was developed, with a fabricated box section of just 500mm depth to span 48m. The cable arch structure is a sort of hybrid structure, adjusting the distribution of forces and restricting structural deformation.

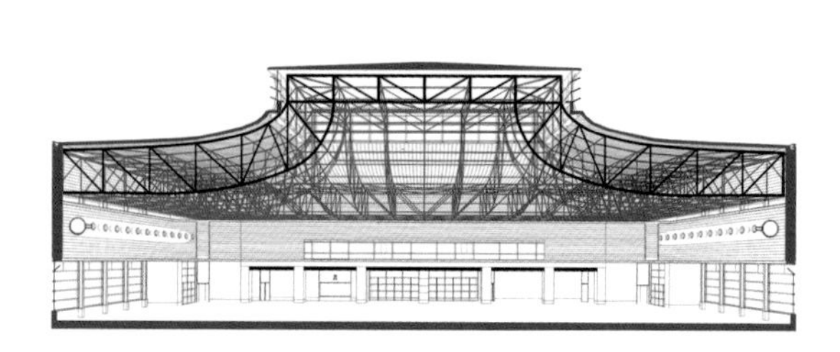

标准展厅剖透视图

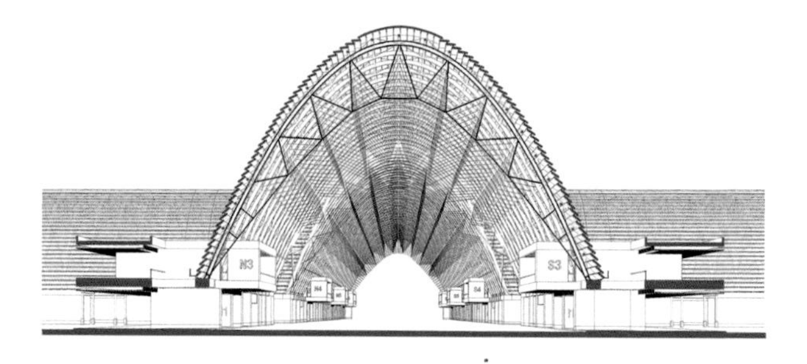

展廊剖透视图

N3
N3

内蒙古广播影视数字传媒中心 Inner Mongolia Radio & Television Digital Media Center

地点 内蒙古自治区呼和浩特市 / 用地面积 226,500m^2 / 建筑面积 149,686m^2 / 高度 97m / 设计时间 2007年 / 建成时间 2015年

方案设计 崔 愷、于海为、陈 宁 刘晏晏、谢 悦、于 玢 王 健
设计主持 崔 愷、于海为、杨益华

建　　筑 谢 悦、刘晏晏 张玉明、于 玢
结　　构 陈文渊、赵剑利
给 排 水 李万华、宋国清
设　　备 劳逸民、王 加
电　　气 陈 琪
总　　图 白红卫
景　　观 史丽秀
室　　内 张 晔、饶 劢

内蒙古广播影视数字传媒中心是一个集广播电视节目制作、播出、传输及行政办公、信息服务于一体的综合性建筑。设计从内蒙古大草原上高低起伏的朵朵白云和奔腾的骏马得到启发，通过错落有致的建筑群体造型表达地域特征和文化内涵，采用半圆形拱建筑顶部结构，创造出别具特色的建筑形象。设计同时从工艺需求和功能使用出发，采用现代建筑语汇和丰富的空间组织手法，建设一个布局合理、功能齐全、设施先进、具有高新技术含量、高效率的广播影视数字传媒中心。

Integrating various functions including production, broadcasting, administrative operation and information service, the media center has an undulating form inspired by the swirling clouds and the galloping horses to present the local characteristics. Semi-circular arches, together with other modern architectural elements and the alignment to processing requirements, have endowed the media center with local features, appropriate layout, high efficiency and cutting-edge technologies, making it stand out as a landmark.

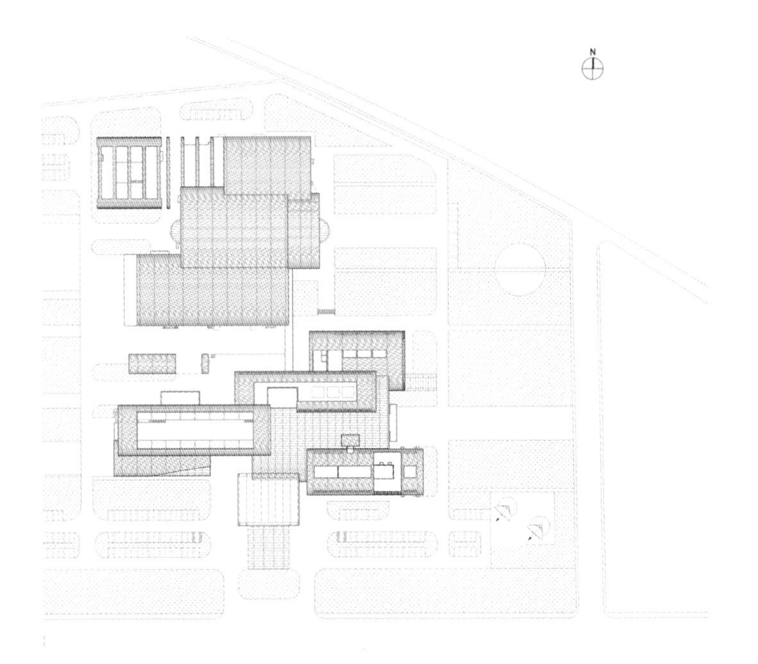

总平面图

极富韵律感的建筑轮廓，既体现了建筑在城市中的标志性地位，又表达出草原人的浪漫辽阔的情怀，

The building's role as a landmark, as well as the profound and romantic characteristics of local people, are reflected through the rhythmical outline.

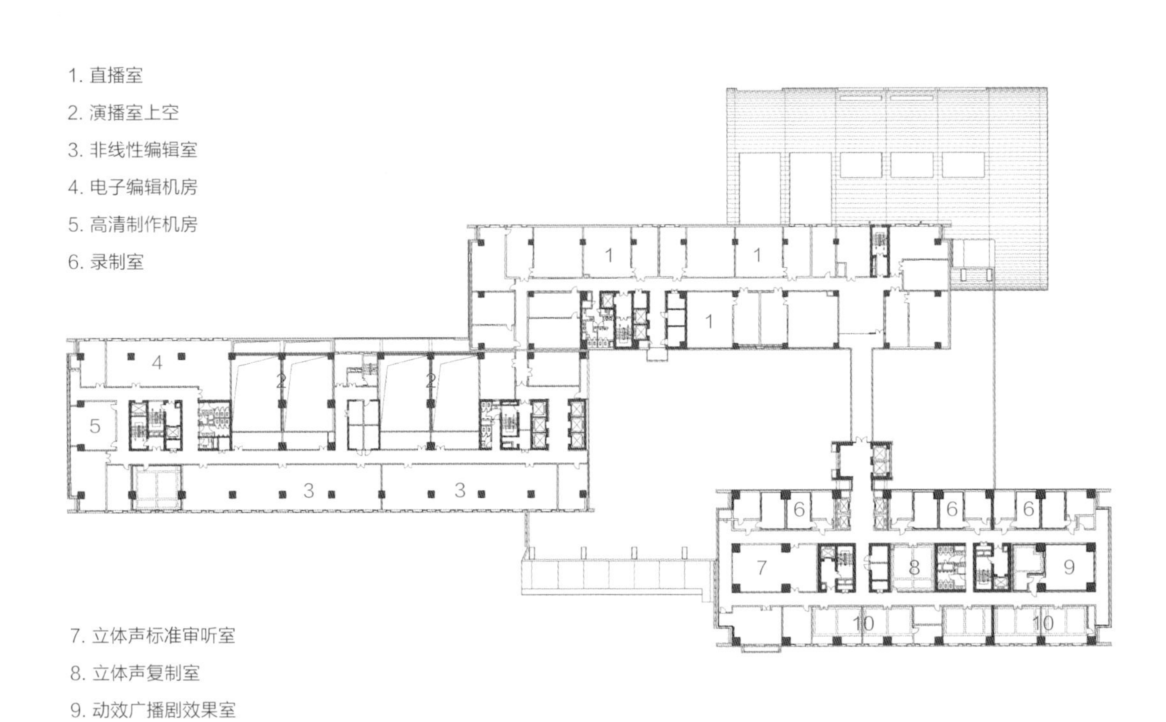
1. 直播室
2. 演播室上空
3. 非线性编辑室
4. 电子编辑机房
5. 高清制作机房
6. 录制室
7. 立体声标准审听室
8. 立体声复制室
9. 动效广播剧效果室
10. 声道混录机房

五层平面图

昌吉州传媒大厦 Changji Media Center

地点 新疆维吾尔自治区昌吉市 / 用地面积 58,029m^2 / 建筑面积 39,470m^2 / 高度 55m / 设计时间 2012年 / 建成时间 2016年

方案设计 于海为、谢 悦、牛 涛
潘天佑、朱起鹏
设计主持 于海为、谢 悦

建 筑 王 静、牛 涛
结 构 赵剑利、许 庆、崔 青
给排水 申 静
设 备 祝秀娟、李 嘉
电 气 程培新、史 敏
电 讯 张 雅
总 图 刘 文

作为容纳功能多样、工艺分工复杂的传媒文化建筑，设计提取传统聚落的空间意向，分类整合各项功能，将普通办公区、广播电台与电视台技术办公区、演播区、职工餐厅等不同功能分别整合为 6 个不同体量的立方体，创造一个既相对独立又方便联系的媒体聚落。建筑单体采用统一的材料和表达语汇，形成具有整体感的形象，同时又以尺度不一的广场、绿地、连桥、台阶分隔、联系，创造出形态丰富的室外空间。连桥保障了广电建筑的必要工艺联系，既为员工提供了通道，也可开放参观展示，成为媒体与受众的互动联系纽带。设于多条重要视觉通廊的交汇点的“信息树”高 40m，兼具地方建筑意象和简洁的现代形式，成为整组建筑的标志性体量。

For a media center with diversified functions and technological processes, the design has adopted the concept of clusters, creating six cubes varied in size to form a media cluster with sectors that are both independent and interconnected. Unified materials and styles have guaranteed the integrity of the cluster, while squares, greening, bridges and stairs of various scales have formed a diversified outdoor space. Bridges between the cubes can also serve as access for staff and an open area for interaction between the media and the audience. An “information tree”, 40 meters in height, stands at the intersection of multiple viewing corridors, becoming the focus of the whole building.

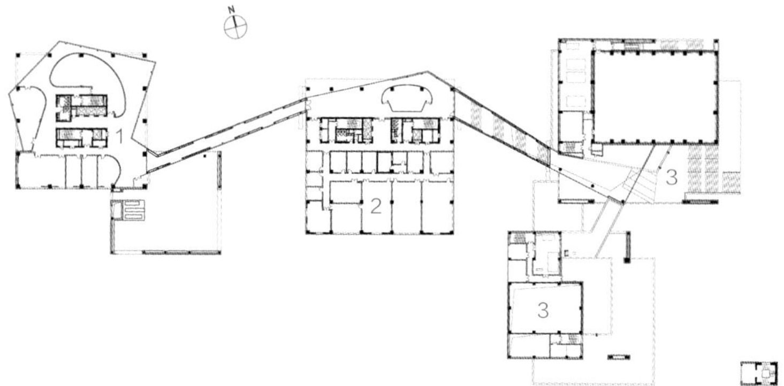

1. 综合办公 2. 综合技术 3. 演播区

三层平面图

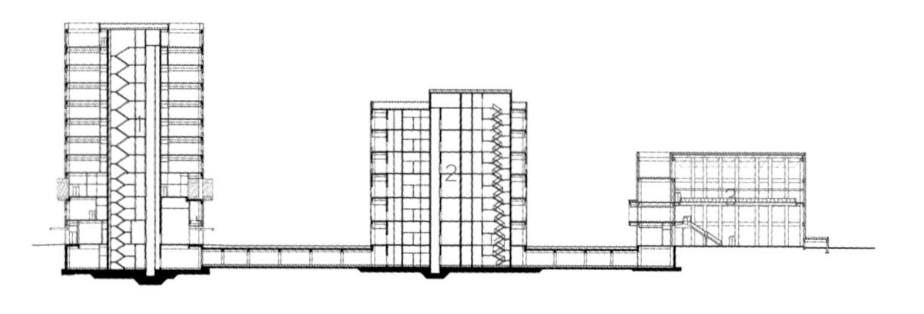

剖面图

传媒大厦
昌吉州艺术剧院
艺术剧院

鄂尔多斯市体育中心 Ordos Sports Center

地点 内蒙古自治区鄂尔多斯市 / 用地面积 900,000m² / 建筑面积 259,000m² / 设计时间 2008年 / 建成时间 2014年 / 座位数 主体育场6万座

方案设计 景 泉、李静威、徐元卿、张小雷、杨 磊
设计指导 崔 愷
设计主持 景 泉、李静威、王更生

建　　筑 徐元卿、黎 靓、张伟成、栗 晗、邵 楠、程 明、郭正同、张月瑶、吴锡嘉、张文娟
结　　构 尤天直、张亚东、施 泓、史 杰、刘文斑、宋文晶
给 排 水 赵 昕、陶 涛、马 明
设　　备 胡建丽、孙淑萍
电　　气 王玉卿、王浩然、李战赠
总　　图 余晓东、王雅萍、李可溯
室　　内 邓雪映、段嘉宾、张 亮、董 岩
景　　观 史丽秀、关午军、王洪涛

鄂尔多斯市体育中心由体育场、体育馆和游泳馆三个场馆组成。建筑形态通过“巨柱”结构的疏密排布使呈现连续、扭结的空间意向，传达出“蒙古摔跤”的喻意，体现了建筑文化的地域性。体育场的入口借鉴了蒙古包“套脑”的空间特色，在上方开了超尺度的采光圆洞。为实现这一空间形式，设计者通过结构技术取消此处的巨柱，并结合仪式性坡道，共同构成了宏大、庄重的仪式性入口空间。座椅以绿色为主，深绿、浅绿合理排布，视觉上延续了绿色的运动场地，形成了一望无际、生机勃勃的草原意象。观景平台高低错落，犹如摔跤时勇士们身上各色飞舞的飘带。主广场上，以“马头琴”抽象变形而成的主体雕塑，如一架连接时空的桥梁，从草原穿梭而来，体现了少数民族豪放、热情、能歌善舞、坚忍不拔的群体特征。

Consisting of a stadium, a gymnasium and a natatorium, the Ordos Sports Center has a form that is both twisted and continuous with its huge columns irregularly arrayed, presenting the dynamic image of Mongolian wrestling. The design of the stadium entrance, with a super-large skylight overhead, was inspired by the features of traditional Mongolian yurt. Dominated by dark and light green colors, the grandstand seems like the extension of the lawn, while viewing platforms of various heights resemble the ribbons on the wrestler's clothes. On the main square, a sculpture with the abstract image of morin khuur, a traditional musical instrument, seems like a bridge between the past and the future, showing the vigor and enthusiasm of the people in Inner Mongolia.

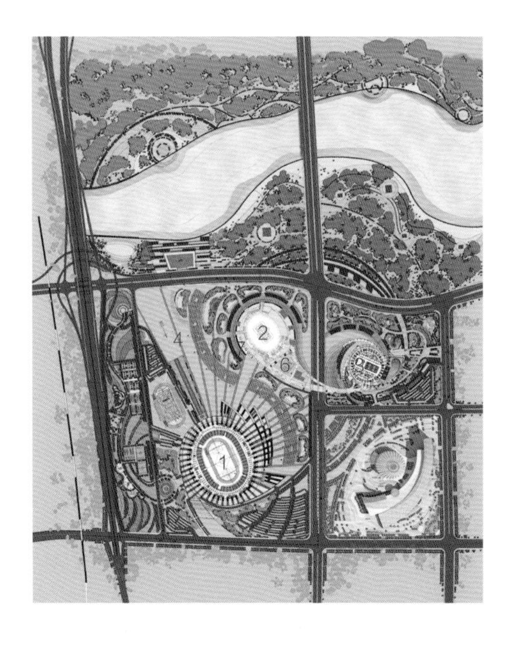

1. 体育场
2. 体育馆
3. 游泳馆
4. 广场
5. 停车场
6. 训练场及健身场
7. 宿舍及培训中心

首层组合平面图

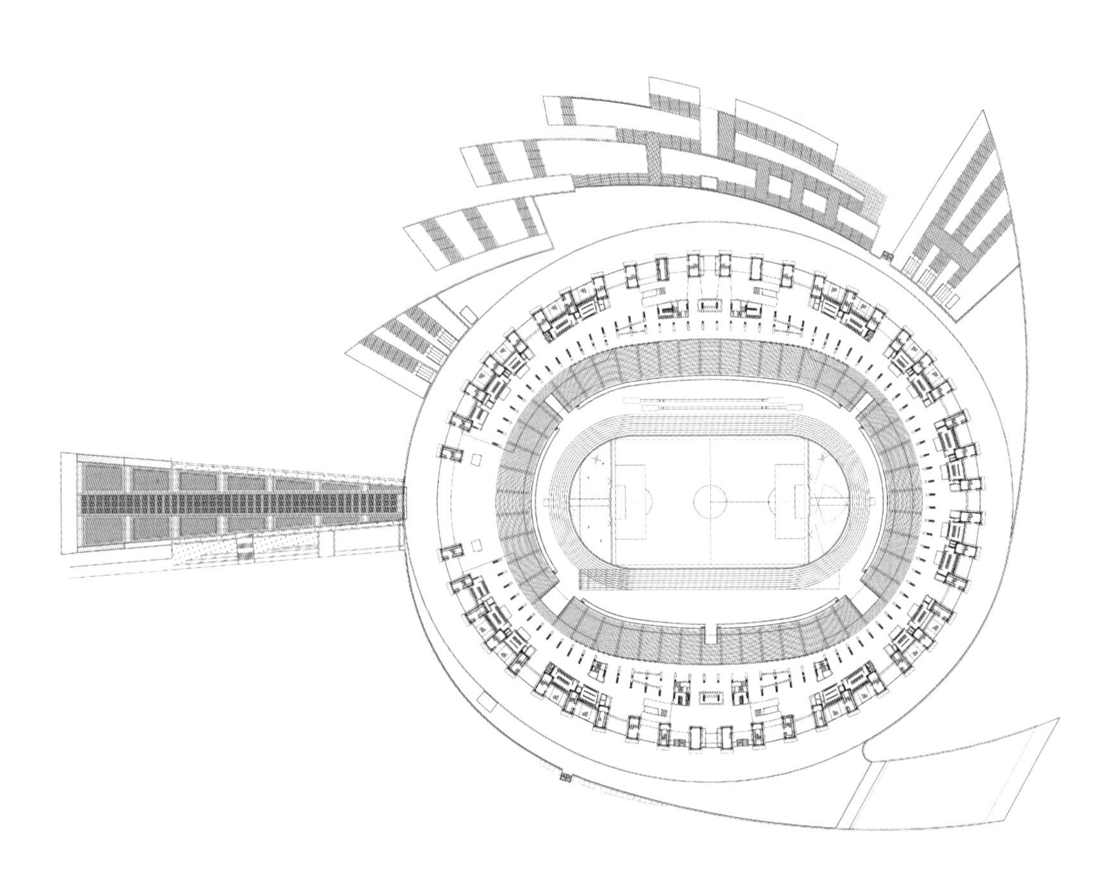

体育场首层平面图

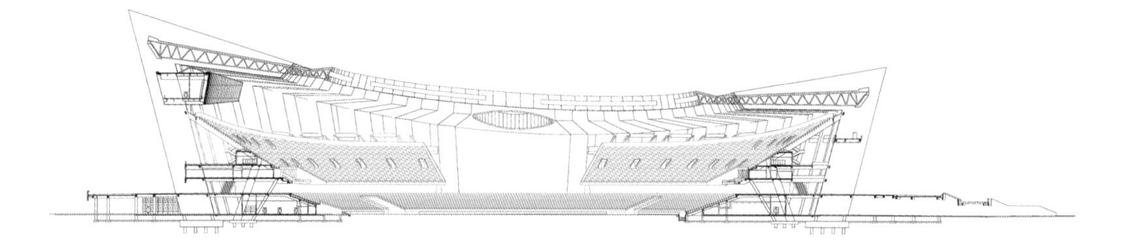

体育场剖面图

体育场巨柱以金色表达“金马鞍”这一设计主题。铝板之间的空隙隐藏了空调洞口、排水管线等，提高了建筑完的完整性。

The gold color of the huge columns implies the symbolic feature “golden saddle”. Openings for air conditioning and pipelines are hidden in the gaps between aluminum panels.

佛山市高明区体育中心 Foshan Gaoming District Sports Center

地点 广东省佛山市 / 用地面积 166,300m² / 建筑面积 45,751m² / 高度 24m / 设计时间 2011年 / 建成时间 2014年 / 座位数 体育场 8000座

方案设计 杨金鹏、李高洁
钱玉斋、王 喆
设计主持 杨金鹏、王 喆

建 筑 李高洁、李建宇
结 构 范 重、胡纯炀
李 丽、张 宇
给 排 水 靳晓红、朱 琳
设 备 汪春华、王春雷
电 气 王玉卿、贺 琳
总 图 吴耀懿

设计将高明区的山、林、水等元素融入建筑之中，形成一个具有当地特色的城市体育公园，一座座如云般连绵的体育场馆展现在人们眼前。波浪状依次展开的拱形钢结构轻轻托起屋顶，造就轻盈欲飞的建筑姿态。建筑金属屋面洁白纯净、起伏有致，强化了“行云”的意境。各个活动广场连接成为整体，景观设计将场地北侧自然水系延续至主入口广场，诠释了“流水”的理念。连续的拱形结构沿主干道依次排开，若连绵的山林在城市中再次呈现，与广场的大地景观共同勾勒了一幅山清水秀的自然长卷。

With the images of mountains, woods and waters of Gaoming Region integrated into the design, the Foshan Gaoming Sports Center has become a city sports park with distinctive local characteristics. The wavy and white metal roof, supported by arched steel structure, has added to the lightness of the building with its resemblance of drifting clouds. Furthermore, the "clouds" are accompanied by "flowing waters", as the waters to the north of the site are extended to the square at the main entrance.

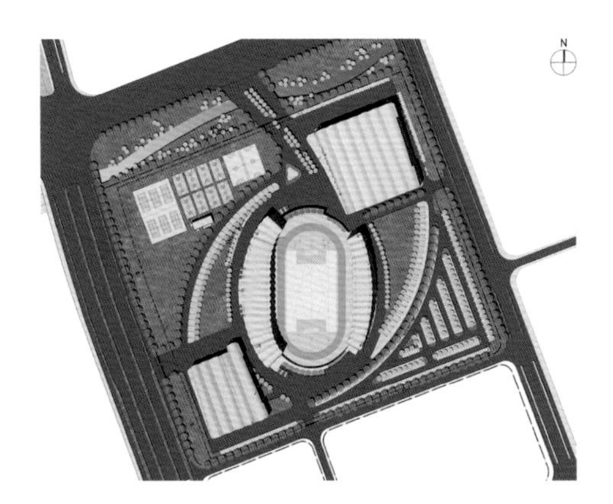

总平面图

项目组供图

扬中市奥体中心 Yangzhong Olympic Sports Center

地点 江苏省扬中市 / 用地面积 109,183m² / 建筑面积 75,598m² / 高度 33m / 设计时间 2010年 / 建成时间 2014年 / 座位数 体育场 10000座

方案设计 杨金鹏、李高洁、闫小兵
设计主持 杨金鹏

施工图设计 上海中森建筑与工程设计顾问有限公司
建　　筑 刘鹏程
结　　构 熊品华
给 排 水 庞志泉
设　　备 焦海涛
电　　气 钟联华
总　　图 郭　欣

扬中市的主体是长江上的一个岛屿，也是仅次于崇明岛的长江第二大岛，其名即“扬子江中”之意。水和岛无疑是最体现扬中地域特点的要素。设计取流水的速度和石头的力量，体现体育的“更快、更高、更强”。连续的屋顶和廊道串起整个建筑群，如同流动的水；体育馆、游泳馆、综合训练馆如同水中的石头，不但形成极具特色的立体景观，更是通过场地的高差，形成自然的分区界限，组织人流和交通。建筑以体育产业为载体，包含竞赛、训练、体育商业，大众健身、大众休闲、商业演出、商业购物等众多功能，改善和丰富了城市功能。

As the region abounds in waters and rocks, the design of the sports center showcases the speed of waters and the strength of rocks to symbolize the spirit of “faster, higher and stronger”. The roof and corridors that connect all the buildings symbolize the flows of water, while the gymnasium, natatorium and the comprehensive training hall take on a solid look like the rocks. The altitude differences of the site serve as natural boundaries between various areas. The fitness, shopping and entertainment facilities have greatly improved the functionality of the city.

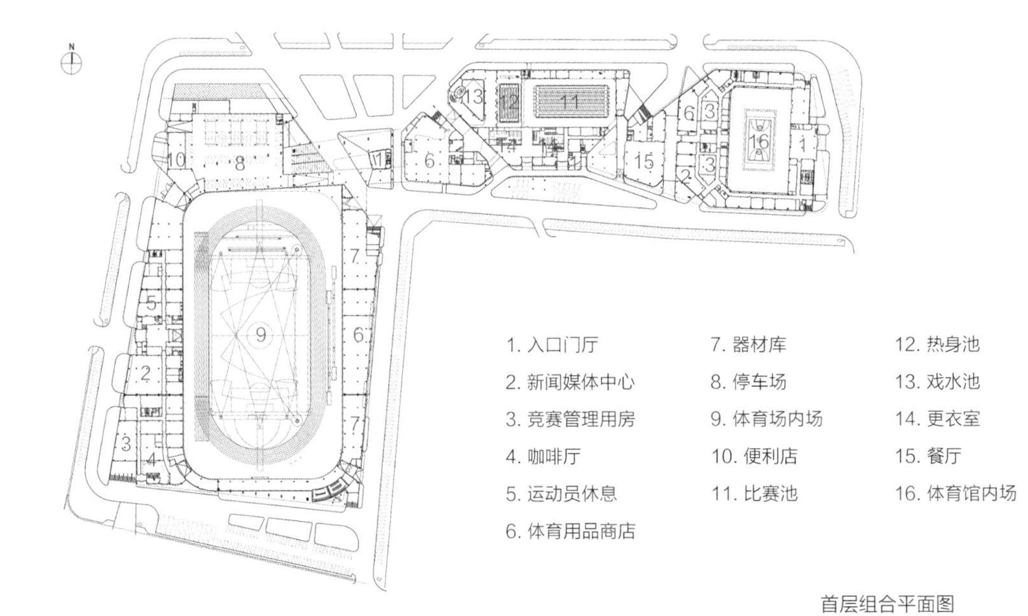

首层组合平面图

展览馆
摄影大展

太原市滨河体育中心改扩建设计 Reconstruction of Taiyuan Binhe Sports Center

地点 山西省太原市 / 用地面积 73,260m² / 建筑面积 73,049m² / 高度 24m / 设计时间 2017年 / 建成时间 2019年 / 座位数 6000座

方案设计 崔 愷、景 泉、徐元卿
张翼南、姚旭元
设计主持 景 泉、徐元卿、李静威

建 筑 张翼南、姚旭元、林贤载
结 构 王 超、陈 越、王 金
给排水 朱 琳、杜 江
设 备 徐 征、唐艳滨、高丽颖
电 气 肖 彦、常立强
总 图 段进兆
室 内 邓雪映、李海波
陆丽如、李 倬
景 观 关午军、朱燕辉
戴 敏、申 韬

合作设计 北京中体建筑工程设计有限公司
北京宁之境照明设计有限公司

滨河体育中心曾经是太原市的地标性建筑，经过多年使用出现功能混杂、停车空间不足、周边风貌杂乱等问题。改造希望以体育中心激活城市，使其融入市民的生活，在充分尊重原有建筑的基础上，巧妙地将新老建筑融为一体，同时与地块东侧的城市滨水绿带和北侧网球中心整合，为城市提供开阔的体育公园。老馆内最大限度地保留了可利用的看台空间，对结构补强加固；保留并延续老馆标志性的造型元素，用统一的屋面将新老建筑融为一体，形成舒展的两翼，条状金属幕墙更强调了形态整体性和动势。平赛结合的体育工艺布置和多种配套功能的引入，保证了场馆运营期间持续的活力。

Originally a landmark of Taiyuan, the Binhe Gymnasium was challenged by a series of problems after decades of use. The project aims to invigorate the city by integrating the center into people's life and present the city with an open sports park that exists in harmony with the original structure. The grandstand of the original building was preserved to the utmost extent, and the structure has been reinforced. The new sector has inherited the elements of the original one, connecting the original building with a wing-like roof with unified elements as the old one.

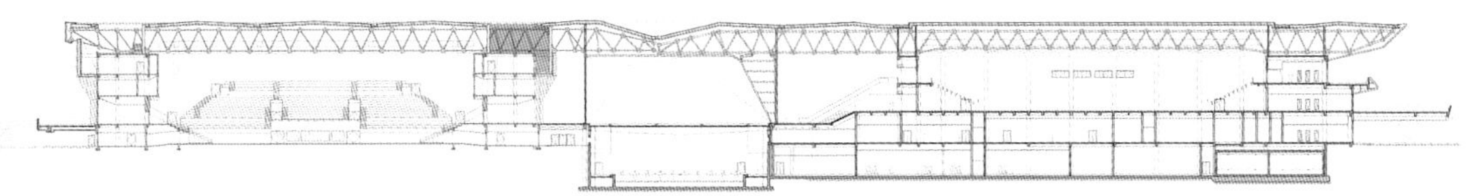

新老两馆剖面组合图

李季 摄

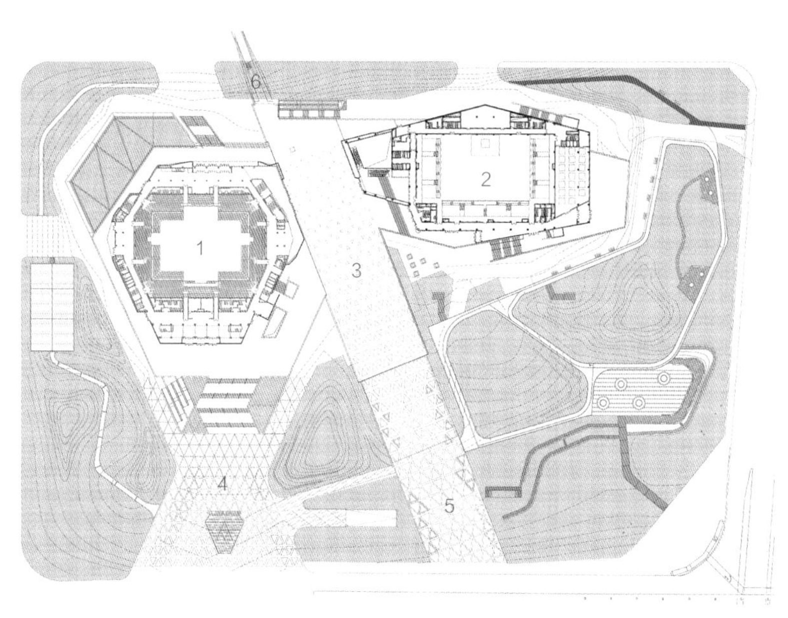

1. 乒乓球比赛馆（老馆）
2. 举重比赛馆（全民健身中心）
3. 景观步道（下方为游泳馆）
4. 喷泉广场
5. 升旗广场
6. 天桥

二层平面图

太原旅游职业学院体育馆 Taiyuan Tourism College Gymnasium

地点 山西省太原市 / 用地面积 20,000m^2 / 建筑面积 17,970m^2 / 高度 24m / 设计时间 2017年 / 建成时间 2019年 / 座位数 2000座

方案设计 景 泉、徐元卿、吴锡嘉
颜 冬、陈 虎
设计主持 景 泉、徐元卿、吴锡嘉

建 筑 颜 冬、陈 虎
结 构 范 重、张 宇、刘家名
给 排 水 张庆康、朱跃云
设 备 胡建丽、陈高峰
电 气 曹 磊、刘征峥
总 图 段进兆
室 内 曹 阳
景 观 关午军、戴 敏、申 韬

太原旅游职业学院体育馆，是第二届全国青年运动会排球项目的决赛场馆。为适应赛时和赛后功能转变的弹性利用，建筑首先被构想为一个简洁的大空间，从一个剖面放样而成，采用最简单有效的形式展现结构美。建筑立面也真实表达了内部空间的秩序与功能，由此强化了建筑形象的识别性。通过利用各类遮光帘、活动座椅、可吊挂舞台灯光设施等，可使场地满足 11 种布置需求，供师生复合使用。屋顶采用金属拱形坡屋面形式，使建筑既具有山西地域文化的凝重，又展现了体育建筑的动势。

The design for the gymnasium aims for the flexible shift between match-time and post-match operation. A simple and authentic large-scale space resembles a volume generated from a section, showcasing the beauty of its structure in a simple but effective way. The function and hierarchy of the interior space is also reflected on the façade, highlighting the identity of the building. The utilization of a series of facilities can meet 11 kinds of demands for the layout. The curved roof implies the dynamic of sports.

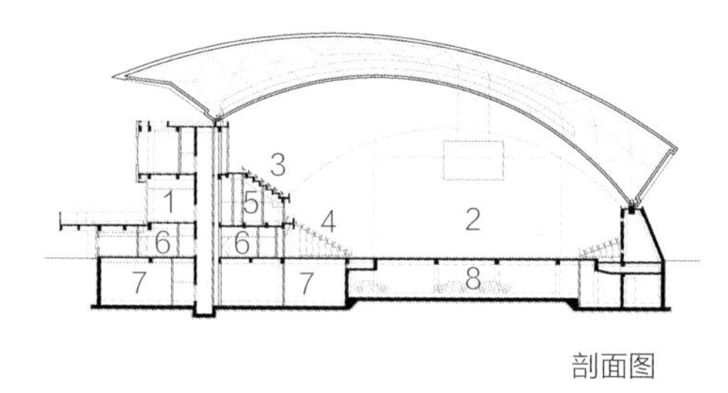

剖面图

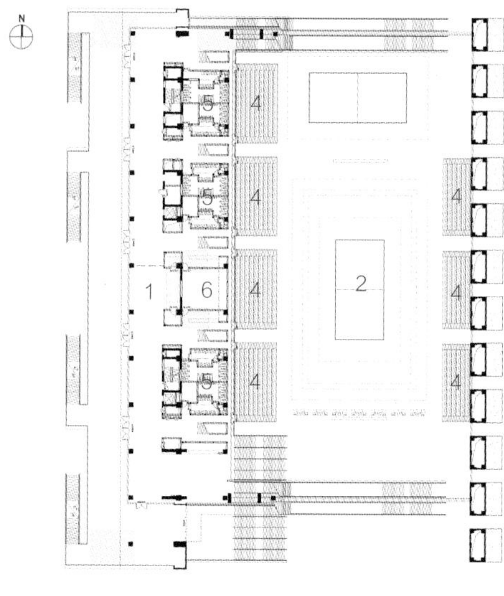

二层平面图

1. 观众集散厅 2. 比赛场地 3. 固定座席 4. 活动座席
5. 卫生间 6. 比赛辅助用房 7. 设备用房 8. 车库

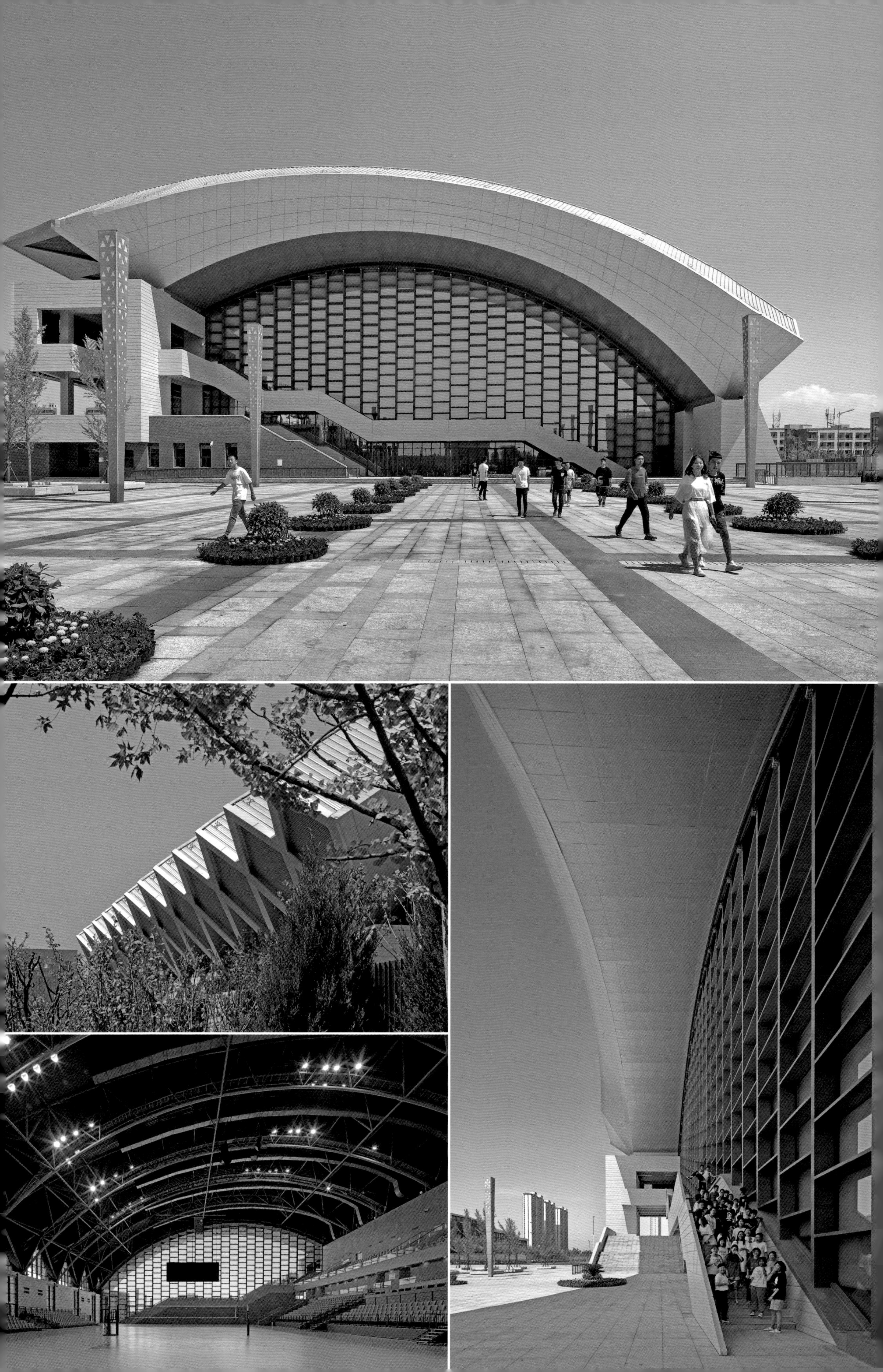

大兴新城北区体育中心 North District Sports Center of Daxing New Town

地点 北京市大兴区 / 用地面积 24,346m² / 建筑面积 31,500m² / 高度 24m / 设计时间 2014年 / 建成时间 2018年 / 座席数 体育馆 3538座

方案设计 李燕云、赵法中、李　卓
设计主持 李燕云、周　玲

建　　筑 赵法中、朱　蕾、殷超杰
结　　构 孔江洪、白晶晶
给 排 水 周　博
设　　备 张志强、牟　璇、刘权熠
电　　气 何　静、丁宗臣
总　　图 刘晓林
室　　内 马萌雪
景　　观 方　威、路　璐

摄　　影 李　季

体育中心是大兴新城的标志性建筑。其以刚强而又富有韵律的体型，大虚大实的形体处理，形成建筑强烈的雕塑感与现代感。浅香槟色直立锁边金属屋面与玻璃幕墙的搭配，在阳光的照射下彰显了建筑品质感与现代感。体育中心由一座乙级体育馆、一座全民健身游泳馆和一处健身中心组成。体育中心日常用作全民健身功能，并可承担全国单项赛事。南北向采光天窗贯穿体育馆比赛大厅、观众休息厅、游泳池区，空间效果丰富，同时为建筑带来了充分的自然采光，从而大大降低了场馆运营费用。

The sports center, consisting of a Grade-B gymnasium, a public natatorium and a fitness center, can host individual sport events while serving the public. The north-south skylights bring natural light into the game hall, the lounge, the swimming pool, reducing the operational costs of the building. The solidness of its form, and the sharp contrast between the transparent and the solid has added to the sculptural beauty of the building.

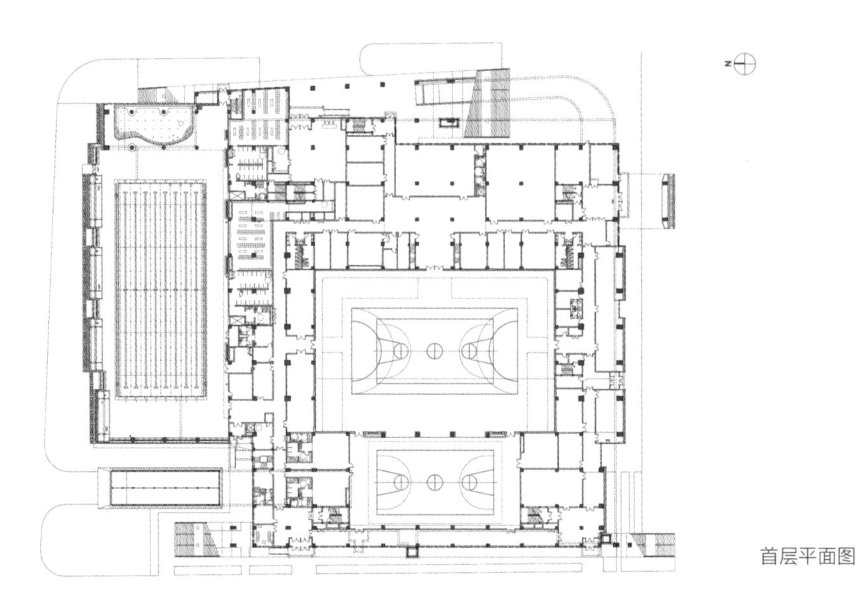

首层平面图

呼伦贝尔海拉尔机场 Hulunbuir Hailar Airport

地点 内蒙古自治区呼伦贝尔市 / 用地面积 10,450m² / 建筑面积 17,000m² / 高度 22m / 设计时间 2015年 / 建成时间 2018年

方案设计 于海为、刘晏晏
设计主持 于海为、刘晏晏

建　　筑 靳哲夫、吕　妍
　　　　 高　超、贺　帆
结　　构 施　泓、王　超
总　　图 李可溯
机　　电 中国民航机场建设集团公司华北机场规划设计院
室　　内 北京市辛迪森装饰设计公司
照　　明 清华大学建筑学院张昕工作室

对海拉尔原有航站楼的改扩建，是在一系列限定条件下进行的。在建筑轮廓、面积、柱网、功能、高度已基本确定的前提下，设计以蒙古包和草原特有的积云、羊群、山峦为意向，运用单元化的设计手法，南透北实的空陆侧策略，使到达机场的使用者无论从云端还是机场道路上都能感受到新航站楼连绵起伏的形态。绵延的雨棚则成为连接新老航站楼的重要元素。双向波浪式单层网壳钢结构屋面由双V形斜杆柱支撑，降低了结构跨度。柱子与吊顶结合的一体化、单元化方式，模糊了水平和垂直的边界，而以蒙古族视为传统高贵的金色、白色为主色调，也让建筑如蒙古大帐般产生连绵、贯通的空间效果。异形龙骨幕墙、TPO屋面、双曲铝板开花柱节点等具有创造性的细部设计，确保了建筑的建成效果。

The expansion and renovation of the original terminal was realized with a series of restrictions as the outline, area, column grid, functionality and height of the terminal were all predetermined. The architectural design, featuring an undulating form, remind people of the clouds, sheep flock and mountains of Inner Mongolia. The canopy, stretching along both the new and original terminals, serves as a crucial element of connection. The wavy roof of single-layered reticulated shell structure is supported by double-V shaped columns. The integration of the columns and the ceiling has blurred the boundary between vertical and horizontal elements.

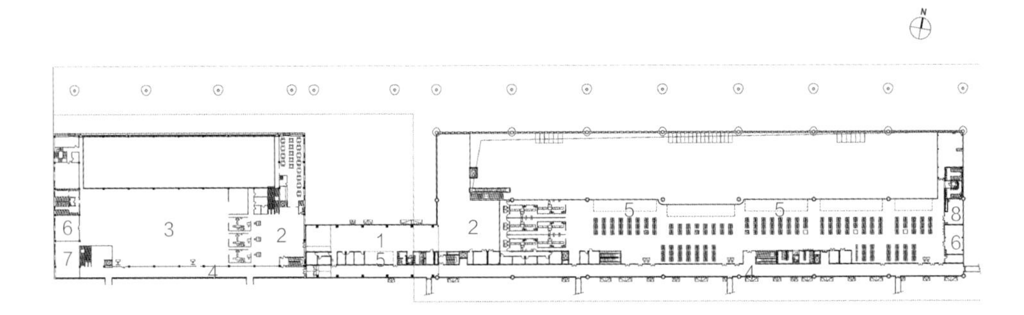

1. 连廊 2. 安检排队区 3. 候机厅 4. 登、离机走廊 5. 商业区 6. 头等舱候机室 7. 咖啡厅 8. 母婴室

二层（出发层）平面图

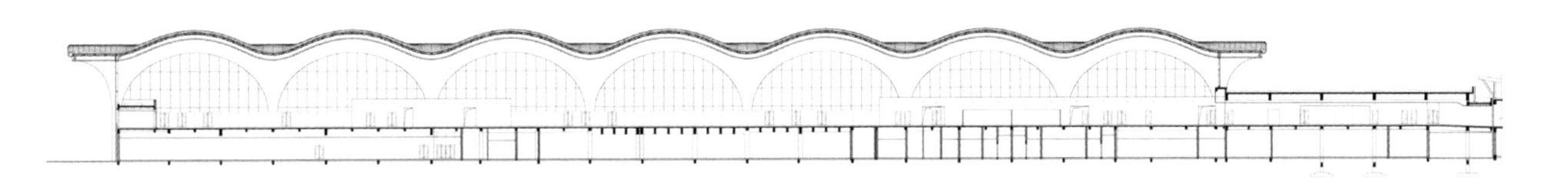

剖面图

09

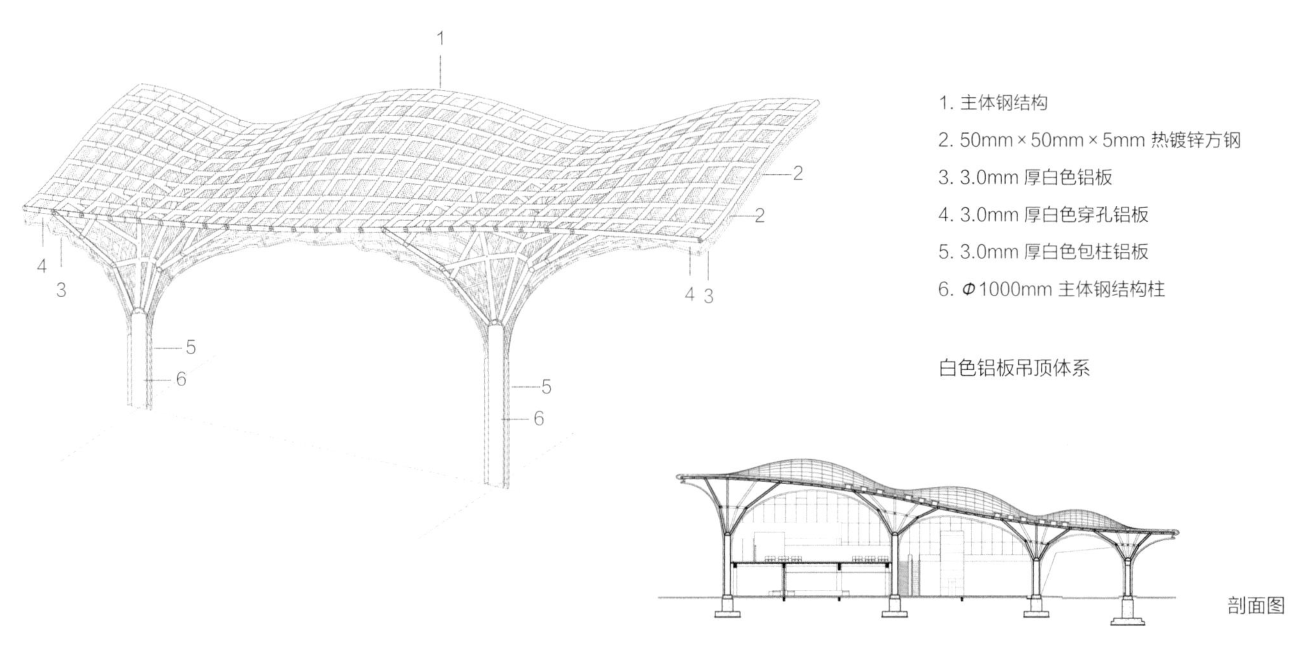

白色铝板吊顶体系

剖面图

腾冲机场T2航站楼 Tengchong Airport Terminal 2

地点 云南省腾冲市 / 用地面积 1,772,640m² / 建筑面积 40,715m² / 高度 22m / 设计时间 2013年 / 建成时间 2018年

方案设计 逄国伟、李慧琴、蒋 鸣
设计主持 逄国伟、李慧琴、陆 静

建　　筑 蒋 鸣
结　　构 鲁 昂、王 磊
给 排 水 吴连荣、高东茂
设　　备 孙淑萍、杨向红
电　　气 李维时、熊小俊
总　　图 余晓东

摄　　影 李 季

腾冲机场 2 号航站楼设计，一方面要与本地民族特色相联系，另一方面要呼应原有的 1 号航站楼。建筑的曲线形源于对飞机机翼的提炼，多条曲线的交织重叠形成了机场的基本形态。而腾冲独特的地质地貌——火山奇观形成的六棱形柱状节理"神柱石"——也被引入其中。一组组长短不一的六棱形曲面簇拥而成的支撑结构，从大地中生长出来，承托着屋面，也同时满足了机场建筑与地域特色的身份表达。两层半的布局方式，巧妙地解决了空侧陆侧大高差的问题，既保证了机场功能流程的通畅性，又做到尊重现有地形，尽量减少了工程土方量。

The design for Terminal 2 at Tengchong Airport was required to be both relevant to local culture and consistent with the existing Terminal 1. Inspired by the form of the aircraft wing, the design features a form determined by the interweaving of various curves. The supporting structure formed by a group of hexagonal curved surfaces, which resemble a sort of stone of the unique landform of Tengchong, seems to have grown out of the ground to support the roof. The building has a 2.5-storied layout to bridge the altitude difference between the two sides of the building.

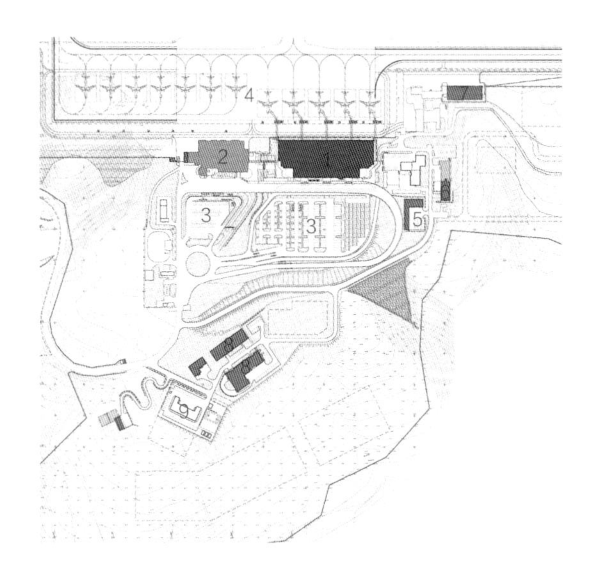

1. 新建 2 号航站楼
2. 原有 1 号航站楼
3. 停车场
4. 飞行区
5. 车库及物资库
6. 公安楼
7. 消防站及场务用房
8. 职工宿舍
9. 规划的联检机关用房

总平面图

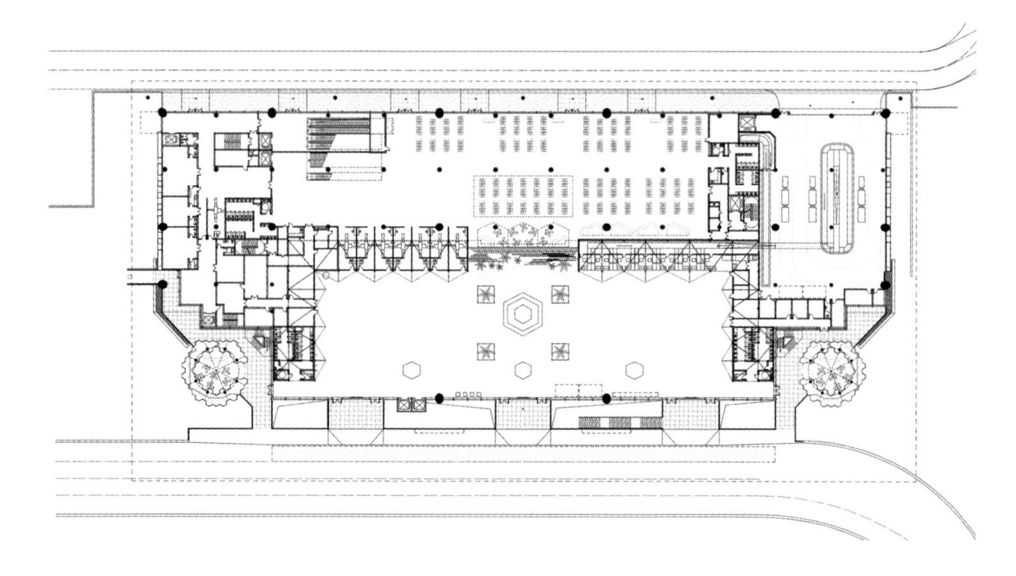

二层（出发层）平面图

TENG
CHONG

重庆VIVO生产基地 VIVO Chongqing Production Base

地点 重庆市南岸区 / 用地面积 172,000m² / 建筑面积 347,000m² / 高度 32m / 设计时间 2015年 / 建成时间 2017年

方案设计 徐　磊、高庆磊
　　　　 谢婧昕、刘　斌
设计主持 徐　磊、高庆磊

建　　筑 谢婧昕、满　乐
　　　　 金　星、赵　迪
结　　构 刘　巍
给 排 水 朱跃云、王仁祐
设　　备 胡建丽、程群英
电　　气 许士骅
总　　图 郑爱龙

本项目力求创造充满科技感和人文气息的生产、生活综合工业园区，依据地势平整出不同标高的台地，以适应各个厂房、宿舍以及人员活动等功能需求。设计遵循各厂房之间的人流物流关系以及场地高差之间的变化，利用全天候的交通联络系统将各栋建筑组合成为紧密、高效的整体。主入口的环形广场是园区的核心，环抱形的空间背山面水，对外传递公司开放的形象，对内满足各主要建筑的内在联系需求。建筑依地势而为，体现了对自然环境的尊重，又将现代化的工业园与重庆的地域特色相融合。

The design aims for a comprehensive industrial park that reveals both scientific and cultural features of VIVO. Terraces varied in altitude are planned, catering to various demands of manufacture, living and other activities. Based on the flow of people and commodities, an all-weather circulation system is established. The circular square at the main entrance serves as the focus of the industrial park, forming a space that showcases the open gesture of the company while meeting various internal demands.

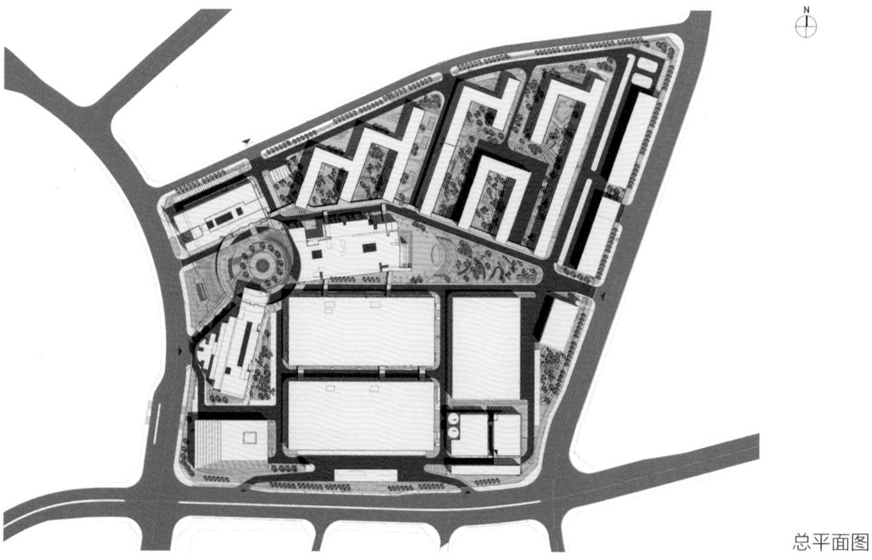

总平面图

项目组供图

不同的功能经过合理的组织，提取其共性，满足功能空间的布局需求，同时针对不同的建筑单体，强调空间形态的可识别性及外观的差异性，形成既和谐统一、又各自鲜明的整体表象。

Common features in various functions were identified to facilitate the arrangement of spaces, while differences in each building's form endow the whole project with diversity.

园区的整体形象简洁、明快，采用统一的立面意象协调各个单体建筑的差异，形成完整的工业园区形象。

Featuring simplicity and vibrancy, the industrial park has obtained an integrated image with a coherent façade style applied to all individual buildings.

北京华为数据通信研发中心 Beijing Huawei Data Communication R&D Center

地点 北京市海淀区 / 用地面积 96,318m² / 建筑面积 147,751m² / 高度 20m / 设计时间 2008年 / 建成时间 2015年

方案设计 GMP 国际建筑设计有限公司
设计主持 陆 静、林 蕾

建　　筑 余 洁
结　　构 曹 清、董 洲
给 排 水 黎 松
设　　备 梁 琳、王春雷
电　　气 胡 桃、崔振辉
电　　讯 张 雅、陈婷婷
总　　图 刘 文
室　　内 邓雪映
景　　观 雷洪强

该中心是华为的大型科研实验办公区，由三栋科研办公楼和一个员工食堂组成。办公楼平面呈 8 字形，围合出主入口庭院和休闲内院，一动一静。食堂采用简洁纯净的立方体形态，建筑空间流畅，阳光通过天窗柔和地射入室内。为保证屋面纯净、平坦的效果，在屋面下局部设置夹层，隐蔽风机、风管等，同时屋面面层采用架空开缝石材设计，结合局部结构找坡和非常规防水节点，使保温、防水和设备通风安全有效。

The center consists of three R&D buildings and a canteen. The canteen has a simple rectangular form, with skylights introducing soft sunlight into the interior. To make the roof as flat as possible, a mezzanine under the roof is inserted to accommodate fans and pipes. Furthermore, the roof covered by stilted and slotted stones has guaranteed insulation, waterproofing and equipment ventilation with structural slopes and unconventional details.

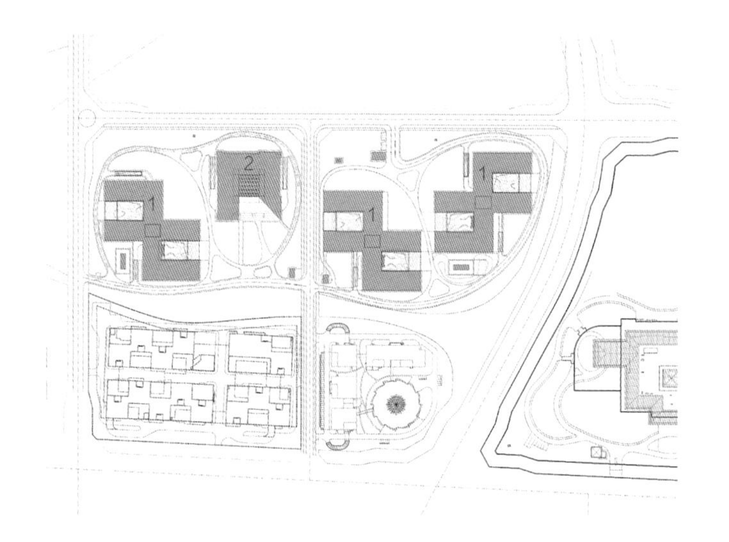

1. 科研办公楼
2. 员工食堂

总平面图

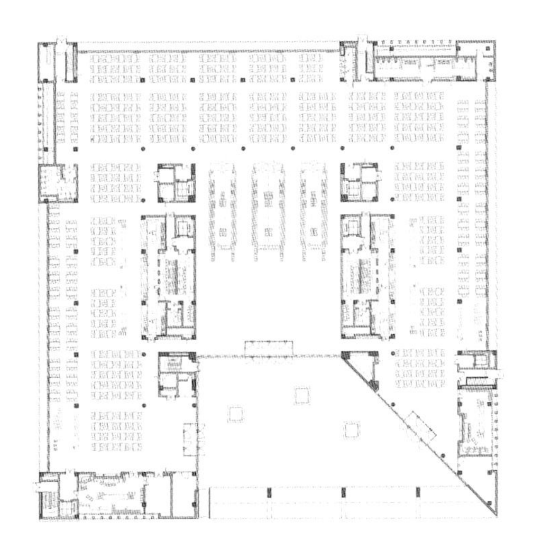

员工食堂首层平面图

科研办公楼立面分为三段：一层为坚实的白色石材勒脚，小尺度的竖直开缝提供采光，并形成百叶式的效果；三、四层为深色竖向玻璃幕墙，以深邃的窗框产生有节奏感的光影变化；相比之下，二层的带状玻璃幕墙则明显内退，强化了上下两部分的对比。玻璃幕墙为内呼吸式幕墙，可开启窗扇外侧采用百叶作为新风口，从而避免破坏幕墙整洁统一的外观。

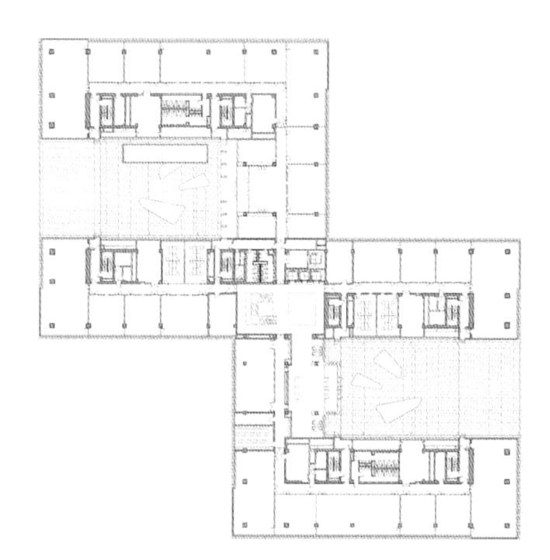

科研办公楼首层平面图

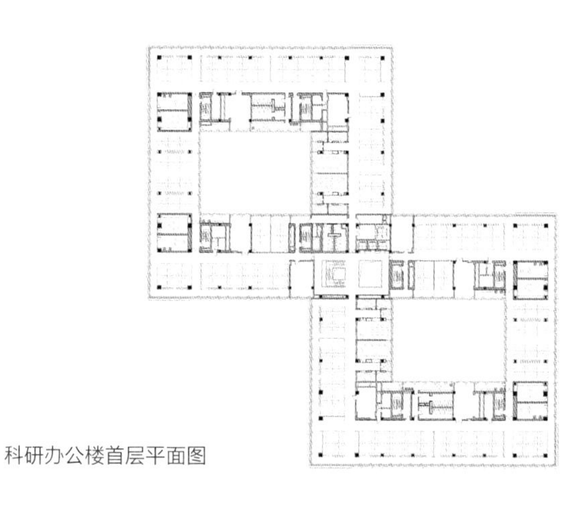

科研办公楼三、四层平面图

On the 1st floor of the R & D building, narrow vertical gaps among the solid and white stone plinths introduce daylight into the 2nd floor the space, while dark colored glazed curtain wall on the 3rd and 4th floors has thick frames which create shadows on the façade; The receded striped curtain wall highlights the contrast between the upper and lower parts of the façade.

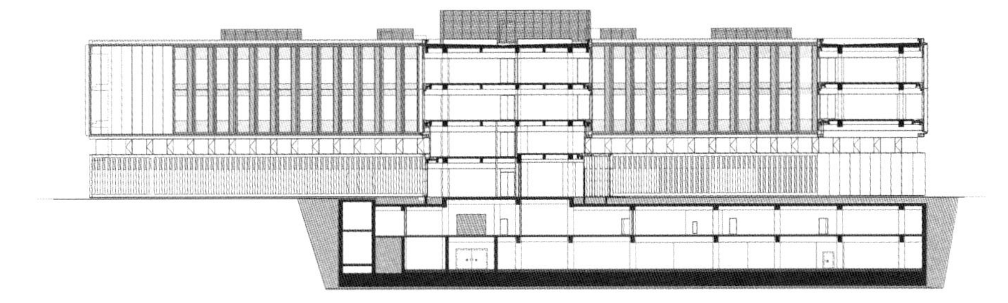

科研办公楼剖面图

中国医学科学院药物研究所新药创制产学研基地 New Drug Base of Institute of Materia Medica, CA

地点 北京市大兴区 / 用地面积 88,279m² / 建筑面积 43,088m² / 高度 24m / 设计时间 2013年 / 建成时间 2019年

方案设计 景 泉、徐元卿、李静威
　　　　 吴锡嘉、张翼南、李雪菲
设计主持 景 泉、徐元卿

建　　筑 吴锡嘉、张翼南、李雪菲
结　　构 王 鑫
给 排 水 李建业
设　　备 祝秀娟、唐艳滨
电　　气 张 辉
总　　图 刘晓琳
室　　内 邓雪映、张全全
景　　观 李 力

药物研究所隶属于中国医学科学院北京协和医学院，设计首先需要满足工艺流程与可持续的理念。基于实验室工艺需求，建筑的进深被控制在三跨柱距内。建筑整体采用回字形布局，并将化学实验室设于下风向的东南侧。为满足药学领域跨学科交流的需要，院落中提供了围合空间、下沉庭院、露台等多种室外交流互动场所，并设置竖向停留空间，形成连贯的立体交往体系。为延续协和医学院的文脉与记忆，建筑适当融入了协和原址的院落、灰砖、绿瓦、垂花门等中西合璧的元素，并延续了药物研究所现有办公楼的立面比例和开窗形式。窗口处设置木色通风百叶，入口处角部处理为回纹形，在呼应传统的基础上加入黄绿相间的彩釉玻璃，既增添了活力，也强化了入口空间的标识性与仪式感，传递出协和东西合璧的特征。

The drug research institute is affiliated to Chinese Academy of Medical Sciences & Peking Union Medical College. Its R & D Base was required to meet demands in both process flow and sustainability. The building has a hollow square-shaped layout with a series of spaces for communication, such as courtyards, sunken yards and terraces, forming a dimensional system together with vertical spaces of the same function. The chemistry laboratories are located at the southeast corner on the leeward. In respect for the history of Union Medical College, a series of building parts of a combination of Chinese and Western elements, such as courtyards, gray bricks, green tiles, floral-pendant gates, are integrated in the newly-built structure.

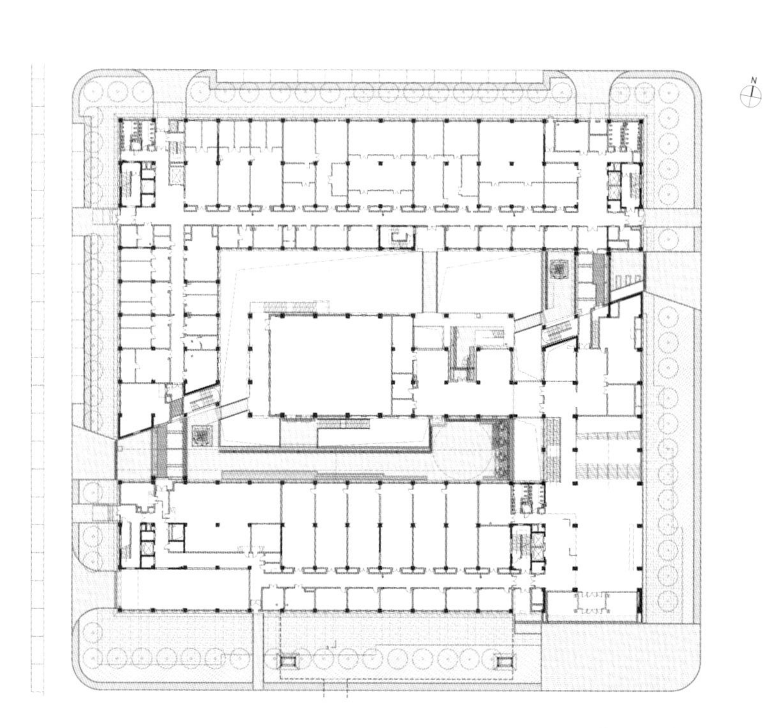

首层平面图

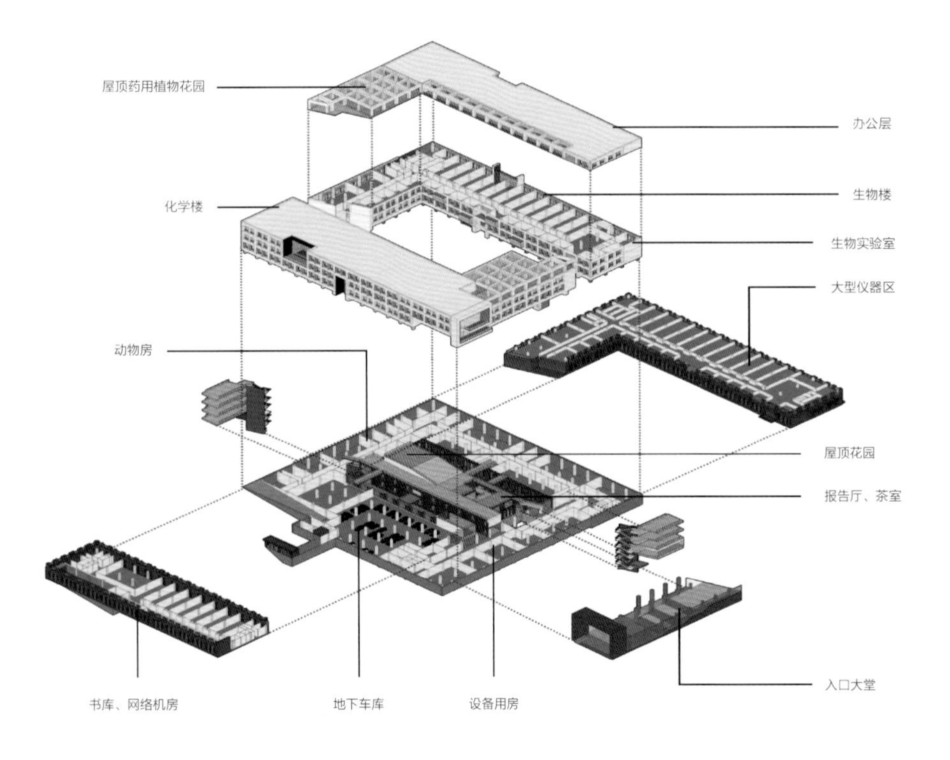

标准实验室单元=
数据分析室 + 走廊 + 实验室

实验室面宽根据实验柜的模数布置，单元间设活动隔墙，可根据不同实验课题组的需要弹性划分，仅针对通风柜中的空气加热或制冷，平衡了使用舒适性与节能两方面需求。

功能分解图

中国建设银行生产基地一期 China Construction Bank Manufacturing Base, Phase I

地点 北京市海淀区 / 用地面积 136,000m² / 建筑面积 284,000m² / 高度 24m / 设计时间 2012年 / 建成时间 2017年

方案设计 徐　磊、高庆磊
　　　　 刘　斌、弓　蒙
设计主持 徐　磊、高庆磊

建　　筑 李　磊、满　乐、张　硕
　　　　 弓　蒙、李宝明
结　　构 王　玮、郑红卫
给 排 水 宋国清、黎　松
设　　备 宋孝春、劳逸民、姜海元
电　　气 郭利群、王玉卿
电　　讯 胡建军、许　静
总　　图 高　治
景　　观 赵文斌、路　璐

园区整体规划采用模块层级化布置，将复杂功能归类整合，明确整个园区的功能逻辑和识别性。设计以具有中国传统“如意”形态的中心开放景观带为园区核心，贯穿左右两个地块，形成内聚型的场地布局方式。场地周边的建筑布局则在保持低调的同时，创造有韵律感的空间。整体空间层级经过有效组织，在形成流畅空间的前提下，保持了各地块自身空间的完整性。

The overall plan is carried out in the form of modules and layers, so that complex functions can be classified and integrated to highlight the functional logic and identity of the area. A central open landscape area, symbolizing “Ruyi ” in Chinese culture, is accessible from two blocks on its both sides, so that the integrity of each plot is preserved.

1. 运维办公楼
2. 监控中心
3. 数据机房
4. 研发中心
5. 预留办公楼

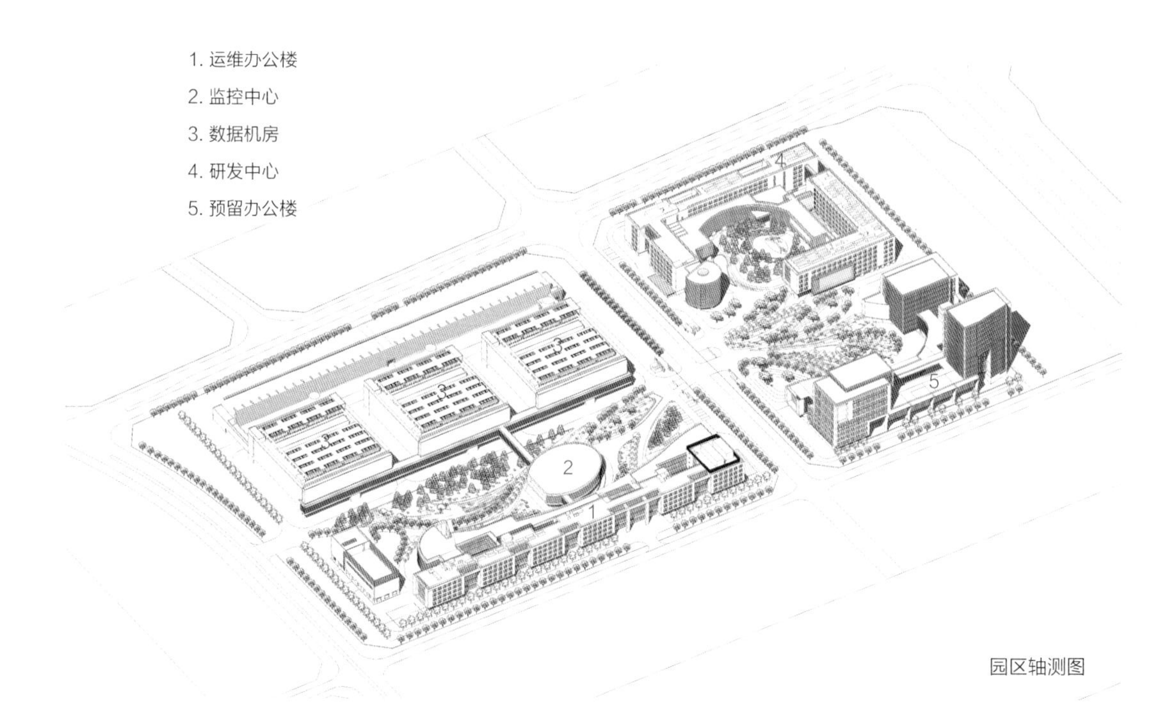

园区轴测图

D

中国农业银行数据中心 Data Center of Agricultural Bank of China

地点 北京市海淀区 / 用地面积 130,000m² / 建筑面积 249,200m² / 高度 18m / 设计时间 2010年 / 建成时间 2018年

方案设计 张 燕、尚 荣、鲍 力
高 明、金 凯
设计主持 张 燕、李衣言

建　　筑 高 明、尚 荣
结　　构 刘 巍、朱炳寅、马玉虎
给 排 水 宋国清、黎 松
高振渊、唐致文
设　　备 劳逸民、娄志亮
姜海元、陈 露
电　　气 胡 桃、崔振辉
电　　讯 任亚武、张 雅
胡建军、赵雨农
总　　图 连 荔
室　　内 张 晖、王 强、张 超

建筑群坐落于北京西山脚下，包含数据机房、研发办公及会议、服务等多组建筑，设有机房区、现场监控区、媒介仓储区、测试区等10个功能区块。设计以“印”为意，取其信物、凭证之意，与数据处理稳定可靠的要求相契合。单体建筑围合布局，外立面饰以石材，与深窗造型共同构成雕刻效果，组成“国字印”的印章笔画，厚重、坚实的形象更强调了中国农业银行作为四大国有银行之一的重要行业地位。设计充分运用轴线，以中轴线广场这一景观纽带连接南北两个地块，建筑均围绕广场四周布置。

The center consists of data building, R&D offices and service amenities, and is divided into 10 functional blocks. Designed with the concept of “stamp”, which symbolizes credit and proof, the building’s image reveals a sense of reliability in data processing. The enclosed layout of each building, together with the stones on the façade and the deep windows, forms the strokes in the Chinese stamp, highlighting the significance of the bank in the industry.

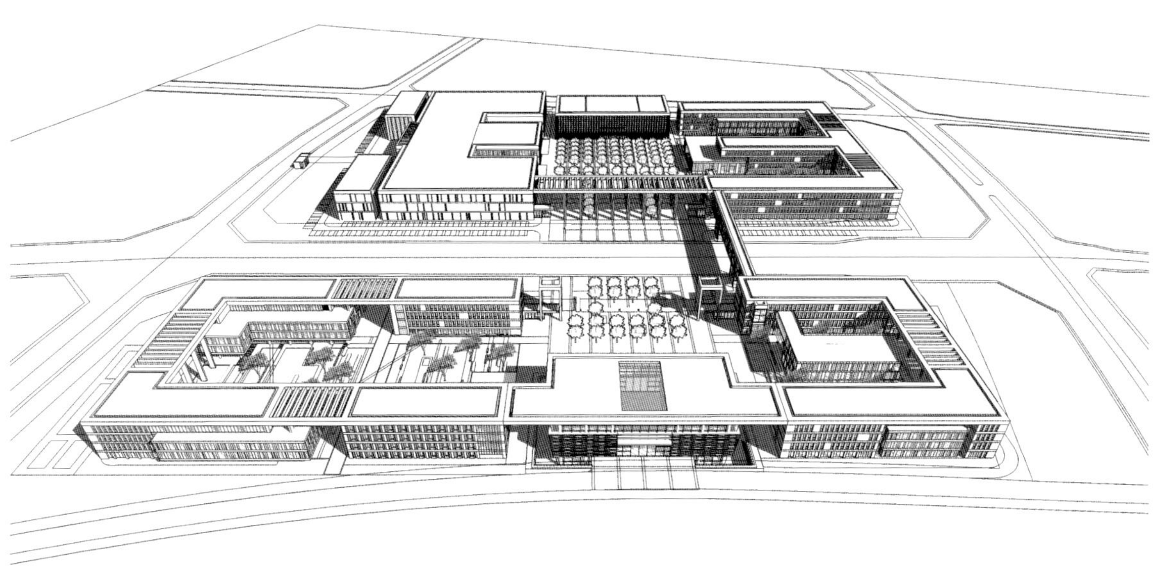

轴测图

中轴两端分别是主楼和会议中心。两座建筑列鼎而建，南低北高。主楼高耸，会议中心下沉，立面庄重沉稳，作为建筑序列和景观轴线的有力收束点。

The main building and the conference center are situated at both ends of the axis of the square, serving as distinct landmarks.

青岛德国企业中心 German Enterprise Center, Qingdao

地点 山东省青岛市 / 用地面积 28,000m² / 建筑面积 75,000m² / 高度 41m / 设计时间 2012－2013年 / 建成时间 2016年

方案设计 德国 SBA 公司
设计主持 马　琴

建　　筑 宋　焱、王　锁
　　　　 郑雪岩、彭　勃
结　　构 谈　敏、李　妍
　　　　 贾　开、那　苓
给 排 水 赵　昕、李盈利
设　　备 何海亮、李　嘉
电　　气 陈　琪、史剑华
电　　讯 张月珍、王　青
总　　图 高　治、李　爽
室　　内 饶　劢、顾大海
景　　观 史丽秀、朱燕辉、关午军

德国企业中心是亚洲地区第一个获得德国 DGNB 铂金奖认证的综合体建筑。作为中德生态园的启动项目，设计致力于贯彻绿色低碳和经济高效的设计原则，采用了多种绿色节能技术手段和大量新型建材。建筑材料多来自当地，既达到了节能减排的目的，也增加了建筑的本土特色。设计将建筑功能归纳为三个类型：办公酒店等相对私密的空间，配套服务等室内交流空间，景观环境等室外交流空间；通过建筑语言转化为南北两组有机融合的建筑，舒展的建筑形态与自然山水紧密结合。上部体量如同崂山山石，错落有致，简洁有力，裙房部分则如水般灵动通透。细部设计与绿色技术充分结合，注重营造优雅和稳重的感受。内部空间则通过丰富和灵活的室内设计，形成高效、开放的氛围。

German Enterprise Center is the first Asian building to obtain Germany DGNB Platinum Certification. As a startup project in Qingdao Sino-German Eco-park, this project implements design principles that are green, low-carbon and cost-efficient. Local materials used in the building has not only added to the local cultural features, but also achieved the purpose of energy conservation and emission reduction. Three main functions – relatively private spaces, indoor communication spaces and outdoor communication spaces are integrated in two groups of buildings. The upper part of the buildings reveals a sense of conciseness and solidity. A modern and elegant atmosphere is presented by the detailed design and green technology.

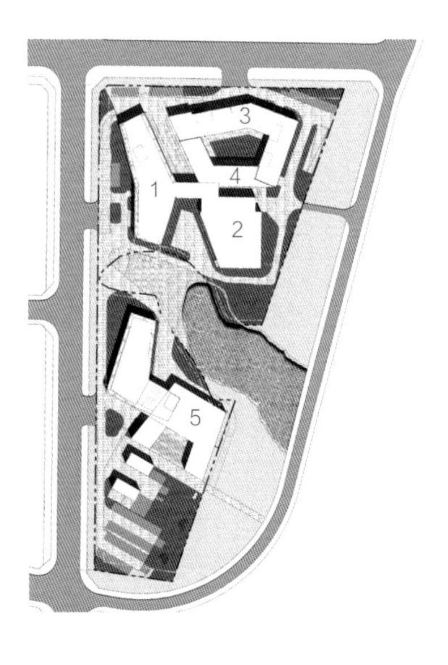

1. 德国中心
2. 培训中心、健身中心
3. 酒店
4. 商业
5. 食堂

首层平面图

ICIC
Sino-GermanEcopark
InterCityHotel

方太厨具研发中心大楼 Fotile Research & Development Center

地点 浙江省宁波市 / 用地面积 269,600m² / 建筑面积 35,257m² / 高度 18m / 设计时间 2013年 / 建成时间 2016年

方案设计 林 朗、邓 烨、祝 贺
梁 叶、王玉廷、罗 荃
李 旸
设计主持 林 朗、邓 烨

建 筑 梁 叶、祝 贺、罗 荃
结 构 朱炳寅、张 路、张夔华
给 排 水 匡 杰、张晋童
设 备 李雯筠、尹奎超
电 气 程培新
总 图 邵守团

项目位于杭州湾新区，建设内容包括研发办公室、方太大学、实验室、制作室和地下车库等，在绿色设计方面达到LEED金级标准。建筑采用方形院落布局，以开放的园林作为底景，围合出郁郁葱葱的中心庭院。首层通过灰空间在东、西、南三面与外部景观相连，并将绿化引入内庭。二层以上向四周层层探出，外围辅以步出式的阳台和玻璃遮阳板的框景；庭院内部环通的室外退台、门厅的弧形楼梯及其上方的小剧场共同构成了丰富的公共空间。设计以清水混凝土奠定空间基调，大量的混凝土梁柱和墙体直接外露，将结构形式和交接关系清晰地体现出来。

Positioned as a high-tech base, the Fotile Research & Development Center, consisting of R&D facilities, Fotile College, laboratories and production rooms, is certificated a LEED Gold-level building. With a layout dominated by a courtyard, the building has introduced landscape into the interior through the buffering zone on the 1st floor. Upper floors stretch out of the outline of the first floor. Terraces, curved staircases and a small theater constitute the diversified public spaces in the courtyard, boosting communication and creation among the staff.

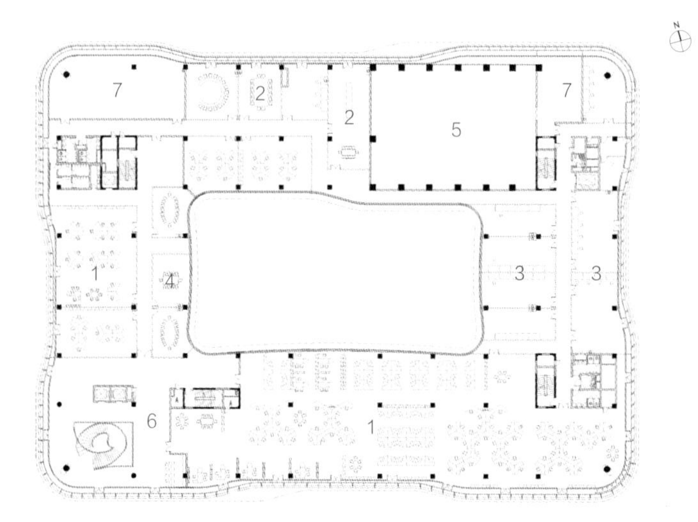

1. 研发办公室
2. 制造操作室
3. 样机试制室
4. 会议室
5. 多功能厅
6. 休息厅
7. 中庭
8. 评审室
9. 门厅
10. 停车库

二层平面图

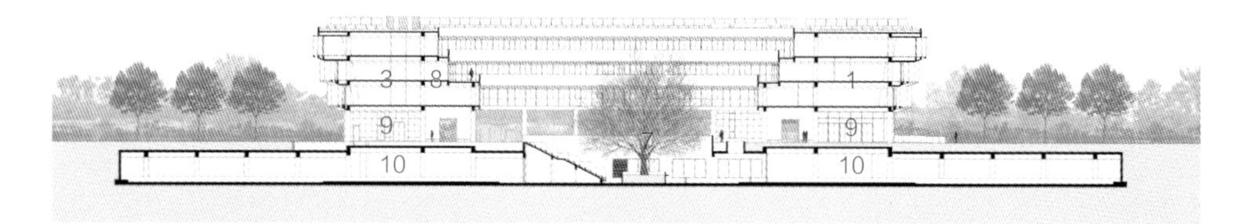

剖面图

北京经开数码科技园 Beijing Jing Kai Digital Science Park

地点 北京市亦庄经济技术开发区 / 用地面积 54,928m² / 建筑面积 83,833m² / 高度 45m / 设计时间 2013年 / 建成时间 2015年

方案设计 崔 愷、郭海鞍
设计主持 崔 愷、沙松杰

建　　筑 郭海鞍、张 婷、刘文轩
结　　构 鲁 昂、何喜明
给 排 水 张庆康、朱跃云
设　　备 金 健
电　　气 许士骅
总　　图 连 荔
景　　观 冯 君

经开数码科技园是一处低密度、绿色精品办公产业园区，也是亦庄地区办公建筑群体中的代表性项目。其规划打破传统办公园区布局方式，将所有小办公楼沿街布置，中间最大限度地围合出城市花园，让每栋办公楼拥有开阔的景观视野，让中心庭院能够服务于城市空间，促进楼宇之间的沟通与交流，兼顾共享与私密。在建筑之间采用了独特的不同角度的凸窗，精心计算让每一个窗户都有良好的景观视野，并避免了小间距办公楼的对视问题。整层通高的蜂窝铝板幕墙实现了幕墙板与板之间的无缝拼接。

As a low-density industrial park, the Jing Kai Digital Science Park has an unconventional layout, where small office buildings are located along the external roads to make space for an open garden in the center. Specifications for bay windows toward various directions are generated through sophisticated calculation, which has guaranteed pleasant views for each window and avoided direct views between offices. Honeycomb-shaped aluminum plates are joint seamlessly to present an integrated and simple image.

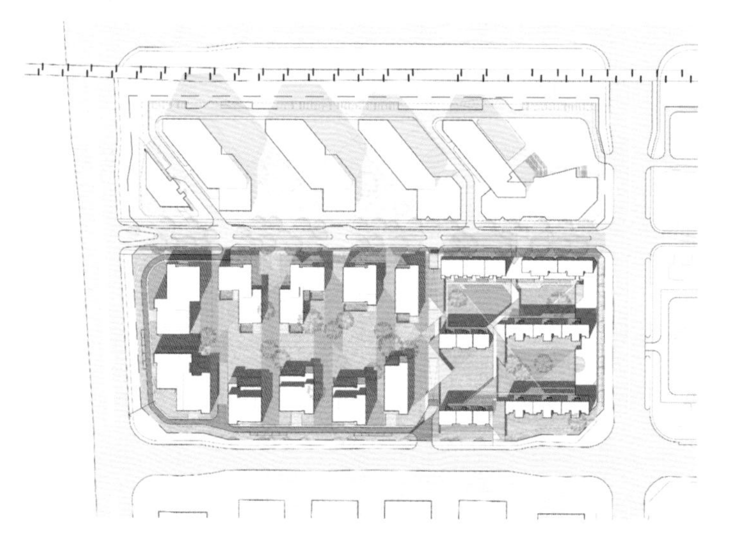

总平面图

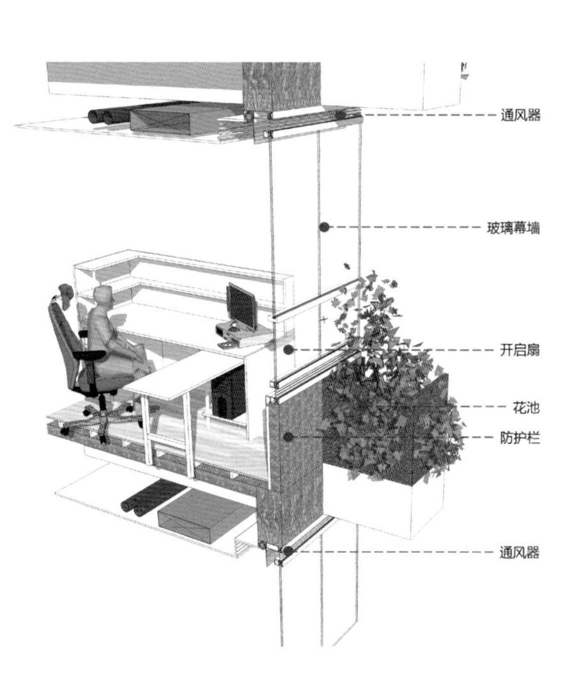

生态幕墙系统

日照文化创意产业园 Rizhao Cultural & Creative Park

地点 山东省日照市 / 用地面积 73,273m² / 建筑面积 106,435m² / 高度 66m / 设计时间 2015年 / 建成时间 2018年

方案设计 于海为、靳哲夫、贺 帆
设计主持 于海为、颜朝昱

建　　筑 靳哲夫、于 玢
　　　　 薛 琪、郝 雨
结　　构 张 猛、徐宏艳
给 排 水 董 超、宋 晶
设　　备 杨向红、吴艺博
电　　气 李沛岩、熊小俊
总　　图 齐海娟

文化创意产业园位于日照城市主轴线北端，毗邻大学城核心区。相对于大尺度的城市环境，园区的设计从构建人性化的小尺度公共空间开始，以形成人文氛围，吸引文化创意人群。设计将艺术馆、酒店、商务办公等大尺度的建筑体量布置在城市界面上，作为对新城周边建筑尺度的回应；创意办公、商业和服务配套、居住等功能空间则转化为较小的建筑体量，由庭院、街巷、广场等外部空间组织在一起。园区内部仅供步行与非机动车交通使用，主要功能区域设有开放的二层交通平台，结合景观设计形成了富有趣味和活力的空间。

Located at the northern end of the city's main axis, the Rizhao Cultural & Creative Park borders the core area of the college town. Situated in such a large-scale neighborhood, the park features small-scale public spaces that appeal to people in cultural and creative industry. Relatively large buildings, including the art gallery, the hotel and the commercial office buildings, line the street to conform to the scale of their surroundings; smaller buildings, such as creative office buildings and service facilities, are connected with courtyards, alleys and squares in the inner area.

总平面图

北京银行保险产业园648地块 No.648 Plot of Beijing Bank & Insurance Industrial Park

地点 北京市石景山区 / 用地面积 27,377m² / 建筑面积 111,449m² / 高度 36m / 设计时间 2015年 / 建成时间 2018年

方案设计 于海为、谢 悦、于 玢
魏亚文、沈若禹
设计主持 于海为

建　　筑 谢 悦、于 玢、魏亚文
沈若禹
结　　构 王 鑫、尹 洋、王 昊
给 排 水 申 静、郝 洁
设　　备 何海亮、李 嘉
电　　气 马霄鹏、高学文
总　　图 吴耀懿

经历了四代发展的金融产业园区，从封闭单一的传统园区，发展为开放混合的多样化园区。北京保险产业园的这一区域，通过公共下沉广场将功能相对独立的办公楼、五星级酒店与会议展示中心整合到一起。地块南北布置五星级酒店和办公楼，东侧为保险园公共建筑群落中的锚点建筑——会议展示中心。会议展示中心无方向的有机形体活泼且具有标志性，作为节点建筑带动整个场地的公共气氛并向城市辐射延续。下沉庭院周边设有配套商业、餐饮、健身等公共功能，使其成为生态与活力相结合的核心空间；创造相互联系、绿色生态的保险工作与服务社区。

The industrial park has been transformed from a traditional and monotonous park into an open and diverse one. A sunken square serves as the hub for a relatively independent office building, a five-star hotel and a convention & exhibition center. The convention & exhibition center at the east of the site becomes the landmark of the park with a vigorous form, boosting the spirit of the place. Retail, catering and fitness services are located along the sunken courtyard, making it a central space that is both sustainable and vigorous in the inter-connected community.

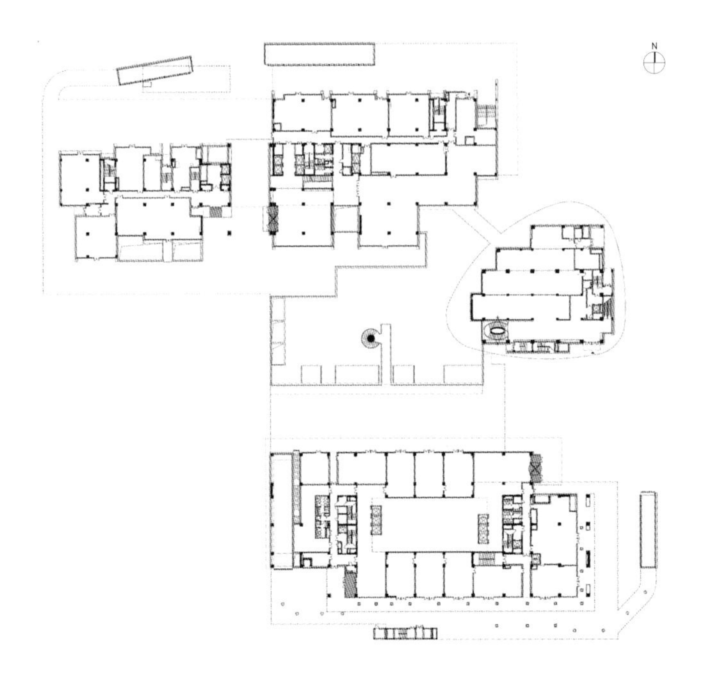

首层平面图

酒店底层面向街道的部分布置配套商业，面向下沉庭院一侧分布有餐厅、健身、活动等较为开放的功能。办公南楼则是大开间灵活分割式办公，底层沿街设置部分配套商业，紧邻下沉庭院的地下一层设置服务园区的美食广场。

The hotel has its commercial spaces lining along the street while its restaurant and fitness center are located along the sunken courtyard. In the office building at the south of the industrial park, large-bay office areas are located on over-ground floors while the food court borders the sunken courtyard.

天地邻枫绿色产业园区 Tiandi Linfeng Green Industrial Park

地点 北京市海淀区 / 用地面积 55,747m² / 建筑面积 98,877m² / 高度 12m / 设计时间 2013年 / 建成时间 2016年

方案设计 崔 愷、丁 峰、杨 凌
胡卫华、张伯榕、张 千
设计主持 崔 愷、丁 峰

建 筑 杨 凌、张 莹、张 峰
结 构 张 猛、肖耀祖、徐宏艳
给排水 董 超、宋 晶
设 备 杨向红、吴艺博
电 气 庞晓霞、熊小俊
总 图 郑爱龙

作为一处绿色、生态的科技型办公园区，建筑群充分利用其靠近国家奥林匹克公园，并与城市道路之间有大片绿地相隔的生态优势，形成高品质、多样性、功能齐全的低层生态组团办公群落，一方面提升了该区块的城市形象，另一方面也提供了自然怡人的建筑氛围。建筑以灰砖坡顶为基调，采用院落式布局，结合步移景异的园林化空间，将新型郊区化办公的空间组织模式与传统建筑精神相融合。

The building cluster avails itself of its adjacency to the Olympic Green and its isolation from the buzz of the urban roads, so that a low-rise office building cluster with top quality is built, improving the image of the block. Dominated by sloped roofs of gray brick, the buildings form courtyards and exterior spaces with features of traditional Chinese gardens, where sceneries vary with changing viewpoints.

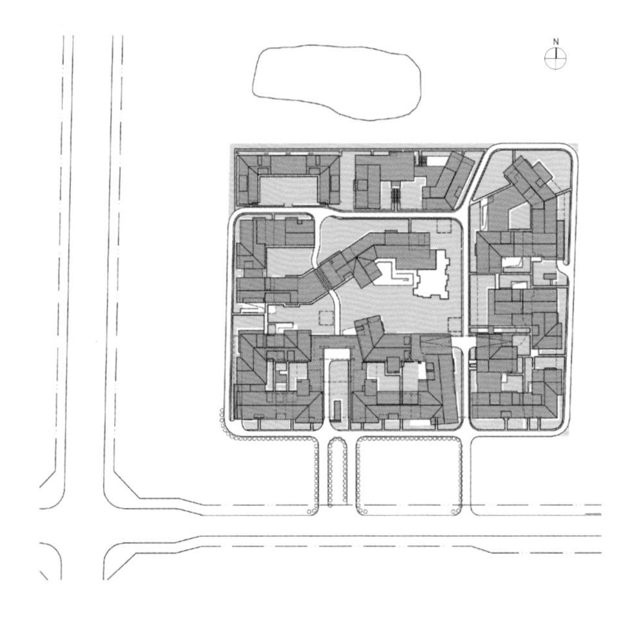

总平面图

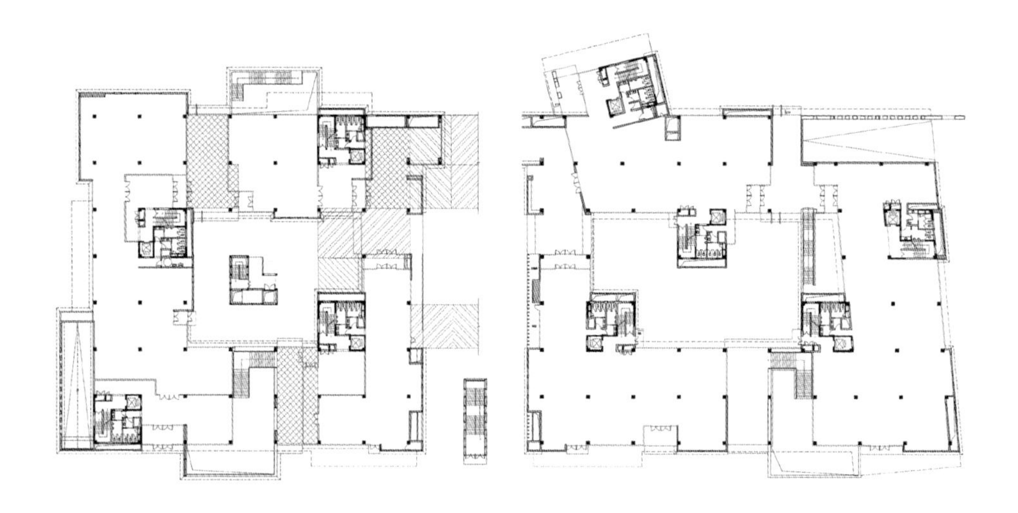

AB栋首层平面图

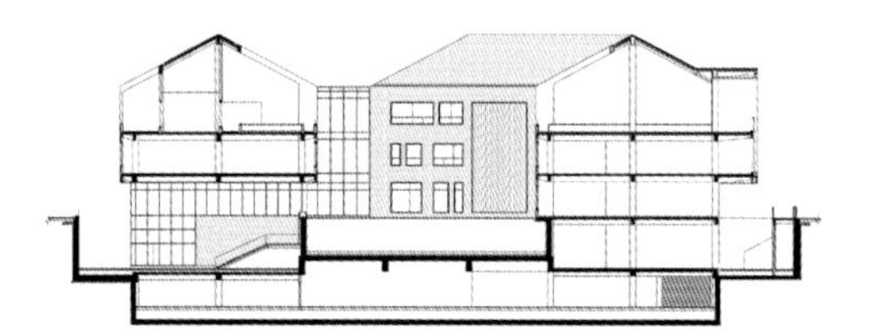

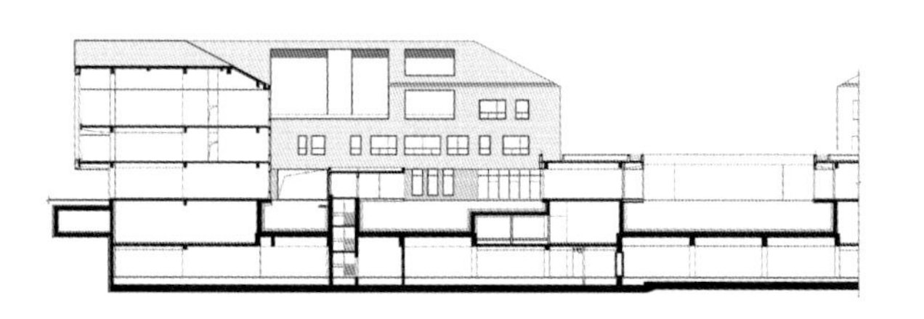

剖面图

华电产业园 Huadian Industrial Park

地点 北京市丰台区 / 用地面积 42,703m² / 建筑面积 248,186m² / 高度 80m / 设计时间 2012年 / 建成时间 2014年

方案设计 邓 烨、段 猛、梁 叶
罗 荃、王永坚、杨 华
郭 韬
设计主持 邓 烨、段 猛

建　　筑 詹柏楠
结　　构 常林润、曾金盛
给 排 水 陈 超、关 维、何 猛
设　　备 张 昕、汪春华
电　　气 王路成、王京生
张晓泉、李战赠
总　　图 邵守团

华电产业园由华电工程集团总部办公楼、商业办公楼、四星级酒店等独立楼栋组成。一街一院的围合式布局，在空间序列上形成南北递进的中轴对称关系。建筑高低错落有序，以浅色石材凸显典雅的气质。北侧的总部办公楼出檐深远且布局对称，采用竖向壁柱，较南侧的商业办公楼更显庄重。总部办公楼中部为开放的公共空间，由下至上设有大堂、报告厅、篮球馆和空中花园等功能，楼内还设有员工餐厅、展厅、档案库、数据中心、健身房等设施，充分满足企业总部办公的需求。园区采用冷热电三联供技术的分布式能源站，可以为整个园区提供电力、制冷、制热、生活热水，提高了能源的利用效率。同时通过对多种主动与被动技术进行整合，使总部办公楼获得国家绿色三星级设计标识和 LEED 金级认证，其他楼栋则获得国家绿色二星级设计标识。

Huadian Industrial Park consists of four independent buildings. The headquarter office building at the north has a symmetrical layout which reveals a sense of solemnness. The central part of the headquarter office building is an open public space incorporating a lounge, a lecture hall, a basketball gym and a garden. Distributed power stations with triple supply technology can meet the demand for electricity, cooling, heating and domestic hot water of the whole industrial park with high efficiency. The integration of active and passive energy-saving approaches has won the headquarter building a national 3-star green certification and a gold LEED certification, while other buildings in the industrial park have won national 2-star green certifications.

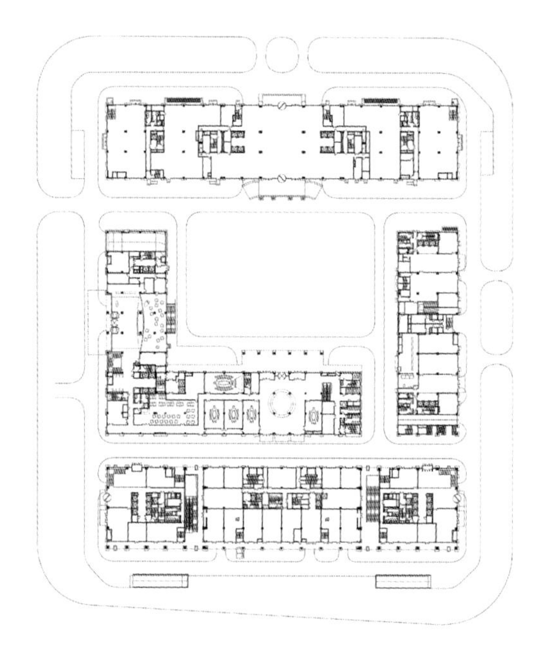

首层平面图

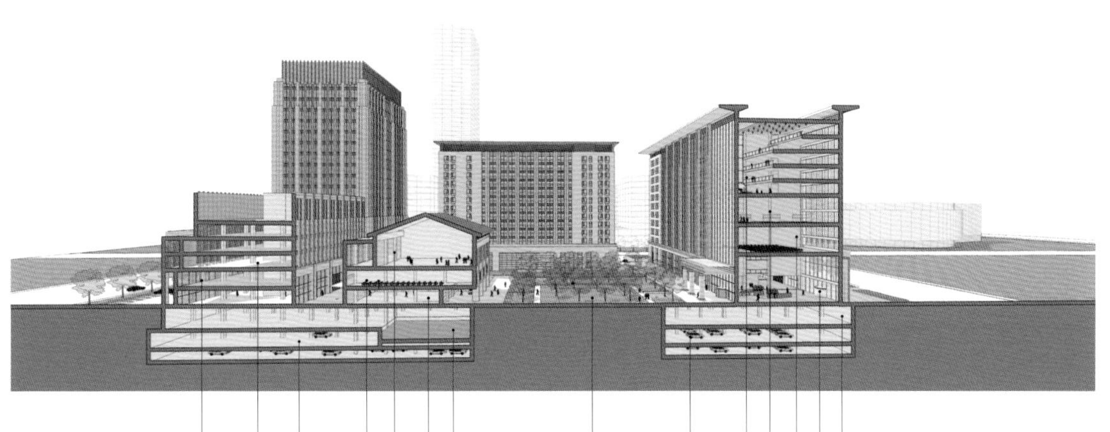

商业
租售办公
汽车库
会议室
大宴会厅
大堂接待
游泳池
中心庭院
汽车库
空中花园
羽毛球馆
大会议室
入口大堂
员工餐厅

剖透视图

中国石油科技国际研发中心 CNPC International Technology Research & Development Center

地点 北京市昌平区 / 用地面积 62,335m² / 建筑面积 161,600m² / 高度 98m / 设计时间 2012年 / 建成时间 2015年

方案设计 徐 磊、丁利群、刘 恒
周绮芸、谢婧昕、王洪跃
设计主持 徐 磊、丁利群

建 筑 刘 恒、周绮芸、孟海港
谢婧昕、金 鼎、金 星
谭寅子、潘 超
结 构 刘建涛、周 昕
给 排 水 朱跃云、张庆康
设 备 马 豫、贺 舒
电 气 许士骅
电 讯 任亚武
总 图 余晓东

建筑位于中国石油科技创新基地核心地块，以会议中心、文化中心、专家公寓、专家工作站等功能为主。其中高层部分设于西南侧，控制总体布局，同时也成为整个园区的标志性建筑。南侧则面向园区中央绿地开放，将景观引入建筑内院。位于三层的公共平台将建筑复杂多样的功能部分串联起来，实现资源共享。建筑立面整体统一在竖向线条中，以石材和玻璃窗有节奏的组合，形成优雅而理性的立面效果，契合企业的形象定位。中心绿地则以“山、水、园、林”为概念打造出景观庭院。

The project consists of a conference center, a cultural center, expert apartments and expert workstations. High-rise buildings is located at the southwestern part of the site, dominating the whole area while serving as the landmark of the whole industrial park. The platform on the 3rd floor serves as a circulation path for the complex that connects different parts and shares public resources. Vertical lines dominate the façade, and the combination of stones and glasses reveals a sense of both elegance and rationality.

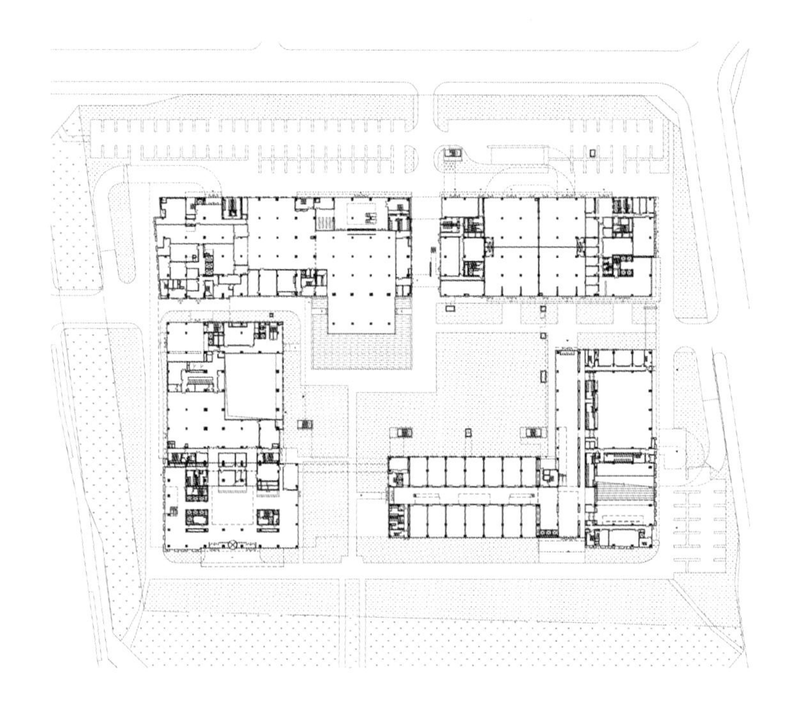

首层平面图

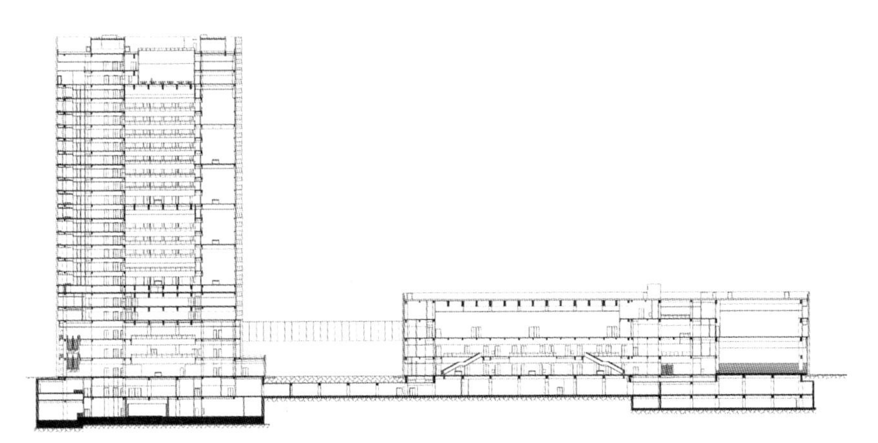

剖面图

龙岩中心城市金融商务中心 Financial Business Center of Longyan Central District

地点 福建省龙岩市 / 用地面积 113,322m² / 建筑面积 530,000m² / 高度 118m / 设计时间 2010年 / 建成时间 2014年

方案设计 李　凌、杨　旭、胡平淳
设计主持 李　凌、胡平淳、杨　旭

建　　筑 马奕昆、徐伯君、林　志
结　　构 曹　清、陈　越
给 排 水 李万华、董新淼、高振渊
设　　备 王　加、李　娟
电　　气 贾京花、凌　劼
电　　讯 孙海龙
总　　图 郑爱龙

龙岩中心城市金融商务中心，位于龙岩行政中心的北侧，形成一条由行政中心向远方山体延伸，沿城市主干道龙岩大道延展的城市脉络。连续的建筑群主要为智能化金融办公建筑，简洁有力的建筑单体在形体上相似，立面则具有丰富的个性，凸显各大金融机构的不同形象，似乎以一条简洁的金属带将所有建筑串联始终。沿街裙房则配备多样的金融营业厅，以及服务于金融办公的餐饮、健身、会议、商业、公寓等多种综合公共服务设施。整体形象既统一又富有变化，呼应着龙岩市山峦环绕、建筑连绵起伏的城市剪影。

The Financial Business Center of Longyan Central District is located to the north of Longyan Administrative Center. The cluster of buildings are mainly intelligent financial office buildings, which are highly identical in volume but varied in the styles of the façade, showcasing the diverse images of various financial institutions. Business halls of the institutions, and service amenities are located in the podium of the buildings. The overall image of the buildings are both coherent and diversified.

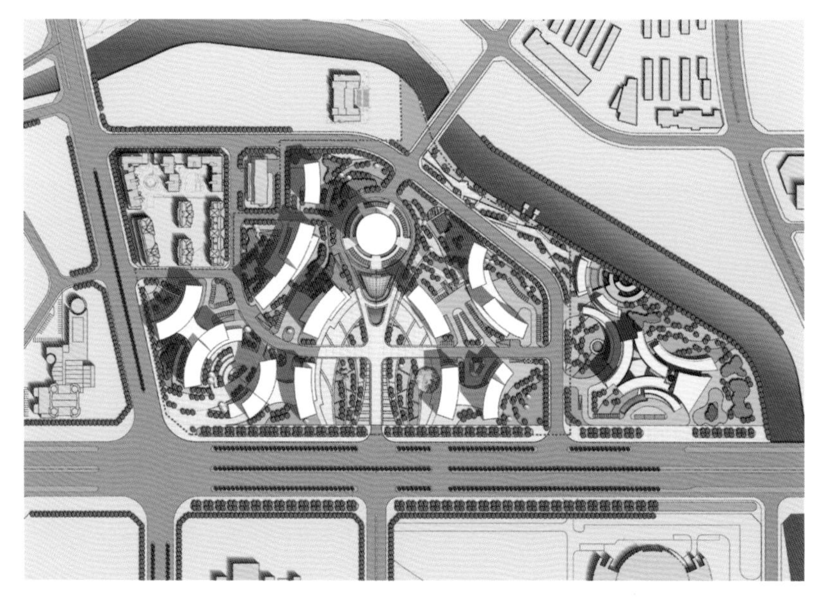

总平面图

商会大厦B

古城创业大厦 Gucheng Entrepreneurship Building

地点 北京市石景山区 / 用地面积 14,367m² / 建筑面积 66,920m² / 高度 80m / 设计时间 2014年 / 建成时间 2017年

方案设计 于海为、刘晏晏、孟 宁
设计主持 于海为

建　　筑 满 欣、郭 睿、赵 鹏
结　　构 郑吉男、辛 宇
给 排 水 张 玉
设　　备 赵建荣
电　　气 丁伟杰
总　　图 郭 睿

合作设计 中国城市建设研究院有限公司

建筑定位为供软件、金融、投资及数据处理类客户使用的中高档研发办公楼。三个建筑体量错落分布在场地上，底层由大堂连接，简洁的建筑造型自成一体，从周边杂乱的城市风貌中脱颖而出。室外绿化延续了极简的特征，为员工提供了一处城市广场；庭院绿化写意精致，成为大堂的点睛之笔。其中，高层塔楼高80m，高层板楼高50m，多层塔楼高21m，三个高低、比例各不相同的塔楼坐落在高12m的裙房之上，裙房楼顶是宽敞的屋顶绿化。三个塔楼中间有一室外庭院，同时为三幢建筑提供了优美的庭院景观。浅灰色石材定义了建筑的主色调，通过铝板开启扇的设计保证了建筑的功能性和立面的纯粹性，石材和玻璃的组合主导了建筑精致而冷峻的气质。

The project is positioned as a medium-to-high class R&D office building for clients in fields including software, finance, investment and data processing. Three towers varied in size are connected by a lounge on the 1st floor, presenting a simple form that stands out from the disorganized neighborhood. A large area of roof greening on the podium, together with an outdoor courtyard among the towers, has added to the beauty of the building. The façade is dominated by light gray stones. Openable aluminum windows have guaranteed the integrity of the glass curtain walls, while the combination of stones and glasses has endowed the building with a sense of both elegance and calmness.

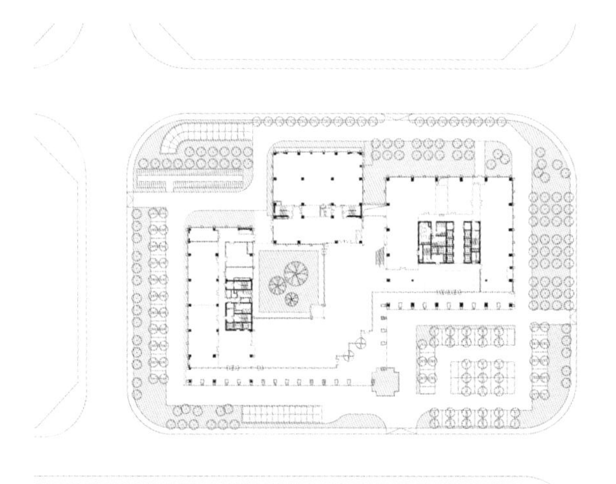

首层平面图

隆福大厦改造 Renovation of Longfu Mansion

地点 北京市东城区 / 用地面积 16,835m² / 建筑面积 58,300m² / 高度 38m / 设计时间 2014年 / 建成时间 2017年

方案设计 崔 愷、柴培根、周 凯
李 赫、戴天行
设计指导 崔 愷
设计主持 柴培根、周 凯

建　　筑 杨文斌、任 重、李 赫
结　　构 任庆英、王 磊、张雄迪
给 排 水 匡 杰、王 松、范改娜
设　　备 宋 枚、雷 博
电　　气 贾京花、刘 畅、陈 游
电　　讯 任亚武、刘 炜、殷 博
总　　图 刘晓琳
室　　内 邓雪映、李 倬、张全全

隆福大厦位于北京东城区隆福寺旧址南段，此地原以隆福寺庙会和东四人民市场知名。原有大厦建于1984年，曾为北京四大商场之一，1993年发生火灾之后扩建，2004年停业。改造设计的定位不再限于商业建筑，而是以文化创意办公为主，在低区和屋顶层配置服务型商业空间等公共功能的办公综合体。新隆福大厦在地面层向建筑内部引入街道，并与外部街巷对接。同时通过拆解低区建筑边界，在尺度上取得与周边住区的融合。中段通过双层幕墙的构造重塑建筑形象，一方面内层灰色实体墙开窗，延续了旧城色彩体系，另外外层设置的玻璃幕墙，通过反射在一定程度上消解了建筑的庞大体量。屋顶仿古建筑作为遗存被保留翻新，重新整理屋顶空间的空间层次，优化屋顶设备机位，形成完整集中的可利用空间，激活了其场所公共性价值。

Longfu Mansion is located at a site once well-known for its temple fair. The former Longfu Mansion, built in 1984 was closed down after a fire. The New Longfu Mansion is an office complex with service-oriented commercial spaces at lower and top floor. Interior streets of the New Longfu Mansion are connected to the streets and alleys of its neighborhood, and the breakdown of its boundaries has integrated the building with its surroundings. Double-layered curtain wall dominates the middle part of the façade. The interior layer is built with solid gray walls with openings, while the exterior layer of glass blurs the hugeness of the building through reflection. The pseudoclassic building on the roof has been renovated to accommodate equipment and form usable public spaces.

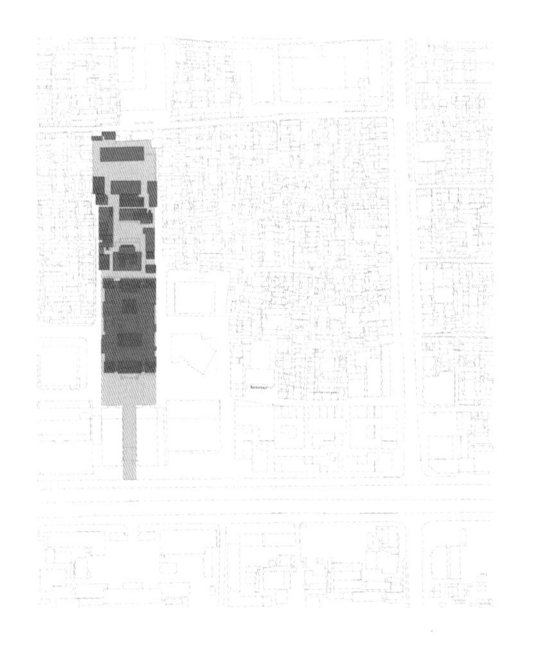

隆福寺地区叠合了各个时代城市发展的痕迹，但又保留着明清以来的街道格局。作为一系列城市更新的起步，隆福大厦的改造强化了原有隆福寺的轴线，随后的改造规划也让原本北侧大大小小随机生长的服务用房获得了新的功能，成为一个开放的文化聚落。

总平面图

ALCHEMICAL
TIMELESS

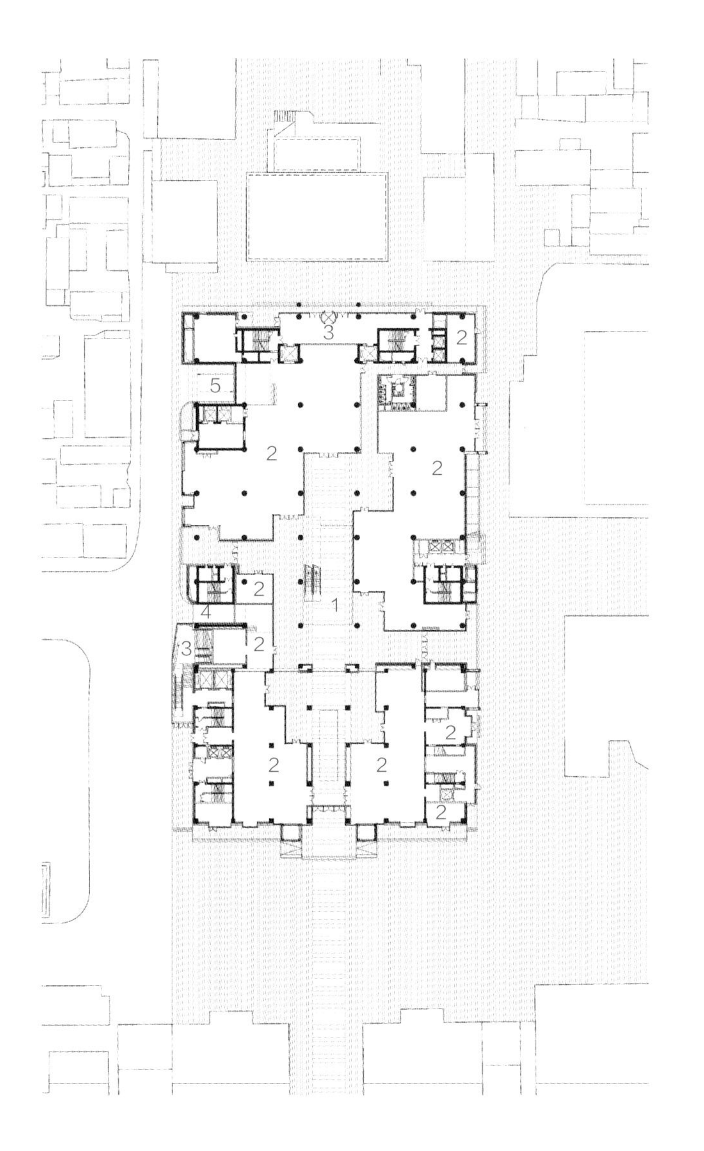

1. 商街
2. 商铺
3. 门厅
4. 地下车库入口
5. 地下车库出口

首层平面图

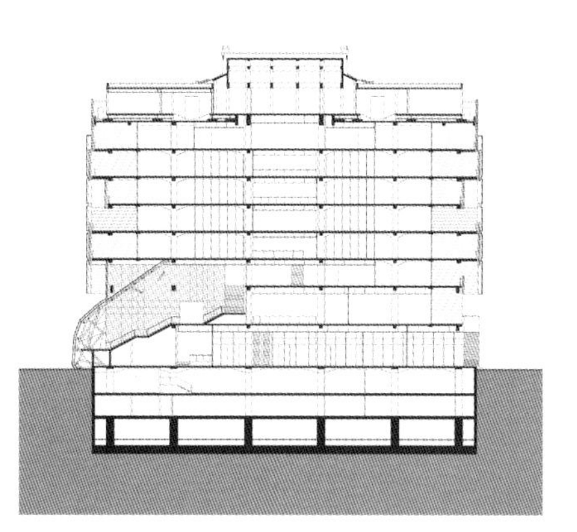

剖面图

西安大华1935 Xi'an Dahua Model

地点 陕西省西安市 / 用地面积 89,922m^2 / 建筑面积 89,050m^2 / 设计时间 2013年 / 建成时间 2014年

方案设计 崔　恺、王可尧、张汝冰
　　　　 陈梦津、冯　君
设计主持 崔　恺
建　　筑 王可尧、陈梦津

施工图设计 中国西北建筑设计研究院

创建于20世纪30年代的大华纱厂，地处西安中心地段，在建厂之初代表着当时国内纺织生产的最高水平；此后的80年间，经历了不断的改造和加建，容纳了各个时期的厂房建筑，也记录了西安这座古都近现代发展的历史段落。同时，工业建筑各时期建造物的高密度共存也有别于民用建筑，如何重新利用这种直白的“密度”成为改造策略的关键。

Dahua Cotton Mill, established in the 1930s, has undergone a series of renovation and extension, with factory buildings of various periods serving as the records for the development of Xi'an. As a result, the high density of factory buildings, different from that of civil buildings, requires the renovation to rediscover the value of its unique texture.

1. 原老南门区 · 休闲餐饮
2. 原老库房区 · 餐饮
3. 原冷东站 · 餐饮
4. 原老医院区 · 餐饮
5. 原动力用房 · 餐饮
6. 原锅炉房 · 当代艺术中心
7. 艺术广场
8. 城市广场
9. 原一期生产厂房 · 商业
10. 原二期生产厂房 · 商业
11. 原老布厂厂房 · 创意商业
12. 原老布厂厂房 · 纺织工业博物馆
13. 原细纱车间 · 商业
14. 原筒并捻车间 · 创意商业
15. 原新布厂厂房 · 创意商业
16. 原新布厂厂房 · 小剧场群落
17. 半室外表演场
18. 原综合办公楼 · 精品酒店

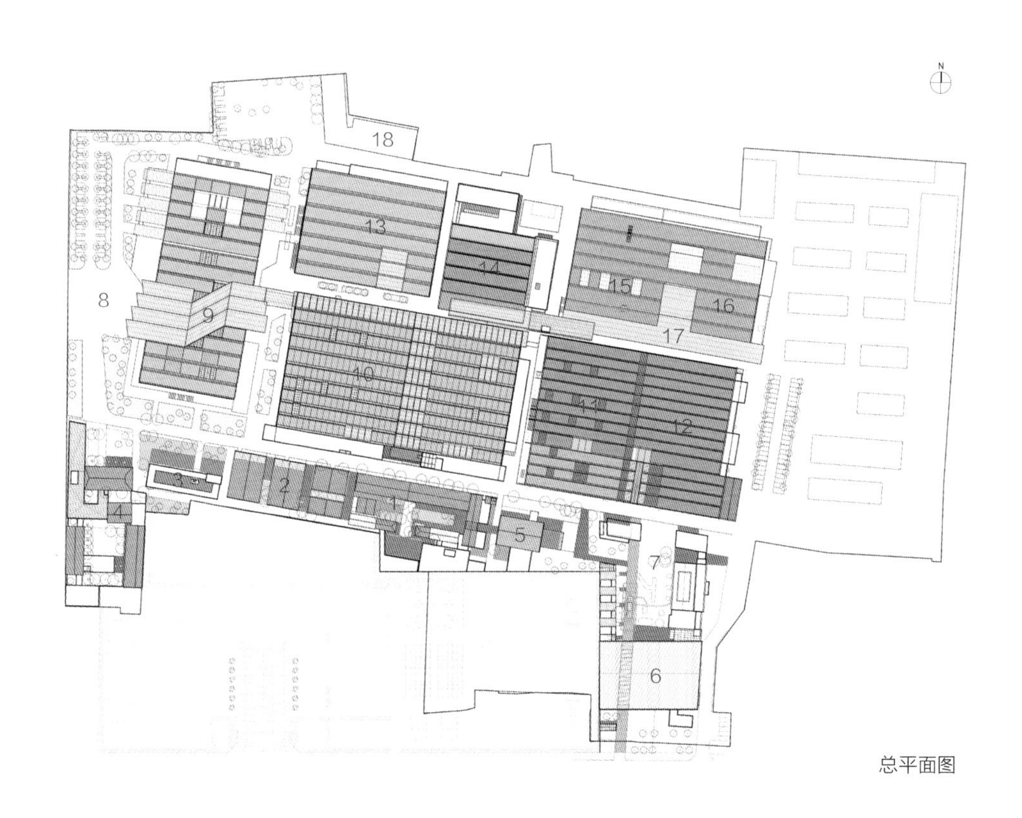

总平面图

早期建成的砖木建筑，围合出尺度宜人的院落空间，设计采取“谨慎的加法”修缮原有建筑，保持材料、空间的本来面目，同时适当增加采用当代建筑语汇的连廊、小品、构筑物，满足餐饮、休闲、文化等新的使用功能，也提示历史记忆和现代生活的共时性。

The brick-timber buildings of early periods were connected by courtyards with appropriate scales, and the renovation design repaired the buildings to maintain their original styles while doing “careful addition” of modern-styled structures to meet new demands and present the co-existence of both the historic remains and the modern living facilities.

中华人民共和国成立后历年建成的厂房多为整齐开敞的车间，屋顶为纺织工业建筑典型的锯齿状天窗，周边设有辅助房间。针对这一区域主要采用“积极的减法”策略，结合城市街区所需空间和尺度，形成新的街道和步行系统，并打开部分结构，产生内部街道和公共空间节点，为产生丰富的城市生活提供了机会。

Structures built after 1949 are mainly plants with zigzagged skylights. Through the strategy of "positive subtraction" , the original auxiliary rooms are displaced by streets and squares, which form a new pedestrian system that invites citizen for culture activities.

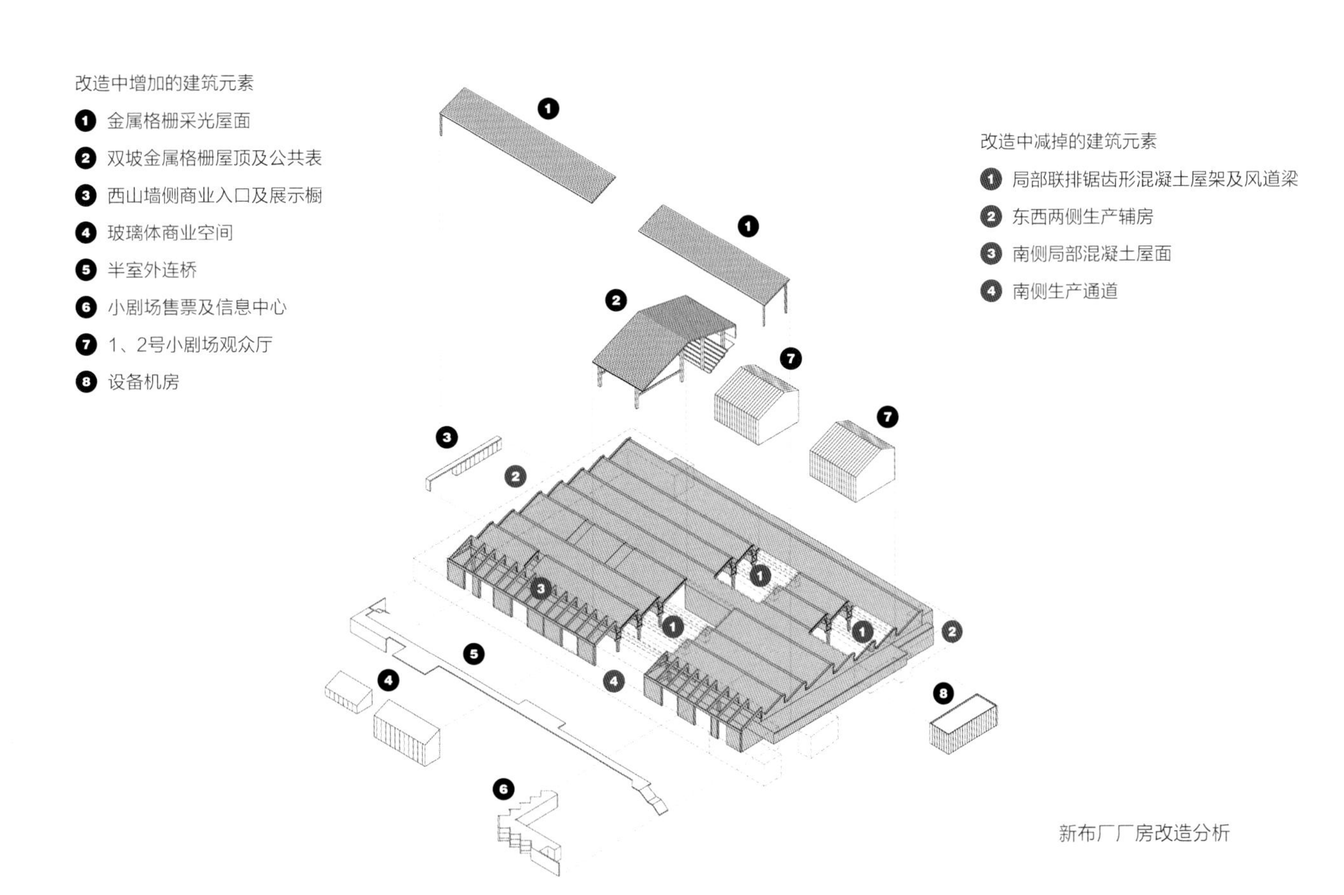

新布厂厂房改造分析

中国大百科全书出版社改造 Renovation of Encyclopedia of China Publishing House

地点 北京市西城区 / 用地面积 4,981m² / 建筑面积 21,312m² / 高度 49m / 设计时间 2013年 / 建成时间 2016年

方案设计 崔 愷、吴 斌
范国杰、杨 帆
设计主持 崔 愷、吴 斌

建　　筑 辛 钰、范国杰、杨 帆
结　　构 魏丽红
给 排 水 裴黎君
设　　备 赵 琪
电　　气 张 罍
总　　图 王雅萍
室　　内 顾建英、张明晓

中国大百科全书出版社大厦位于北京西二环路外，原有建筑于1987年建成，其入口位于西侧的内街，道路狭窄，交通不便，小开间的办公空间显得低矮局促。主楼和东侧裙房围合而成的内部庭院未得到充分利用，毗邻二环路的良好景观和形象价值也未得到有效挖掘。改造设计将入口调整至二环一侧，让建筑回归主要交通路线，为原有内院覆盖玻璃屋顶，用作室内中庭，兼具展览、接待、礼仪功能。为了释放内部空间提供开敞办公，主要设备管线打破传统方式沿外墙布置，并利用窗下墙空间进行整合。在外立面上用金属铝板将其包裹起来，铝板同时也起到遮阳作用，并使得功能和立面形象统一起来，增强了建筑的标志性。

The office building of Encyclopedia of China Publishing House is located along the West 2nd Ring Road of Beijing. The renovation design relocated the original entrance so that the building can be accessed from the 2nd Ring Road. The former courtyard has been covered with a glass roof, so that it has been transformed into an interior atrium for exhibition and reception. Main pipelines are located along exterior walls, integrated behind walls under the windows and covered by aluminum panels serving as sun shades, highlighting the building's identity as a landmark.

改造前原状

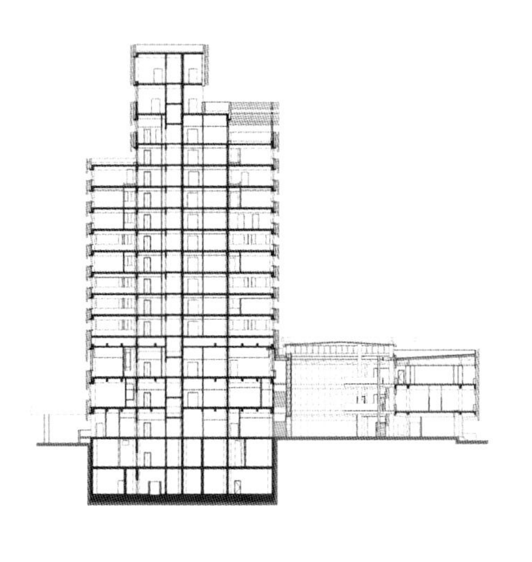

剖面图

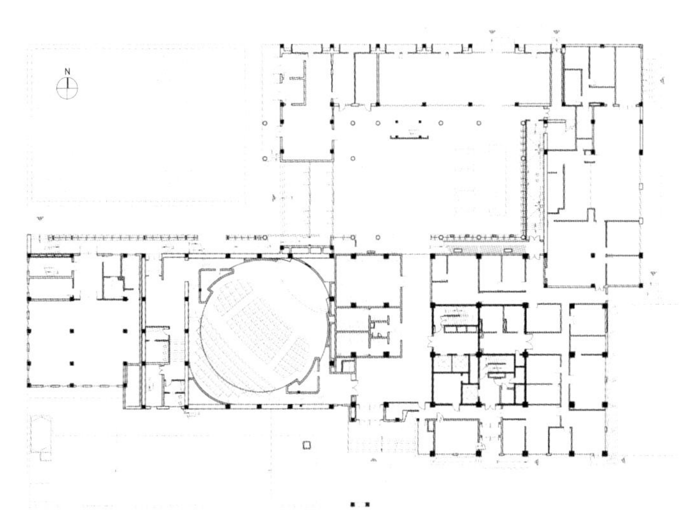

首层平面图

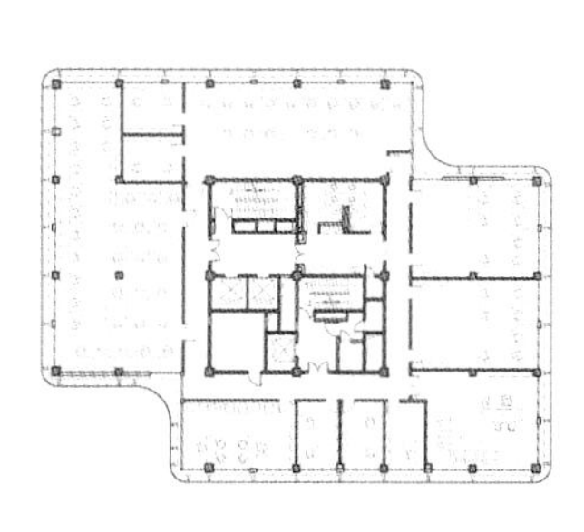

九层平面图

首钢二通厂房改造 Reconstruction of Two-way Factory, Shougang Group

地点 北京市石景山区 / 建筑面积 5,492m² / 高度 17m / 设计时间 2012年 / 建成时间 2015年

方案设计 徐 磊、张 华
金 星、禹 航

设计主持 徐 磊

建 筑 金 星、张 华

合作设计 首钢建筑设计院

基地位于废弃的首钢二通厂区内，周边经过改造已成为环境优美的创意产业园，作为改造对象的旧厂房处于基地中央。一组高低错落、形状各异的实体砖楼被插入场地西侧高大的杨树之间，在“让”的同时呈现出一种建筑的“重”。对于原本粗陋而居场地中部的旧厂房，没有通过装饰美化，而是直接加入一排单层小砖房，让厂房在空间上退居其后，让围绕树木的建筑体量在尺度和材料上统一，巧妙化解了原来场地中的局促。老建筑曾经的一切都真实地得到保留，体现了另一种尊重的方式。为防止雨水迸溅而设置的连接落水口和地面的铁链、窗台和屋檐的黑色石板压顶、白色的卵石散水，在细节上为建筑的品质提供了保障。

The original workshop to be renovated was located in the center of the site, beside which brick buildings with various volumes were scattered among the tall trees, presenting the solidness of the buildings. The original workshop was not decorated or refurbished; instead, a row of single-story brick volumes were built in front of the workshop, so that the workshop seemed to have retreated to make way for the brick volumes of unified material and scale. All the elements of the original structure have been preserved in pursuit of respect and authenticity. Furthermore, detailed design for various building parts has guaranteed high quality of the building.

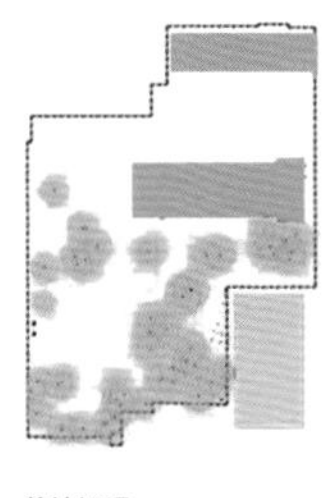

基地概况

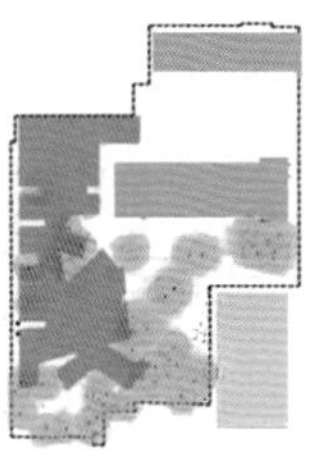

新建中餐厅退让树

原有厂房

内部新建退让厂房

设计概念图

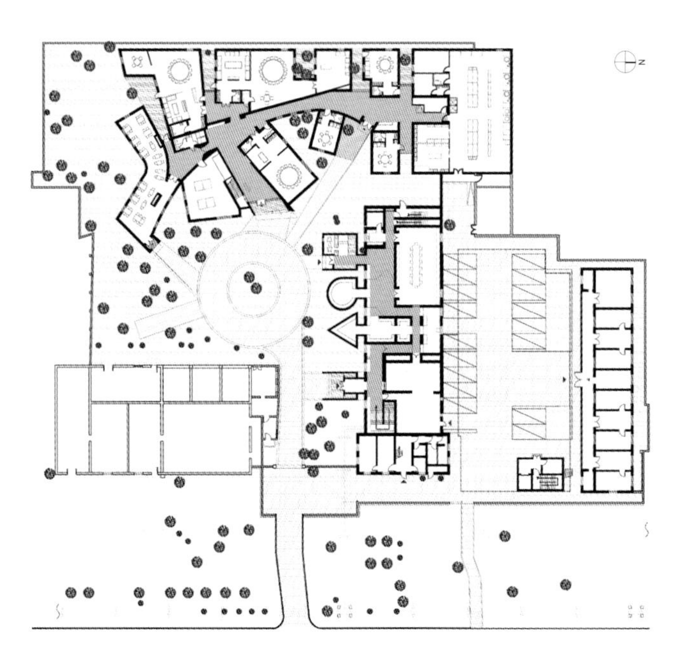

首层平面图

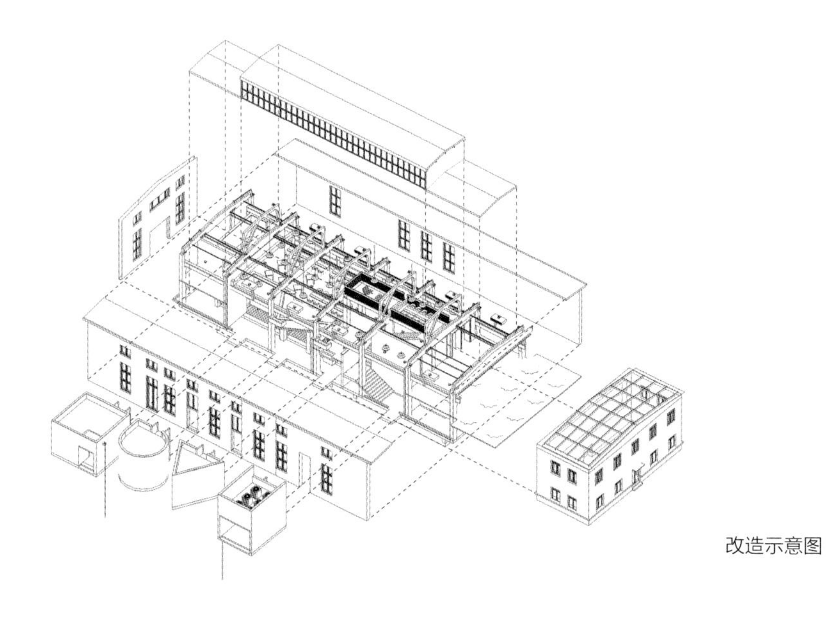

改造示意图

北京工业大学校医院立面改造 Facade Renovation for Hospital of Beijing University of Technology

地点 北京市朝阳区 / 建筑面积 3,000m^2 / 高度 22m / 设计时间 2016年 / 建成时间 2016年

方案设计 潘天佑、冯 晴
设计主持 于海为、张玉明

建　　筑 潘天佑、冯 晴

北京工业大学校医院的改建，是一处通过立面改造恢复建筑活力并改善校园空间的实例。校医院地处校园北门，紧邻重要的城市交通干线。原有建筑建于 20 世纪 90 年代，场地布局缺乏规划，造成校园入口处公共空间的混乱。立面改造将建筑的外部空间清晰地划分为三个层次，解决了医务人员、学校师生和市民等多股人流相互交叉的问题：第一层次的雨篷，被用作与城市对话的窗口，成为市民活动的发生地；第二层次的砖廊，内部用作停车棚，整合了底层空间，也缓解了杂乱无章的自行车停放问题；第三层次增加种植幕墙，提升了医院与校园北门的外在形象。

The renovation has both revived the building and improved the quality of campus. It has divided the exterior space of the hospital into three tiers to separate the various people flows, such as medical staff, students and citizens. The first-tier space under the awning serves as a hub to the city. The second-tier space in the brick corridor is used as a bicycle parking place, which has reorganized the space and solved the problem of scattered bicycles. The third-tier space is a curtain wall with vegetation, which improves the image of both the hospital and the area around campus gate.

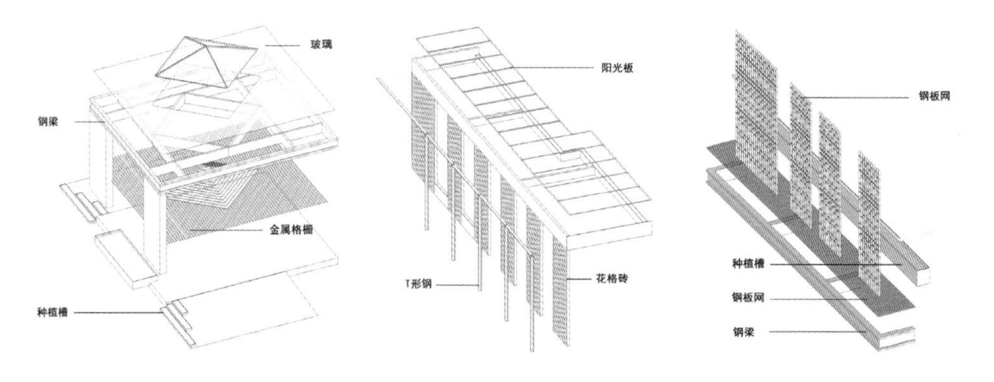

各层次材料做法

上地·元中心附属展示中心 Exhibition Hall of Shangdi Yuan Center

地点 北京市海淀区 / 用地面积 11,555m² / 建筑面积 2,400m² / 高度 10m / 设计时间 2015年 / 建成时间 2015年

方案设计 徐 磊、弓 蒙
设计主持 徐 磊

建 筑 弓 蒙
结 构 张淮湧
给排水 朱跃云
设 备 金 健
电 气 曹 磊
总 图 高 治

作为原三元乳品厂II、III段老厂房的局部改造项目，设计的着眼点是在原有的工业特质基础上，让新功能置入后的空间仍具有充分的灵活性。设计保留了II段厂房原有的排架结构，仅对局部结构进行改造。首层中心部位为带有演出功能的多功能展示厅，两侧利用空间高度设置夹层，上部设置办公，下为服务多功能展示厅的辅助空间。III段厂房保留原有排架结构形式，去除内部隔墙后获得开敞的展示大厅。两段厂房交接处去除部分排架，改造为采光庭院及卫生间。室外场地设计简洁明快，根据建筑、绿地、小品的外轮廓衍生出若干条“控制线”，形成网格，内部填充彩色透水混凝土路面，形成类似蒙德里安画作的艺术效果。

The design focuses on the preservation of the original industrial features so that the spaces remain flexible after the introduction of new functions. The existing bent structure was preserved, and the original interior walls were dismantled to obtain an open exhibition space and a small theatre space. At the joint of the two workshops, the bent structure was partially replaced by a courtyard and washrooms. The boundaries of buildings, greeneries and landscape elements form grids filled with colored permeable concrete pavements, just like the classical works of Mondrian.

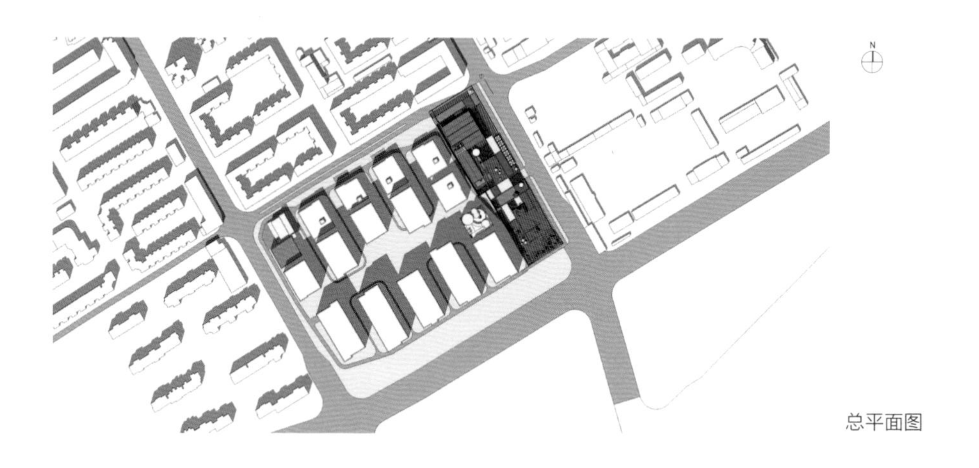

总平面图

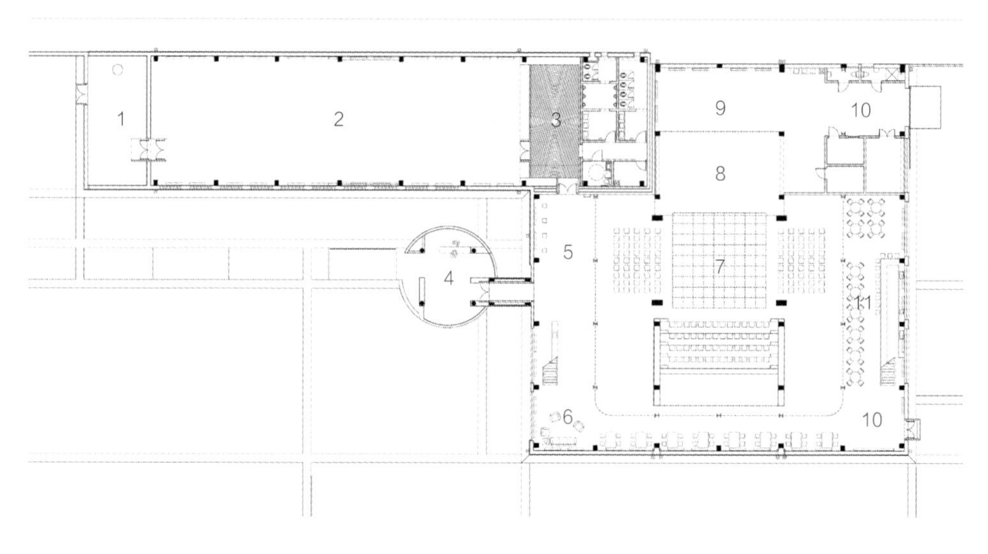

1. 展厅小院 2. 展厅 3. 采光庭院 4. 圆厅 5. 多功能厅入口区
6. 休息区 7. 主舞台区 8. 后台区 9. 化妆区 10. 次入口 11. 茶座区

首层平面图

昆山锦溪祝家甸砖厂改造及民宿学校 Kunshan Jinxi Zhujiadian Brickyard Reconstruction & B&B Scho

地点 江苏省昆山市 / 建筑面积 5,154m² / 高度 9m / 设计时间 2014年 / 建成时间 2017年

砖厂改造

方案设计 崔 愷、郭海鞍
张 笛、沈一婷

设计主持 崔 愷、郭海鞍

建 筑 张 笛
结 构 陈文渊、冯启磊
给 排 水 安明阳
设 备 王 加
电 气 胡思宇
总 图 李可溯、吴永宝
景 观 无界工作室
照 明 北京宁之境照明设计有限公司

民宿学校

方案设计 崔 愷、郭海鞍、宁昭伟
设计主持 崔 愷、郭海鞍、刘 勤

建 筑 苏易平、孟 杰
结 构 何相宇、曹永超
给 排 水 安明阳
设 备 王 加
电 气 胡思宇
总 图 吴永宝
结构顾问 常民建筑

祝家甸是古代皇家金砖产地，至今仍有古窑在继续使用，但村落已破旧不堪。乡村复兴计划采用“微介入”的方式，首先将一座建于 20 世纪 80 年代的废弃砖厂改造为古砖博物馆，容纳展厅、手工坊、咖啡吧等功能。阳光从残破的屋顶投入室内的动人场景被设计者以透明瓦延续下来。室内空间则采取模块化设计，所有地板单元、家具单元、设备单元、镂空单元均可移动和替换，为小建筑提供更多的可能性。为了将底层结构加固的钢拱融入老窑的氛围，在钢拱上增加精美的砖拱，如同形成一条时空隧道，可以回味当年热火朝天的生产场景。入口处独立基础的钢楼梯将老坡道保护其中，新老交叠而存。建筑山墙上“淀西砖瓦二厂”的旧水泥字，则保留着近半个世纪的乡愁和回忆。

Zhujiadian Village was once a place producing gold bricks for the imperial family, with several kilns in use till today. Through “micro-intervention”, a deserted brickyard was renovated into a museum with an exhibition hall, a DIY workshop and a coffee shop. The transparent tiles have preserved the appealing scene of sunlight pouring in through the cracks of the roof. All the components of the interior space are designed with modules that allows for mobility and replacement. A brick arch was built over the steel arch, reminding people of the history. A steel staircase at the entrance covers the old ramp to present the co-existence of the old and the new.

改造后砖窑内举办的民俗活动

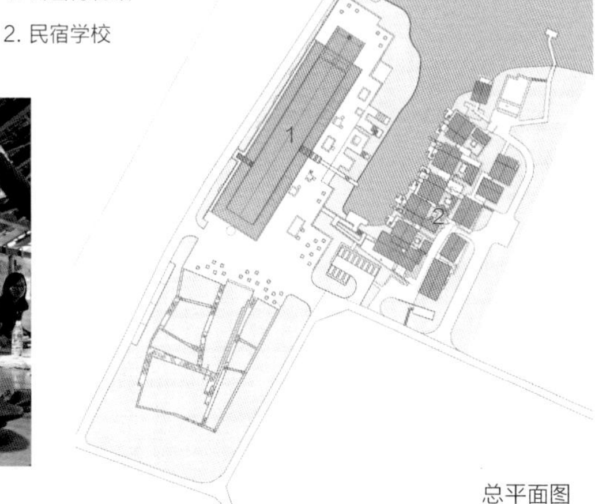

1. 砖窑博物馆
2. 民宿学校

总平面图

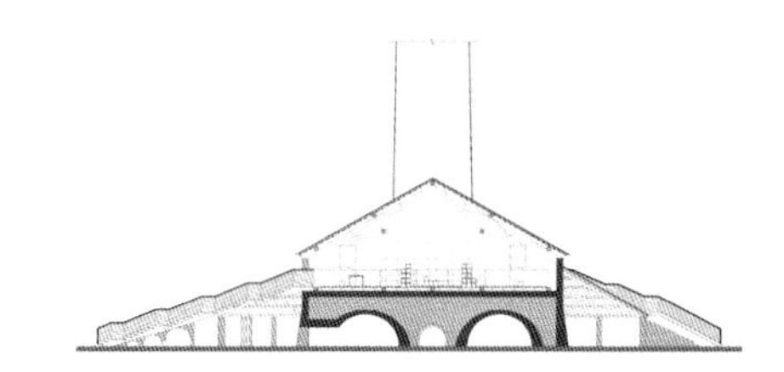

砖厂改造横向剖面图

1. 砖艺手工坊
2. 辅助间
3. 会场
4. 文化展厅
5. 观景平台
6. 原砖窑地面
7. 烟囱

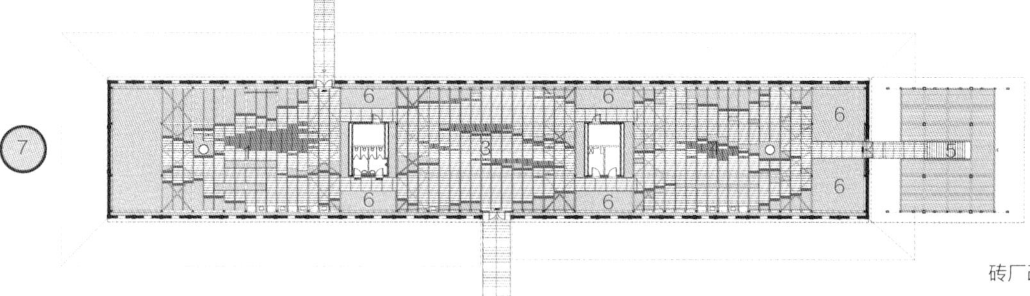

砖厂改造首层平面图

随后建造的民宿学校与砖厂隔河湾而望，水埠头、亲水平台、白墙黛瓦呈现经典的江南水乡面貌。一条小街将访客带入乡村，街的一边是小商铺，另一侧是一家一户的院落。建筑采用了操作安装方便的轻钢框架结构，铝锰合金的金属瓦降低了屋面自重，草泥墙和竹木板墙组成的外墙保持了生态。这些现代措施同样能产生传统意味，并教会村民如何建造轻盈而对自然环境无害的房子。

The B&B school built afterwards faces the brickyard across the river bay. The structure of the school, featuring light-weight steel framework, is easy for the local workers to build. The roof of alloy tiles is light-weighted, and the exterior walls of mud, bamboo and wood have maintained the ecological features of local buildings. These measures have all presented traditional styles while enabling the villagers to build environmental friendly houses.

民宿学校首层平面图

墙里的院子共有四座，每三个客房可以共享一座，客人可以分房而居，亦可以包下一间小院，享受江南水岸的宁静与清新。每个院子里有一个公共小客厅，可以用来接待亲友，小聚盘桓。

Four courtyards, each equipped with a small public lounge and shared by three guest rooms, are enclosed by the walls. Guests can book separate rooms or a whole courtyard to enjoy the serenity of the water town.

昆山西浜村昆曲学社 Kunshan Xibang Village Kun Opera School

地点 江苏省昆山市 / 建筑面积 2,868m^2 / 高度 7m / 设计时间 2014年 / 建成时间 2016年

方案设计 崔 愷、郭海鞍、沈一婷
设计主持 崔 愷、郭海鞍

建 筑 沈一婷
结 构 何相宇、曹永超
给 排 水 安明阳
设 备 王 加
电 气 胡思宇
总 图 李可溯
景 观 冯 君

昆山绰墩山北的西浜村，地处昆曲发源地，却早已无昆曲之音。设计从研究昆曲发展脉络入手，避免大规模重建的行列式做法或涂脂抹粉式的乡村改造，以微介入的有机更新策略，选取村口四座已坍塌农园，通过原址重建和改建，将缺失的乡村肌理重新织补完整，设计成昆曲学社，让村落再有昆曲流传，通过文化引导乡村复兴。粉墙与竹墙组合形成梅兰竹菊四院，又结合水系设计了戏台，两层游廊的穿插，共同组成一个空间丰富、光影交错的昆曲研习之场所。

The school is located in Xibang Village, the birthplace of traditional Kun Opera. The design was initiated by tracing back the Kun Opera's history, and carried out through "micro-intervention". Through reconstruction of collapsed plantations, the context of the village was repaired by the school, invigorating the village through cultural promotion. With a stage anchoring the river and a two-story corridor, it has become a school with diversified spaces.

总平面图

首层平面图

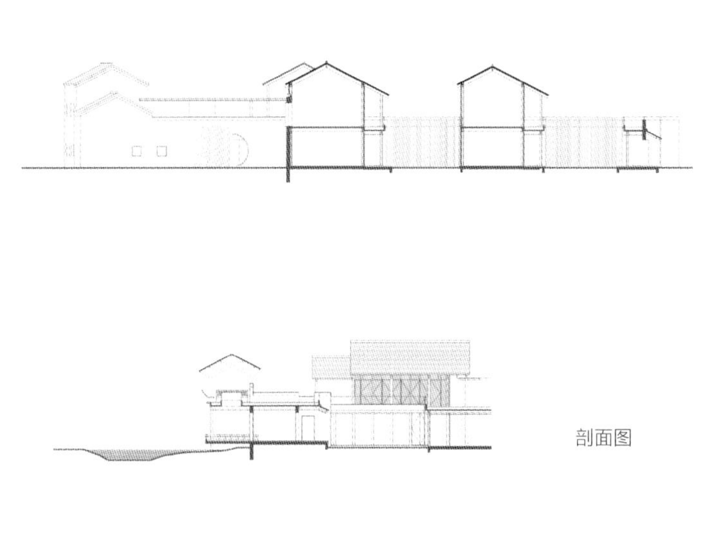

剖面图

1. 门房 2. 梅院 3. 多功能厅 4. 菊院 5. 戏台 6. 序厅 7. 竹院 8. 舞蹈教室 9. 化妆间 10. 教室 11. 教室办公室 12. 兰院 13. 食堂 14. 练声台

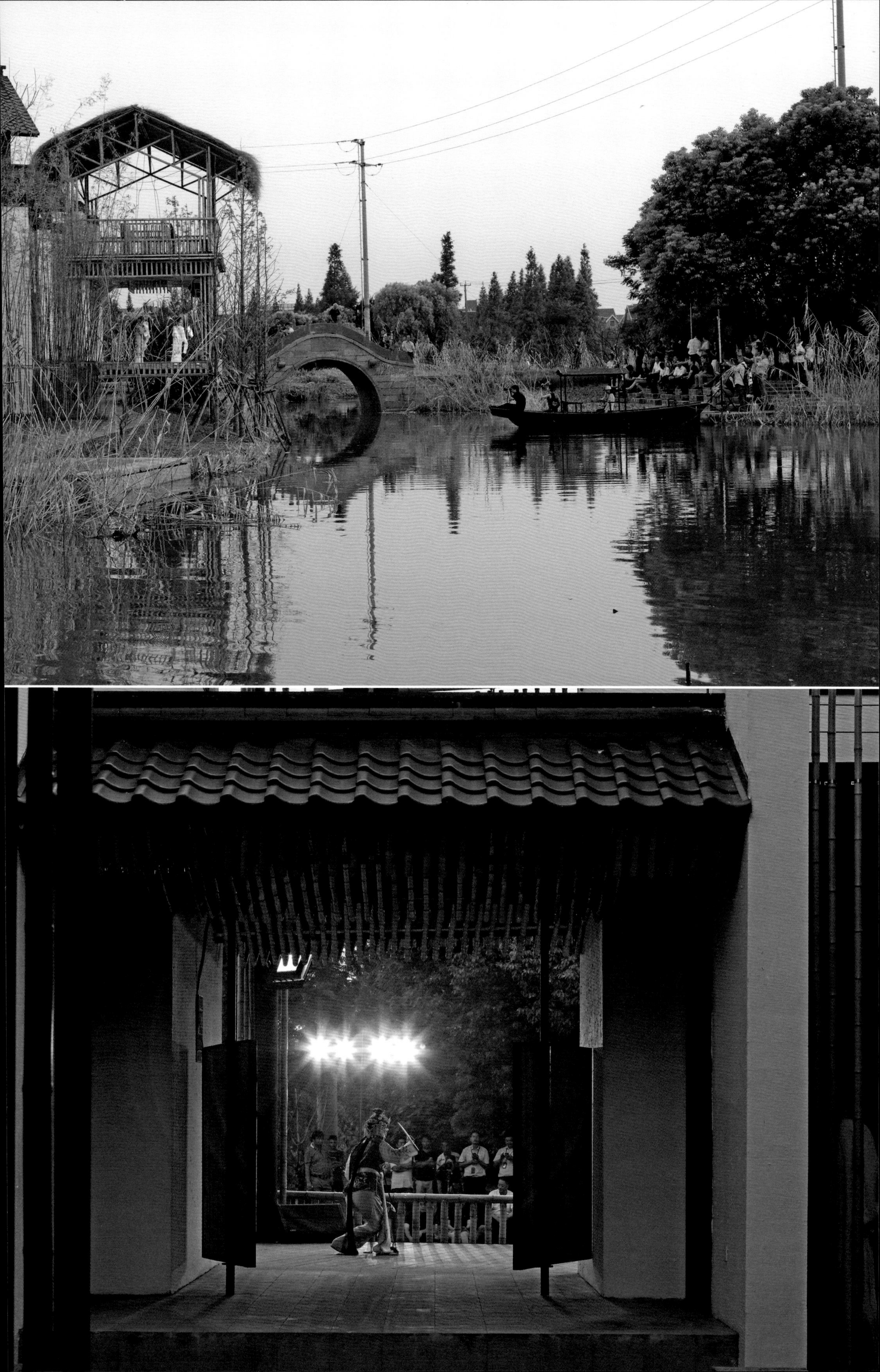

天水街亭古镇示范院落改造 Courtyard Renovation of Tianshui Jieting Ancient Town

地点 甘肃省天水市 / 建筑面积 43号院387m² 167号院416m² / 高度 8m / 设计时间 2015年 / 建成时间 2016年

方案设计 苏　童
设计主持 苏　童

建　　筑 窦　通
结　　构 刘文珽
给 排 水 高东茂
电　　气 熊小俊
总　　图 刘　文

街亭古村，是国家历史文化名村，但近年来私搭乱建现象严重，需要通过整治激活村落并适度发展旅游。改造规划首先在自愿的原则下选取若干典型民宅作为示范性启动项目，其中43号院作为自营农家乐的代表，由居民与政府共同出资改造。设计除了增加经营面积、整理外部空间，还提供了具有简单绿色技术的阳光房，减少了冬季采暖费用。增设的街边凉亭，既能增加商业吸引力，也为古镇街道提供了公共休闲场所。167号院则是位于重点风貌控制区的典型自用商业民居。原有残破的木结构建筑经过落架大修与功能提升，维持了古朴的风貌，同时增建了一定的居住和商业空间。新旧之间的联系与对比仍保持了古镇的完整性，也为后续的民居改扩建提供了参照范本。

Jieting Town is a national historic and cultural town. The renovation of the No.43 courtyard, which was conceived to be a B&B, has been funded by both the government and the residents. Besides additional business area and reorganized exterior spaces, a sunlight room with simple green technologies is built. The No.167 courtyard is a small shop and a house for the owner's private use. Overhaul of dilapidated wooden structure has preserved the modest look of the houses while extensions were made. The harmony and contrast between the new and old has retained the integrity of the town while making the place a model for future renovation.

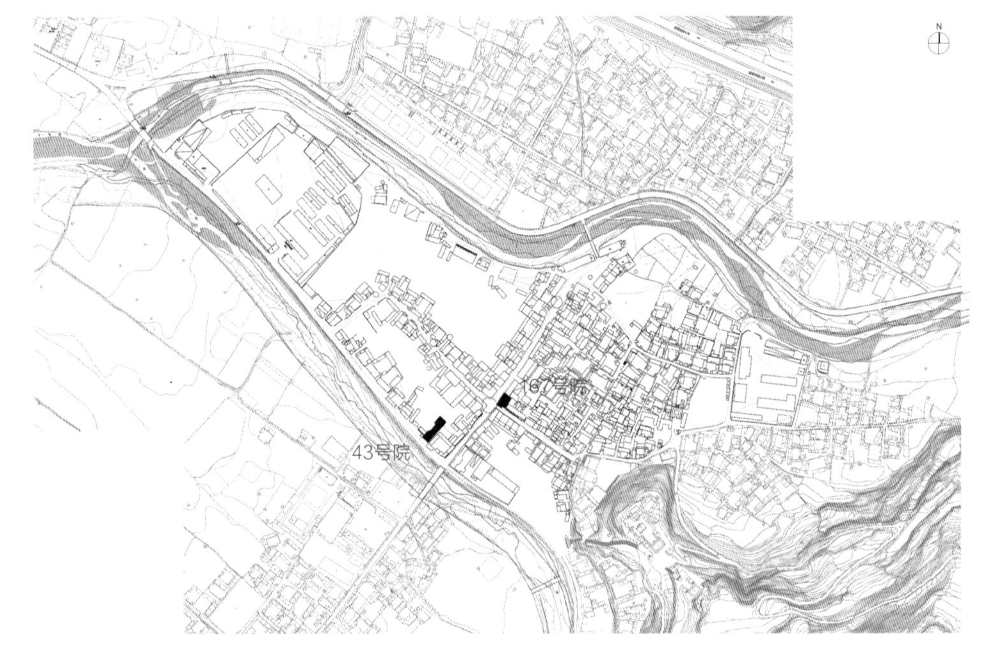

改造院落在村中位置图

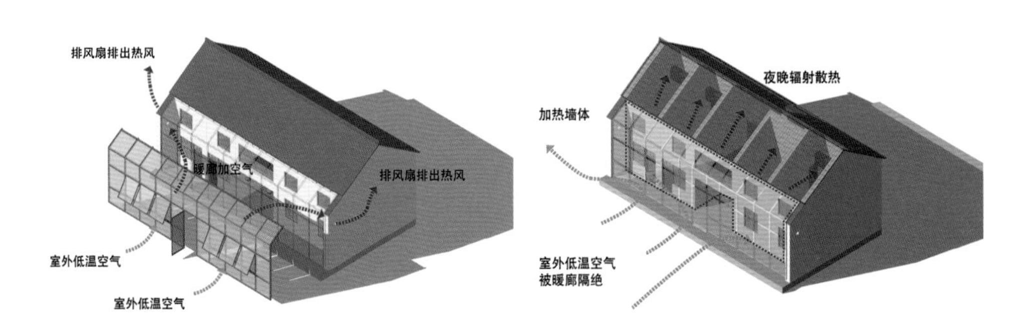

43号院夏季节能措施分析图　　43号院冬季节能措施分析图

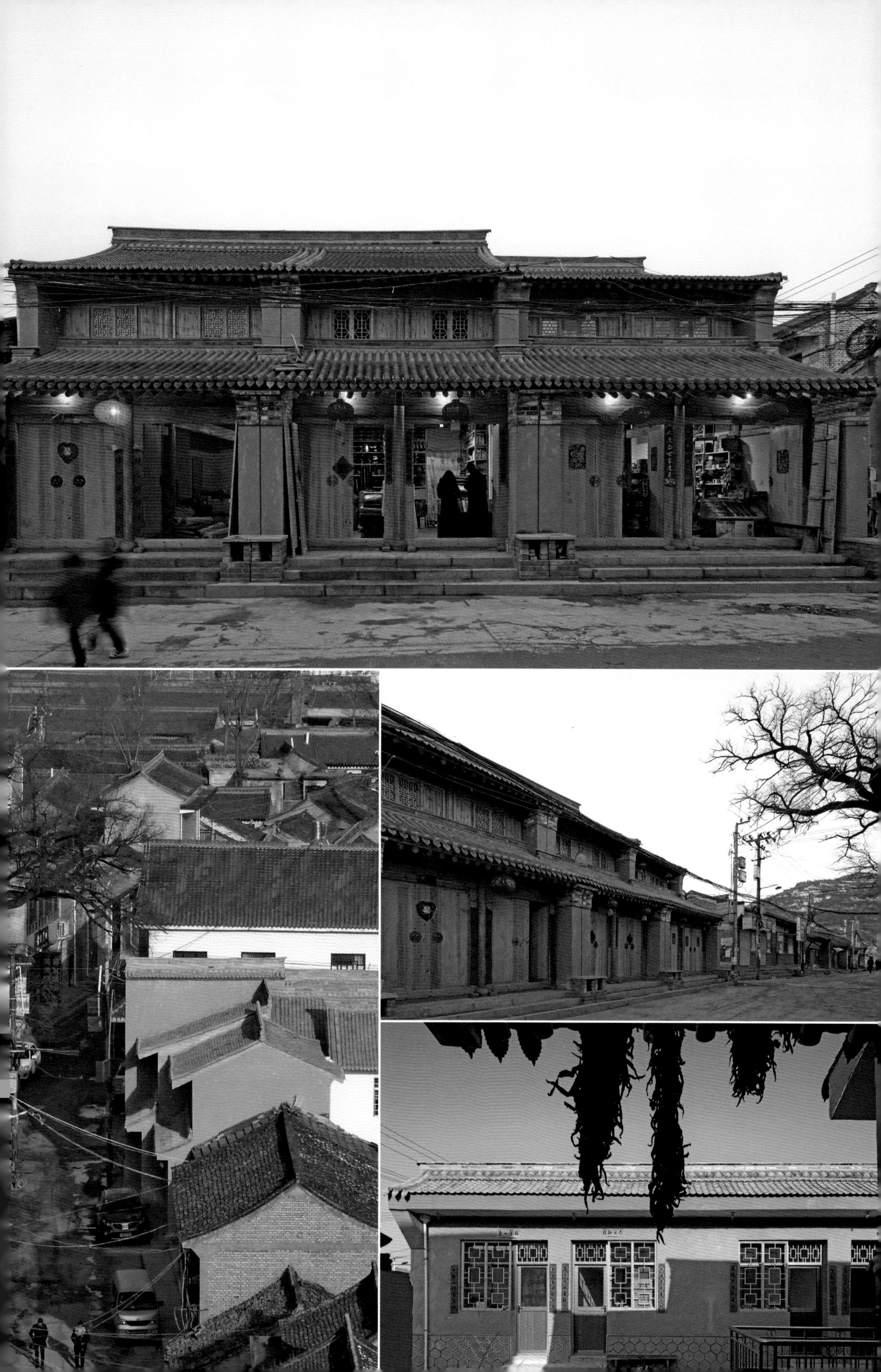

鄂尔多斯折家梁村改造更新 Ordos Shejialiang Village Renovation

地点 内蒙古自治区鄂尔多斯市 / 建筑面积 2,800m² / 高度 4m / 设计时间 2015年 / 建成时间 2016年

方案设计 苏 童、杜戎文、王 宇
设计主持 苏 童

建 筑 杜戎文、王 宇
结 构 马晓雷
给 排 水 高东茂
设 备 刘 娜
电 气 熊小俊

折家梁村位于鄂尔多斯市乡村，村民30余户，与周边众多村落一样存在空心化的问题。改造从村落发展和民生问题出发，采用系统性乡建策略，深入挖掘本村特色，提出以“羊家乐”为主题的发展策划，打造与羊互动的体验模式，依托现有产业提供配套服务，结合网络提供线上线下的农产品渠道，并专为农产品设计系列标识与包装。改造更新结合现状问题、策划定位制定规划，解决民生基础设施等，将村内建筑分为三个等级：焦点建筑——村文化活动中心；重点建筑——其他公共配套建筑；背景建筑——民宅、产业建筑等，不同的类型设置差异化的设计原则和深度。景观设计对节点、绿化、标示系统、村庄家具等制定导则进行引导设计。同时，在羊家乐特色住宿体验中探索利用地域沙土材料筑构沙包土屋，生态循环，降低成本，体现趣味特色；在村公共服务中心的设计中应用被动式生态技术，目前服务中心已投入使用，村民反馈建筑较为舒适，冬暖夏凉。

Located in the rural area of Ordos, Shejialiang Village, with over 30 households, was challenged with the problem of vacancy. The renovation, based on the villages' development and people's well-being, features a systematic rural construction strategy. A scheme featuring “sheep tourism” was proposed, including interaction with sheep, marketing channels for agricultural products and brand & package design for the products. Furthermore, the buildings in the village were classified into three groups so that they would be treated with various renovation strategies. Landscape design ranges from greeneries and signs to public furniture. Houses built with local sand and soil provide accommodation for the tourists with both fun and cost effectiveness. Passive energy-saving technologies were applied in the design for the public service center, and have received positive feedback from the users.

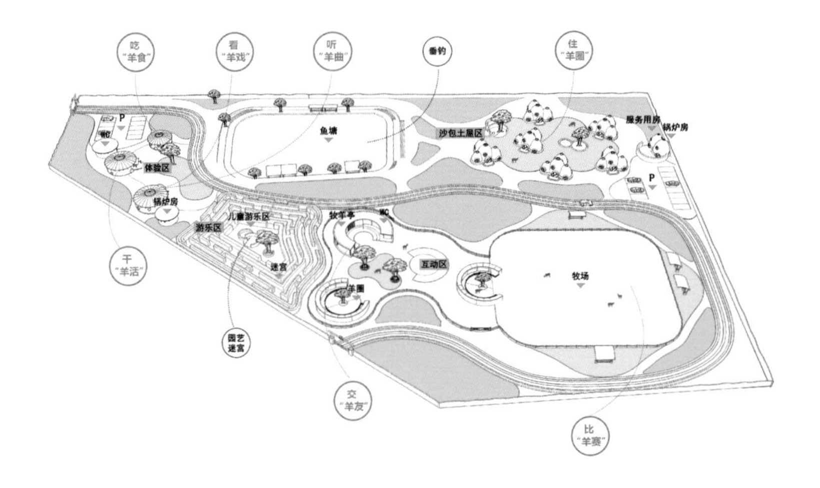

总体布局图

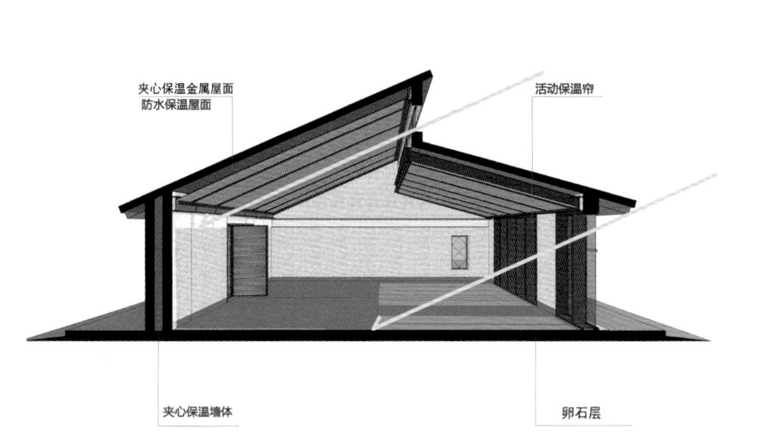

综合利用被动式技术实现节能保温

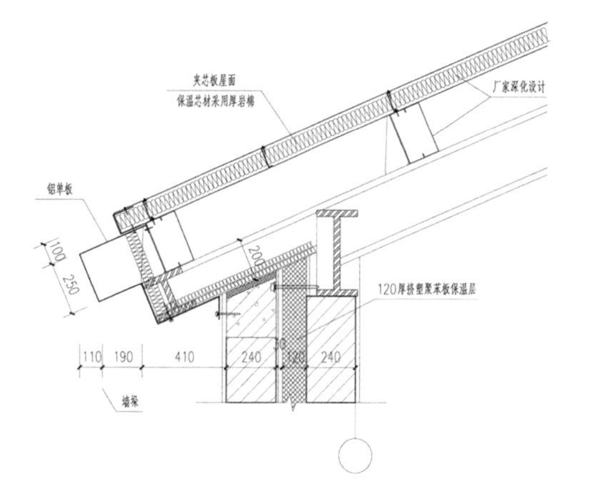

夹心保温双层墙体墙身构造详图及模型示意

华为荔枝园员工宿舍 Huawei Staff Dormitory of Litchi Orchard

地点 广东省深圳市 / 用地面积 198,972m² / 建筑面积 460,205m² / 高度 99m / 设计时间 2010年 / 建成时间 2016年

方案设计 徐 磊、李 淼、王洪跃
设计主持 徐 磊、李 淼、蒋成钢

建　　筑 王绍刚、李宝明、王洪跃
结　　构 王 载、张雄迪
　　　　 贾文颖、张晓宇
给 排 水 吴连荣
设　　备 李京沙
电　　气 丁志强、李 磊
电　　讯 许 静
总　　图 连 荔
室　　内 赵 虹
景　　观 关午军、孟 妍

荔枝园距华为总部约 3km，能够为华为内部员工提供约 4600 套租赁宿舍，其中近 1000 套为短租宿舍，供外地的华为员工来深圳出差使用。用地范围内最大高差达 60 余米，起伏变化很多。设计通过对场地的优化整理，移山填壑，尽可能减少封闭的背阴和背风区域，打通主导风向穿行场地的路径，构建优质的区域小气候；而后将三十几栋建筑单体环绕基地错落有致地依山而建，根据地形、朝向及与周围的关系权衡选择每一栋建筑物的布置方式；4 栋短租宿舍与会所融为一体，数层曲线体量依山水平展开，勾勒并呼应了山的意象。

Located 3 kilometers from Huawei Headquarter, Huawei Staff Dormitory of Litchi Orchard offers 4,600 staff apartments for renting, 1,000 of which provide short-term accommodation for staff on business trips to Shenzhen. Challenged by a site with a maximum altitude difference of over 60 meters, the design has made optimizations to the terrain for a pleasant micro climate. The layout of each of the buildings, located on the hillside and encircling the site, has taken the orientation and its relationship with its surroundings into consideration.

总平面图

丰富的景观植被和多元的活动场所，进一步烘托了舒适而兼具趣味性的建筑群体景观。

Rich vegetation and diverse activity spaces are also planned in the community.

华为南方工厂员工公寓 Huawei Southern Factory Dormitory

地点 广东省东莞市 / 用地面积 144,644m² / 建筑面积 236,783m² / 高度 38m / 设计时间 2008年 / 建成时间 2015年

方案设计 张 燕、李衣言、孙 雷
王 冠、曹 莹
设计主持 张 燕、李衣言

建 筑 沈晓雷、王 冠
结 构 张晓宇
给排水 杨兰兰、何 猛
设 备 宋 玫、王 佳
电 气 王苏阳
总 图 余晓东
室 内 赵 红
景 观 关午军、孟 研

这一项目是为华为南方工厂员工的技术与管理人员的单元式员工宿舍，分为南北两区，共有23栋9～11层的员工公寓和4栋配套公共建筑。单体布局通过长短结合、点线相间的方式适应曲线形基地，围合形成错落有致、活泼自然的建筑群体。结合南方地区气候特点，通过共享的景观带和曲线的院落空间，将内外空间相互融合，中部形成员工活动和服务配套中心区，通过悬挑结构、格栅、木饰挂板等形成标志性空间。规划及建筑单体设计强调兼顾朝向、日照、通风、遮阳的综合处理手法，如利用开敞楼梯间形使楼梯户型实现南北通风，以及利用场地高程设置半地下开敞车库，改善采光通风，降低造价，同时提供丰富的立体景观。

Provided for the staff in Huawei Southern Factory, the community is divided into a northern and a southern districts, which consist of 23 apartment buildings of 9-11 stories and 4 public buildings. A vigorous and diverse layout of the buildings conforms to the fan-shaped site, where landscape belts and courtyards have blurred the boundary between exterior and interior spaces. The center of the community is an area for leisure and services, where buildings with cantilever structures and wooden boards endow the area with distinctive features. The planning has taken various factors into consideration to achieve high quality and cost effectiveness.

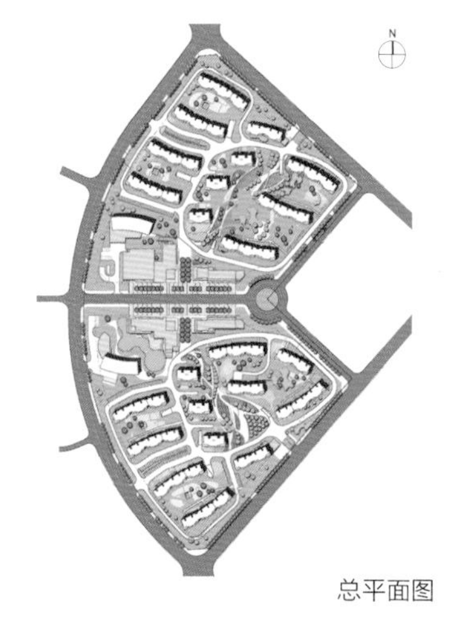

总平面图

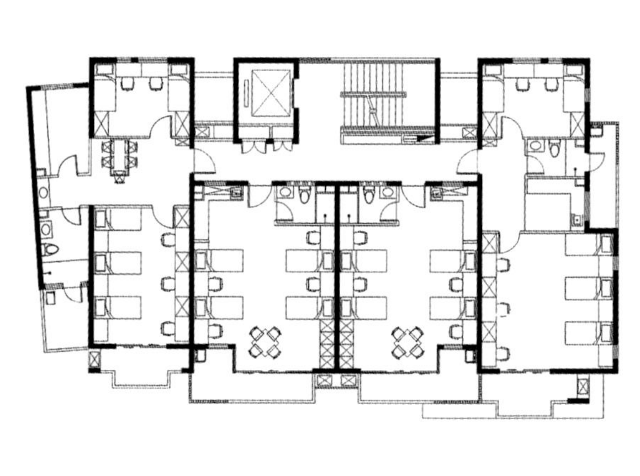

标准层平面图

唐山第三空间综合体 Tangshan No.3 Space

地点 河北省唐山市 / 用地面积 4,991m^2 / 建筑面积 88,011m^2 / 高度 99m / 设计时间 2009年 / 建成时间 2015年

方案设计 李兴钢、谭泽阳、付邦保
孙 鹏、赵小雨、张一婷
张玉婷、闫 昱
设计主持 李兴钢、付邦保

建 筑 孙 鹏
结 构 张付奎
给 排 水 赵 昕、李建业
设 备 胡建丽、王微微
电 气 许冬梅
总 图 郝雯雯

"第三空间"位处唐山繁华的建设北路，其用地东侧紧邻一片南北向平行排列的工人住宅，建筑朝向、布局和塔楼及裙房的体量、形状几乎完全以满足日照规范要求计算得出；两栋平行的百米板状高楼顺着西南阳光的入射方向旋转了一个角度，朝向东南方向，裙房的屋顶也被"阳光通道"切成了锯齿形状，其东侧留出一个带状的花园空地。由此形成一个向高空延伸的立体城市聚落，城市中垂直叠摞的76套"别业"宅园。"标准层"中惯常平直的楼板以错层结构的方式层层堆叠，形成每个单元中连续抬升的地面标高，犹如几何化的人工台地，容纳从公共渐到私密的使用功能，在不断的空间转换中形成静谧的氛围。收藏及影音空间被塑造成"坡地上的小屋"形态，大小、形态、朝向各异的"亭台小屋"被移植于立面，以收纳城市风景，就像敞开于都市的一个个生动的生活舞台，成为密集分布的垂直"城市聚落"的象征。顶层单元则借屋顶之便，引入真正的庭院。所有复式单元既收纳城市及自然景观，自身也成为城市中的新景观。

Tangshan No.3 Space is located along a busy road in the city. Bordering a residential district on the east, the layout, orientation and size of this apartment building are all determined by daylighting regulations. As a result, two paralleled towers face the southeast for effective daylighting, and the roof of the podium has a cut to make way for sunlight. As a multi-dimensional "urban cluster", the property consists of 76 "villas". In each unit, the unique combination of flat floor slabs has formed a group of artificial "terraces", which serve for both public and private functions. Spaces for art collection and film projection are made into the form of "small houses on slopes", and "small pavilion houses" with various sizes and orientations are embedded on the façade as symbols of "urban clusters", which introduce the landscape into the interior spaces. All the duplex apartments enjoy urban and natural sceneries while becoming sceneries on their own right.

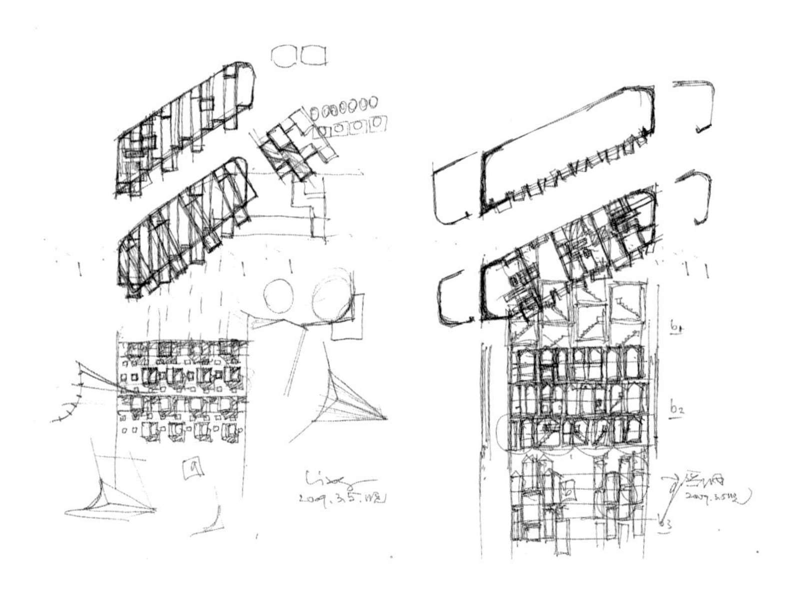

设计草图

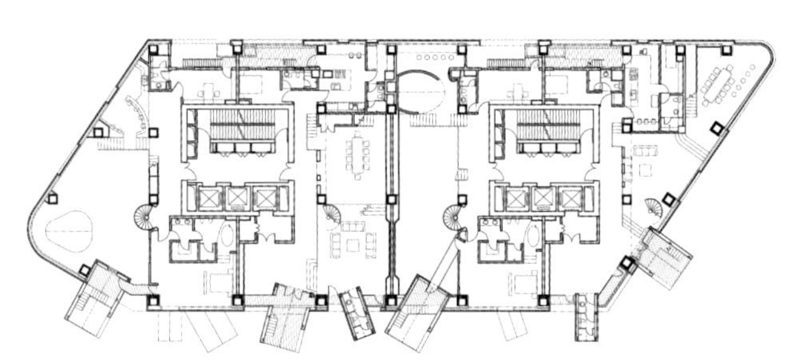

南楼七层平面图

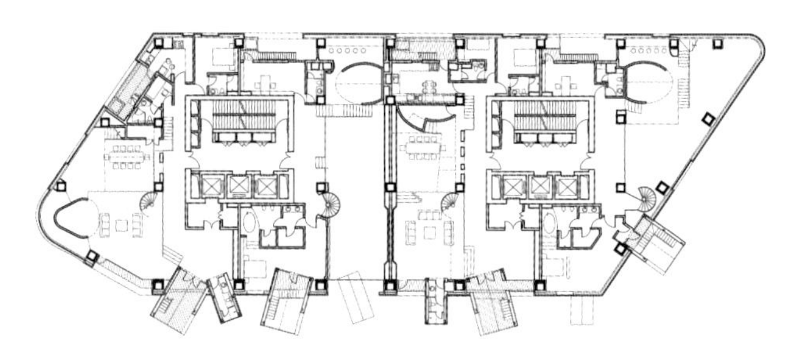

南楼八层平面图

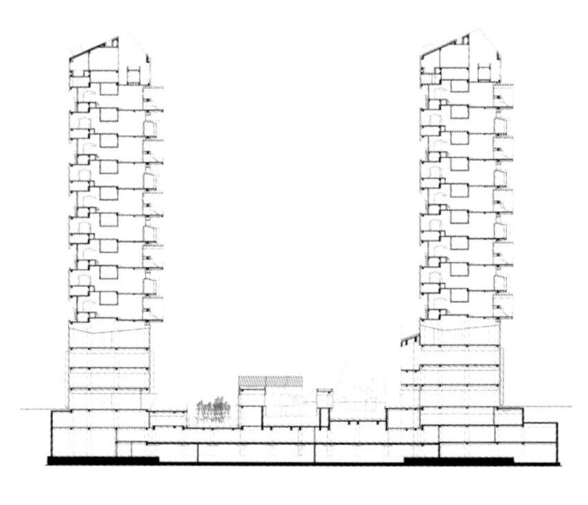

剖面图

亦庄金茂悦 BDA Jinmao Residence

地点 北京市亦庄经济技术开发区 / 用地面积 134,800m² / 建筑面积 414,700m² / 高度 60m / 设计时间 2013年 / 建成时间 2016年

设计指导 刘燕辉、李佳玲
设计主持 谷德庆、詹柏楠

建　　筑 王永坚、张　岳
　　　　 赵庭珂、李　超
结　　构 常林润、曾金盛、包梓彤
　　　　 朱正洋、杨　勇、李　芳
给 排 水 陈　超、关　维
　　　　 李仁杰、唐　安
设　　备 张　昕、王代兵
　　　　 彭　博、丁　聪
电　　气 张晓泉、王京生
　　　　 贾志丹、邵　楠
总　　图 高　治、路建旗、朱庚鑫

合作设计 维思平（WSP）建筑设计咨询有限公司

亦庄金茂悦毗邻北京南郊环境优美的南海子公园，以高层住宅为主。设计以“台地花园”为切入点，采用轴线秩序、工整大气的布局方式。住宅单体南北通透，全明设计，中央储藏空间结合设备集成设计形成套型核心。此外还采用了一系列的先进技术，营造“恒湿、恒温、恒氧”的绿色科技住宅。顶棚设置毛细管辐射末端加单元集中新风的温湿度独立控制空调系统，确保室内温度分布均匀，易于控制温湿度，有效控制空气污染物。中水系统、太阳能热水系统、地源热泵系统等技术措施也进一步强化了住宅的绿色设计品质。

Dominated with high-rise buildings and built into a “terraced garden” with hierarchical axes and regular layout, the community boasts apartments with both southern and northern orientations. A series of cutting-edge technologies have been applied to achieve “constant humidity, constant temperature and constant oxygen level” and facilitate the balanced distribution of heat and easy control of humidity, temperature and air pollution.

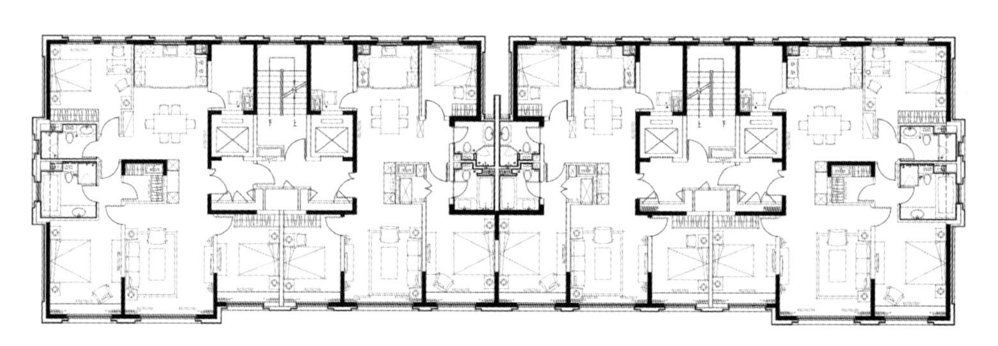

标准层平面图

龙湖·长楹天街 Longfor Changying Paradise Walk

地点 北京市朝阳区 / 用地面积 175,700m² / 建筑面积 678,600m² / 高度 80m / 设计时间 2010年 / 建成时间 2014年

方案设计 柴培根、童英姿
孙博怡、李 颖
设计主持 柴培根、童英姿

建 筑 孙博怡、李 颖
结 构 孙洪波、郝国龙、余 蕾
给排水 侯远见、俞剑锋
设 备 张 斌、姜海元
电 气 陈 红、熊小俊
电 讯 张月珍
总 图 齐海娟

长楹天街是龙湖地产投资开发的城市商业、住宅综合体项目，位于北京城区至通州的城市快速路南侧一块狭长的东西向用地内，地块中间还有一条南北向的市政道路通过。为了将商业价值最大化，集中商业沿主干路展开，并以一座超过百米的廊桥跨过市政路，将东西两部分连接起来，形成总长度 700m 的大型商业体。商业部分内部设置了多组中庭形成既有差异又相互联系的共享空间。地下二层与地铁站厅相衔接，采用非常规的做法，在地铁站和商业空间之间增设了贯穿东西的地铁出站大厅。这样一方面改善了地铁人流的出行体验，缓解了地铁瞬间人流对城市地面交通的冲击；另一方面也连通了东西地块地下商业空间，使集中商业和地铁人流在多个标高上有了衔接机会。

The urban complex is built on a narrow site bordering an urban expressway. The retailing part is arranged along the road to maximize its value, and a bridge over 100 meters in length has connected the eastern and western parts of the complex, forming a 700-meter-long shopping mall with a series of shared public spaces. The retailing part is connected to the subway, and an extension of the subway station's hall is built, merging the western and eastern parts while buffering the impact on the over-ground transportation brought by outbound crowds from the subway. Furthermore, people's access to the commercial spaces are established at various altitudes.

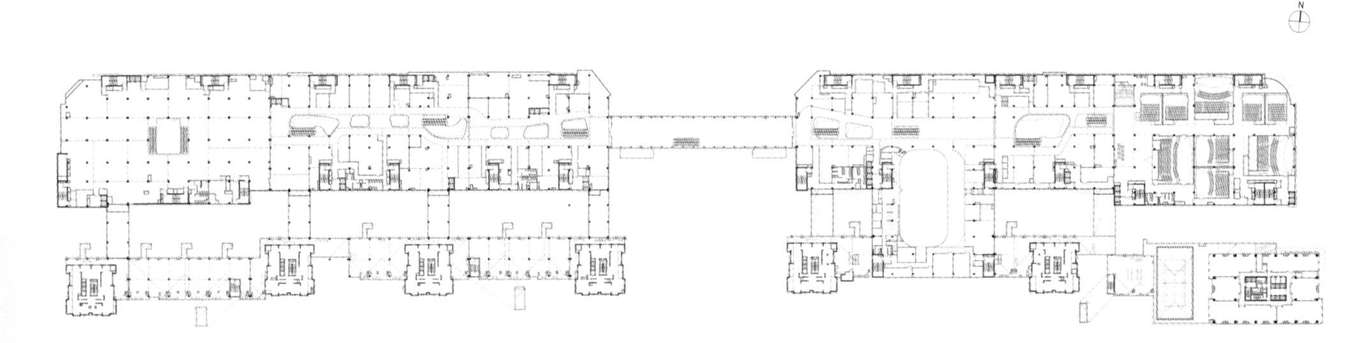

三层（廊桥层）平面图

立面设计穿插使用了石材幕墙、透明玻璃幕墙、彩釉玻璃幕墙、金属隔栅等元素，光厅、橱窗、广告位、楼梯间、出入口、进出风口各个分散的元素被整合在了一个大设计体系之下。项目东端影院区弧形玻璃幕墙，采用透明玻璃、磨砂玻璃和彩色玻璃肋结合的材料组合方式，配以夜晚的灯光处理，在面对城市大道时，获得了炫动、耀眼的标识性商业效果。

A rich combination of materials, consisting of stone curtain walls, transparent and colored glazed curtain walls, metal grilling, dominate the façade while elements such as show windows, advertising spaces and air inlets are integrated into a system. The combined use of various kinds of glazed walls has made the movie theater of the complex stand out with its dazzling look.

龙湖·时代天街 Longfor Times Paradise Walk

地点 北京市大兴区 / 用地面积 250,000m² / 建筑面积 670,000m² / 高度 80m / 设计时间 2011年 / 建成时间 2016年

方案设计 柴培根、李佳玲、李 楠
刘 旻、赵国璆
设计主持 柴培根、李佳玲

建　　筑 李 楠、刘 旻
赵国璆、杨文斌
结　　构 史 杰、王树乐
杨 飞、文 欣
给 排 水 王耀堂、王则慧、周 博
设　　备 宋孝春、韦 航、张 斌
电　　气 许冬梅、王 铮、庞晓霞
电　　讯 张月珍、张 雅、高爱云
总　　图 白红卫、李可溯

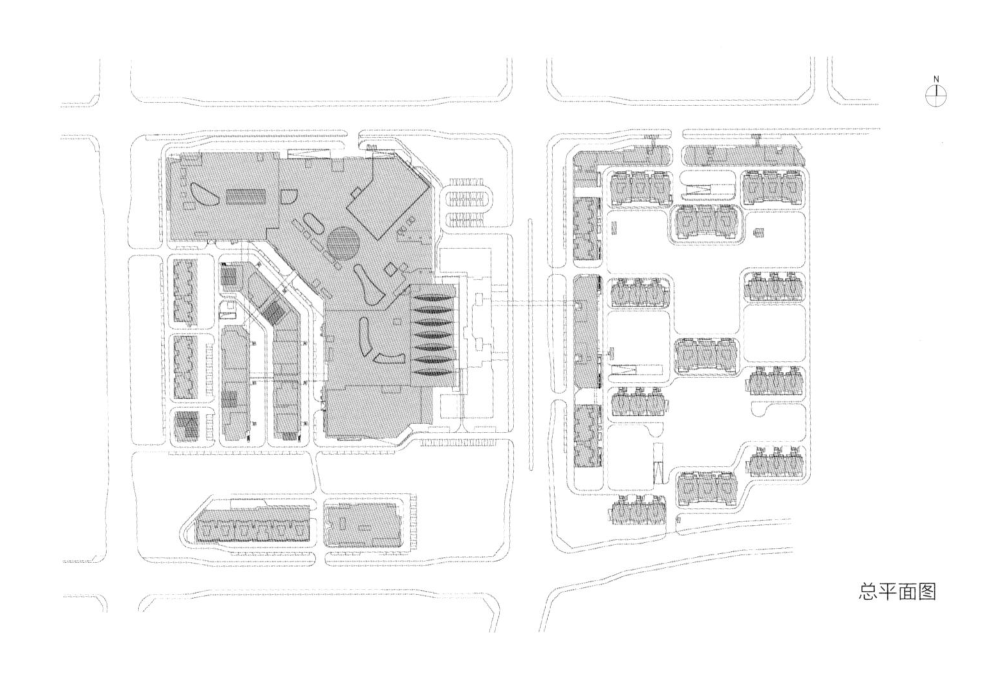

总平面图

龙湖·时代天街位于北京大兴新城核心区域，紧邻地铁四号线站点，是一处以商业综合体为特色的大型商业、居住复合社区。整组建筑分为东西两部分，西区主要是大型一站式集中商业及商业步行街，结合现状地铁站厅进行轨道交通一体化设计，实现商业与站点的无缝衔接，形成多维、立体的城市交通流线和公共空间。东区以住宅为主，沿街配置少量商业服务设施，通过人性化的公共空间和简洁的建筑设计，塑造高品质的宜居社区。

Longfor Times Paradise Walk is a large-scale community of both commercial and residential buildings. The property is divided into two parts. The western part mainly consists of a centralized one-stop shopping mall and a pedestrian zone. The eastern part is dominated by residential buildings with a few commercial spaces. The simple and people-oriented design has made the property a high-quality community.

嘉都会所 Jia Du Clubhouse

地点 河北省燕郊 / 建筑面积 2,500m^2 / 设计时间 2013年 / 建成时间 2014年

方案设计 于海为、刘晏晏
　　　　 魏　伟、刘滨洋
设计主持 于海为

建　　筑 刘晏晏、魏　伟、刘滨洋
结　　构 周　岩
给 排 水 陈　静、李梦辕
设　　备 韦　航、张　干
电　　气 贾京花、刘　畅
BIM 支持 石　磊

建筑是一个大型居住社区的启动项目，前期作为项目展示接待场所，待社区建成后作会所使用。设计的初衷是让建筑不破坏现有的高大乔木，不遮挡优美的自然环境。建筑形态由躲避多株乔木的根系半径而确定，经过抻拉柔化，不规则的环形更为契合场地的自然属性。金属屋顶根据遮阳需要，在西侧出挑最多，形成充足的廊下空间沟通室内外。为衬托高大的树木，建筑净高和屋面构造厚度经过了刻意控制。为实现空间的纯净，垂直构件尽量与外表皮结合，“树杈”形的梁各向交错，形成一个整体，共同传递荷载，尽可能减小了结构构件的截面尺寸。

As the starting project of a large residential community, the building, initially used for demonstration and reception, serves as a clubhouse. The design aims for the preservation of existing huge trees, so the form of the building was determined by the distribution of roots, presenting an irregular ring shape that conforms to the natural quality of the site. The metal roof has long eaves at the west side, forming a buffering zone between the exterior and interior. Vertical building parts are mostly integrated with the façade, and the crossed beams, like branches, have formed a whole system that minimizes the section dimension of the beams.

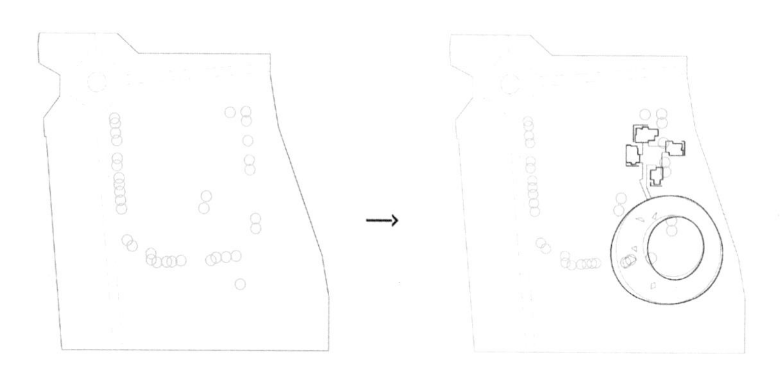

概念形成图

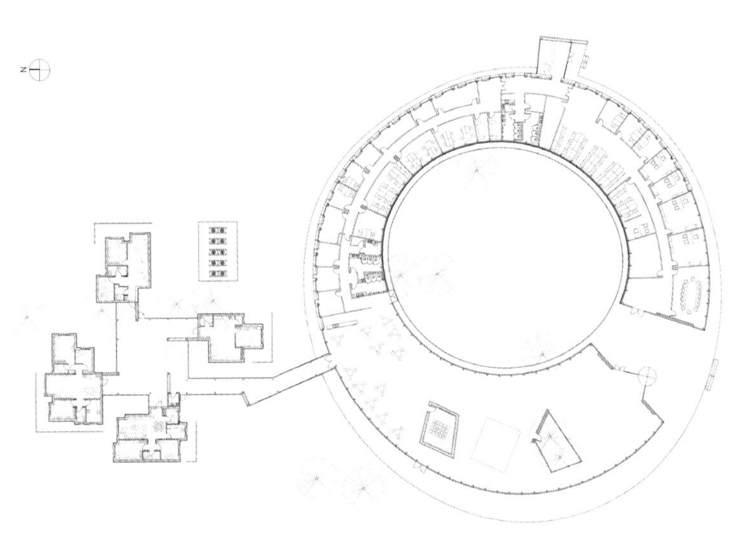

平面图

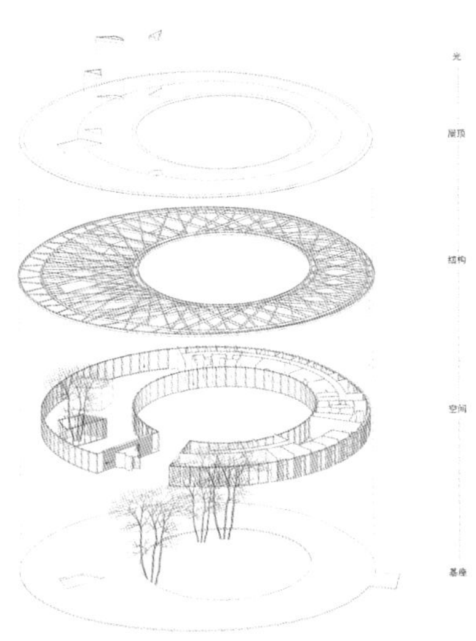

形体构成分层图

环形进深较大的部分是公共活动区域，平面内部实体的影音室和玻璃树院为建筑加入了空间的变化和层次感。光线在不同方向的玻璃间折射出璀璨质感。

Public functions are located at spaces with larger depths, and the diversity of the space is achieved through the contrast between solid and transparent parts.

朝阳区生活垃圾综合处理厂焚烧中心 Incineration Center of Waste Treatment Plant at Chaoyang Dist

地点 北京市朝阳区 / 用地面积 50,000m² / 建筑面积 36,000m² / 高度 48m / 设计时间 2011年 / 建成时间 2016年

方案设计 李兴钢、张音玄、李 喆
梁 旭、李 欢

设计主持 李兴钢

建　　筑 张音玄、梁 旭
李 喆、李 欢

合作设计 中国城市建设研究院有限公司

垃圾是人类城市生活的必然产物，垃圾处理中心也是城市重要的市政公用设施。焚烧中心主厂房是厂区内体量最大的单体建筑，通过对内部工艺和外部围护系统的整合设计，建筑作为一个整体，将垃圾运卸、存放、焚烧、净化、发电的工艺流程及空间特征表达出来，并在保证工艺要求和使用功能的前提下，强调体量的完整性和工业建筑的力量感，选用反映工业建筑特征的建材，形成简洁、高效的外观。焚烧中心还具有展示、教育的社会服务功能，在厂房内部空间设置了一条生动、安全的教育展示通廊和垃圾处理参观廊道，参观和展示区域与内部生产区域流线分开，互不干扰。

The incineration center undertakes the processes of waste transportation, storage, incineration, decontamination and power generation. Through the integrated design of both the internal process and the external image of the building under the premise of smooth operation, the integrity of the form and the strength of an industrial building are manifested. Serving as an educational space, the incineration center has a specially designed corridor for garbage disposal education, which is independent form the production lines in the plant.

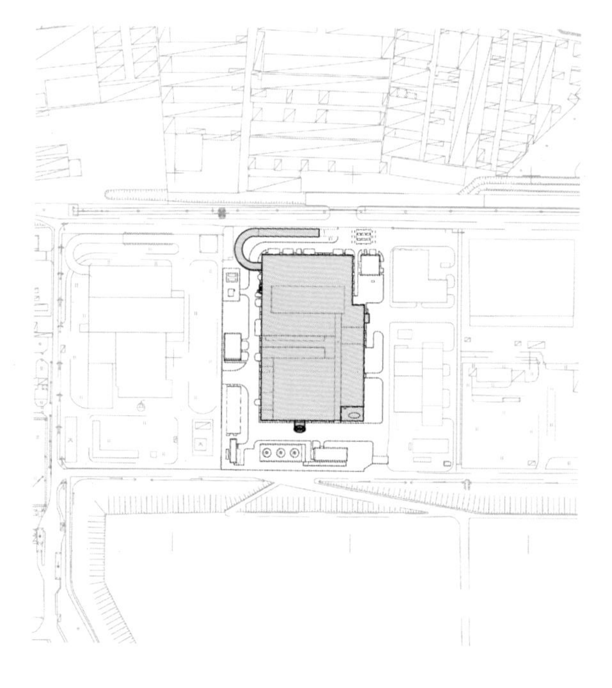

总平面图

设计在“方整”的基本形态基础上，对建筑转角、主要出入口、屋顶等位置进行了细化处理，并重点处理了办公、参观、展示和教育空间的立面表达。

On the basis of a regular and square form, detailed modifications are made to corners, entrances, roofs, as well as the exterior envelope of offices, exhibition spaces and educational spaces.

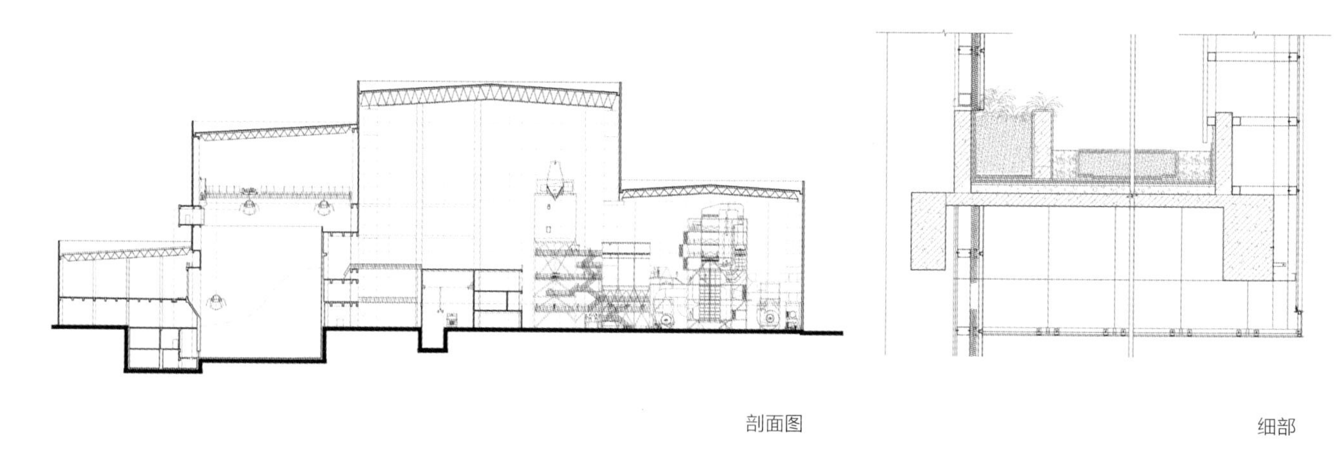

剖面图

细部

#1炉烟气回流风机出口

咸宁仙鹤湖鹤鸣居 Crane House of Red-crowned Crane Lake in Xianning

地点 湖北省咸宁市 / 用地面积 25,416m² / 建筑面积 10,614m² / 高度 19m / 设计时间 2015年 / 建成时间 2018年

方案设计 郑世伟、黄鹤鸣
设计主持 郑世伟

建　　筑 黄鹤鸣、王 浙
结　　构 王 超、周 岩
给 排 水 董 超、刘园园
设　　备 杨向红、马任远
电　　气 熊小俊、崔家玮
室　　内 黄 新
景　　观 黄 新、卢 青、于楠楠
照　　明 马 戈

鹤鸣居作为泰康纪念陵园中的综合服务中心，位于咸宁梓山湖新城一片湿地公园之中，用以接待、展览和举办纪念仪式。设计从中国传统文化提炼出“生于自然，归隐山水”的哲学理念，建筑取意于抽象的山峦，形势舒缓安详，薄壳结构轻盈飘逸，素雅的色彩以及在光线下微妙的光影变化，使其更显超脱。设计着力于与环境、气候、微气候的匹配——巨大的屋盖为其下的几个独立的体量提供遮蔽和阴凉，同时也创造出很多通透的灰空间。北侧为仪式空间，从主入口至骨灰堂，由东西向的轴线串联起来；南侧则为休息、纪念空间，休息厅、餐厅、礼堂等五个独立功能体块覆盖不同材质，轻盈地漂浮于湖面之上，面向湖面的一侧均为完整的落地窗，室内明亮、舒适，使得空间氛围庄重而不沉重，神圣而不阴郁。

As the comprehensive service center of the cemetery, the Crane House is located in a wetland park of Xianning. Inspired by the traditional concept of “born in the nature and recess into the mountains and waters”, the design, featuring a light shell structure, is an abstraction of mountains. The modest colors, as well as the delicate effects of light and shadow, have added to the serenity of the building. The design conforms to not only its surroundings, but also the macro and micro climate. The huge roof shades independent volumes underneath while creating gray spaces. The northern part of the building are mainly used as ritual spaces while the southern part mainly serves for memorial ceremonies and breaks.

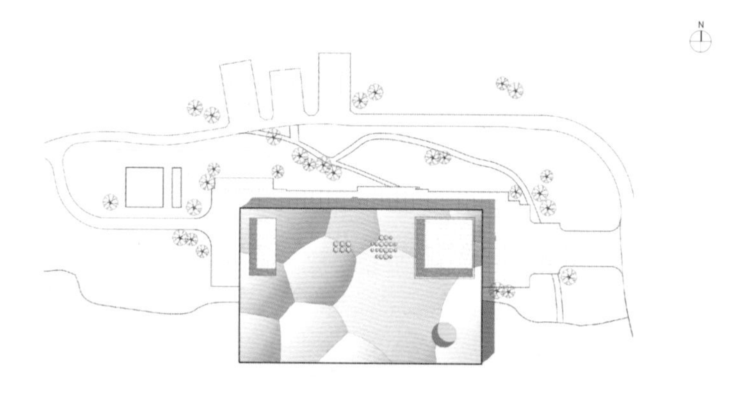

总平面图

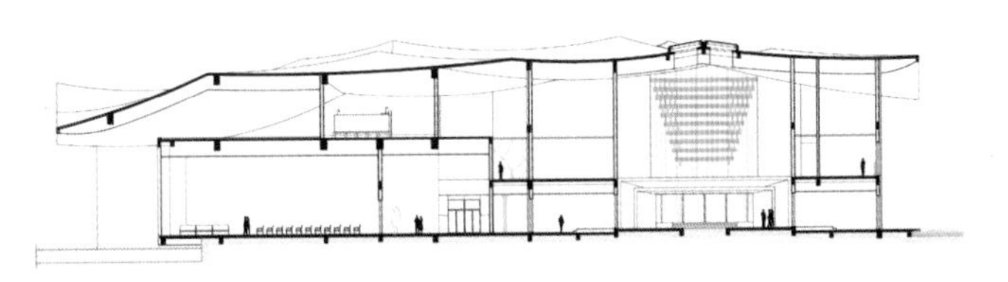

冥想礼堂剖面图

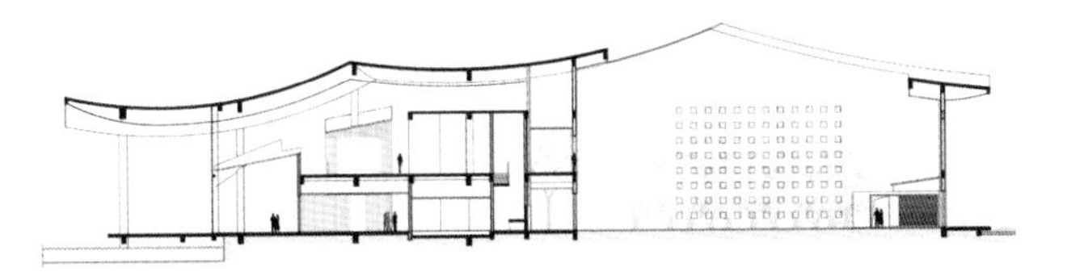

餐厅剖面图

仙鹤湖·鹤鸣居

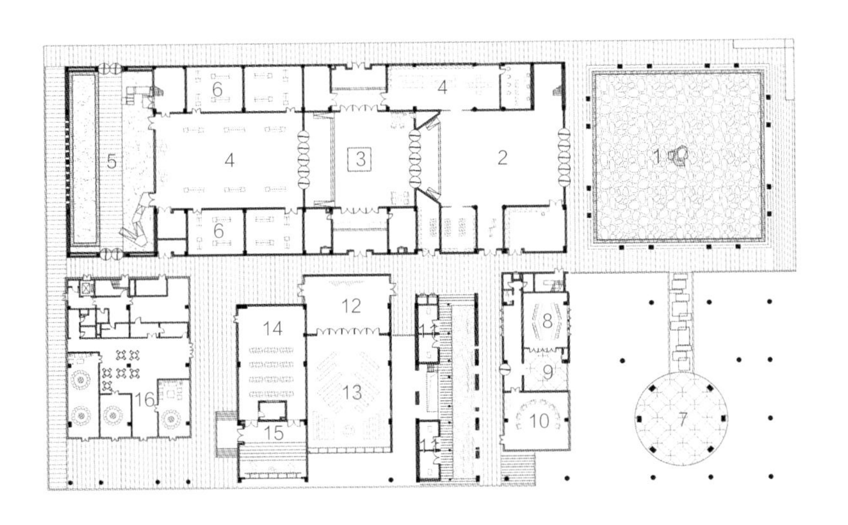

首层平面图

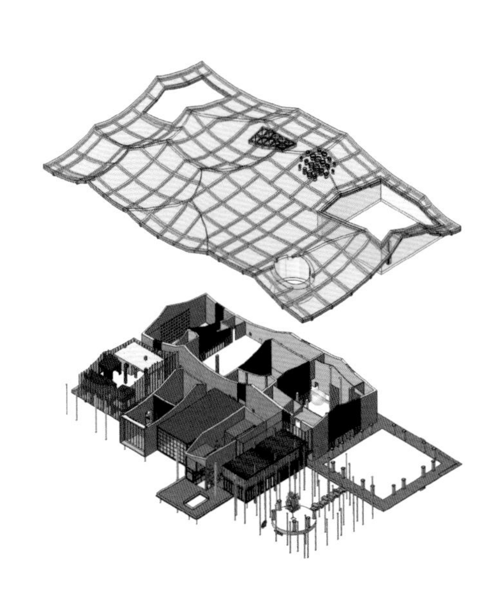

爆炸轴测图

1. 松庭 2. 斗转星移厅 3. 时光沙漏厅 4. 展示陈列区 5. 竹园 6. 祠堂 7. 苍穹厅
8. 多媒体放映室 9. 磐石厅 10. 湖景议事厅 11. 洽谈室 12. 极地雪莲厅 13. 冥想礼堂
14. 生命体验馆 15. 休息厅 16. 餐厅

中粮营养健康研究院一期 COFCO Nutrition & Health Research Center, Phase I

地点 北京市昌平区 / 用地面积 132,934m² / 建筑面积 49,075m² / 高度 49m / 设计时间 2011年 / 建成时间 2014年

方案设计 杨金鹏、李健宇、王霖硕
设计主持 杨金鹏

建　　筑 李健宇、王霖硕、罗　颖
结　　构 张淮勇、周　岩
给 排 水 王耀堂、周　博
设　　备 刘玉春、刘　伟
电　　气 蒋佃刚
电　　讯 任亚武
总　　图 刘　文

中粮营养健康研究院位于北京未来科技城内，由研发、检测设施、办公、粮油食品文化博物馆、食品体验中心组成。建筑环境将生活和记忆融入科研设施和室外空间，在参观路线中设置了多个陈列展示空间，并重点设计了休息区域、运动设施、餐厅和景观系统。

Situated in the Beijing Future Science & Technology City, the COFCO Nutrition & Health Research Center consists of R&D areas, examining facilities, office spaces, a food culture museum and a food exploration center. The sense of daily life and memories are reflected in the scientific research facilities and the outdoor spaces. Great efforts have been made in the design of resting area, sports facilities, canteens and the landscape system.

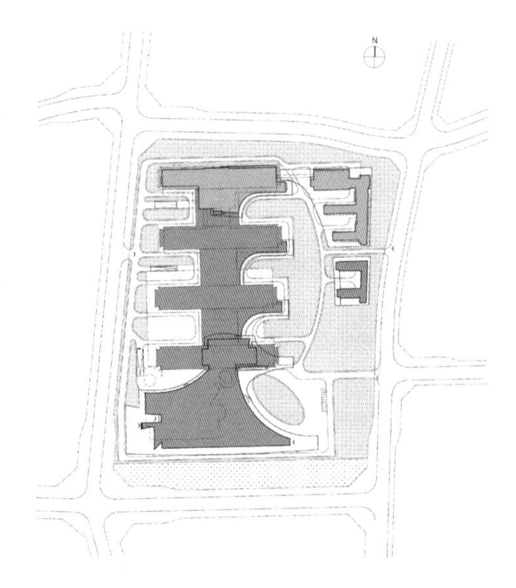

总平面图

中国人民解放军第八十八医院 No.88 Hospital of the PLA

地点 山东省泰安市 / 用地面积 40,000m² / 建筑面积 81,000m² / 高度 39m / 设计时间 2009年 / 建成时间 2014年

方案设计 宦洪桥、杨 光
设计主持 陈一峰、赵 强

建 筑 宦洪桥
结 构 张亚东、潘敏华
给排水 赵 昕、贾 鑫
设 备 徐 征、樊 燕
电 气 李战赠、何 穆
总 图 连 荔

医院位于泰山脚下现有院区内，场地高差变化大，且建造施工不能影响医院正常运营。新建工程通过整体规划有效整合原有散落的区功能、医疗流线及室外空间，完成了由传统医院向新兴现代化医院的转换。在满足医院复杂工艺流程的基础上，建筑形态采用退台及形体穿插的手法，以适应山地地形。形体的拆解削弱了医院巨大体量的压迫感，同时形成高低错落的山地建筑特色。

The design of the hospital at the foot of Mount Tai was challenged by the big altitude difference. Through the integration of functions, routes and outdoor spaces, a conventional hospital has been transformed to a modernized one. The terraced form of the building conforms to the terrain of the mountain, and the segmentation of the building's volume has alleviated the sense of pressure of a large building.

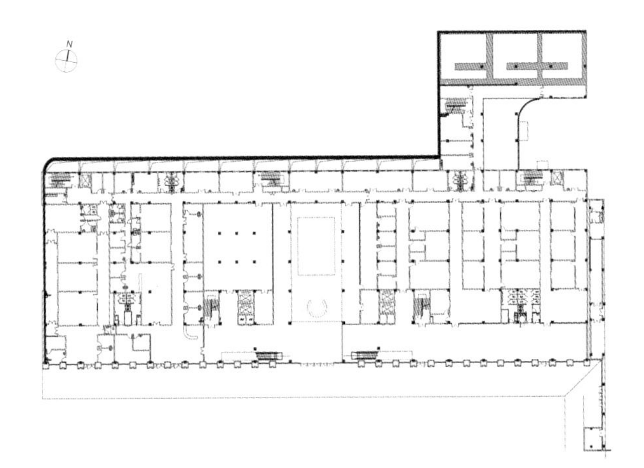

门诊医技楼首层平面图

图书在版编目（CIP）数据

作品 2019 : 中国建筑设计研究院作品选 = SELECTED WORKS 2019 OF CHINA ARCHITECTURE DESIGN & RESEARCH GROUP / 中国建筑设计研究院编 . — 北京 : 中国建筑工业出版社 , 2020.12
ISBN 978-7-112-25605-1

Ⅰ . ①作… Ⅱ . ①中… Ⅲ . ①建筑设计－作品集－中国－现代 Ⅳ . ① TU206

中国版本图书馆 CIP 数据核字 (2020) 第 232001 号

责任编辑：徐晓飞 张 明
责任校对：王 烨

主　　编：崔 愷
执行主编：张广源 任 浩
文字编辑：任 浩
美术编辑：徐乐乐
英文翻译：谭雅宁

建筑摄影：张广源（署名者除外）

参编人员：李 季 周 萱 朱荷蒂

作品 2019：中国建筑设计研究院作品选
SELECTED WORKS 2019 OF CHINA ARCHITECTURE DESIGN & RESEARCH GROUP
中国建筑设计研究院 编
*
中国建筑工业出版社出版、发行（北京海淀三里河路 9 号）
各地新华书店、建筑书店经销
北京雅昌艺术印刷有限公司制版印刷
*
开本：965 毫米 ×1270 毫米 1/16 印张：25 插页：1 字数：400 千字
2020 年 12 月第一版 2020 年 12 月第一次印刷
定价：228.00 元
ISBN 978-7-112-25605-1
(36687)